Energy Metropolis

HISTORY OF THE URBAN ENVIRONMENT

Martin V. Melosi and Joel A. Tarr, Editors

ENERGY METROPOLIS

An Environmental History of Houston and the Gulf Coast

Edited by MARTIN V. MELOSI
and JOSEPH A. PRATT

University of Pittsburgh Press

To Carolyn and Suzy, with love

Published by the University of Pittsburgh Press, Pittsburgh, Pa., 15260

Manufactured in the United States of America
Printed on acid-free paper
10 9 8 7 6 5 4 3 2 1

Library of Congress Cataloging-in-Publication Data

Energy metropolis: an environmental history of Houston and the Gulf Coast / edited by Martin V. Melosi and Joseph A. Pratt.

p. cm. — (History of the urban environment)

Includes bibliographical references and index.

ISBN-13: 978-0-8229-4335-8 (alk. paper)
ISBN-10: 0-8229-4335-2 (alk. paper)
ISBN-13: 978-0-8229-5963-2 (pbk.: alk. paper)
ISBN-10: 0-8229-5963-1 (pbk.: alk. paper)

1. Houston (Tex.)—Environmental conditions—History. 2. Gulf Coast (Tex.)—Environmental conditions—History. I. Melosi, Martin V., 1947- II. Pratt, Joseph A.

GE155.T4E54 2007
363.7009764'1411—dc22 2007006107

CONTENTS

ACKNOWLEDGMENTS

This book grew out of "An Environmental History of Houston and the Gulf Coast" conference held at the University of Houston in March 2003. We would like to thank those who helped stimulate interest in the Bayou City's and the Island City's environmental history through their participation in the conference, including Barry Moore, Gary Garrett, Janice Harper, David Snyder, Steven Klineberg, David Todd, Alecya Gallaway, James McSwain, and several authors in this volume. Thanks to Joseph Pratt and the Cullen Chair in History and Business, the Tenneco Lecture Series, and the Environmental Institute of Houston for their generous support in helping sponsor the conference and the book. Thanks also to the outside readers of the book manuscript and to the able staff at the University of Pittsburgh Press for their valuable assessments in helping make the project successful.

Introduction

Cities are by their very nature energy intensive. The concentration of human and material resources for purposes of survival, construction of infrastructure for transportation and communication, and the production and consumption of goods and services are essential characteristics of communal living. As William Cronon and others have demonstrated, the connection of city and hinterland often extends the impact of urban development beyond the city's political borders while creating interdependence between the built and natural world.[1] The production and use of energy have been potent forces in driving this process of change.

The historical impact of energy reaches beyond the conversion of resources to stationary and motive power to make machines run; to illuminate streets and interiors; to move trains, cars, buses, and trucks; and to generate heat and refrigerated air. Broadly understood, energy encompasses all processes of production and consumption that allow people to function in the physical world. A key challenge in urban and environmental history is to identify and analyze the central impacts of energy production and use on the evolution of cities.

One general impact is clear: changes in energy supply and demand have greatly affected the economic context within which cities have grown. New technologies using new sources of energy have shaped the transportation and communication revolutions that have transformed the world economy in the last two centuries. The impact of new sources of energy on the American economy has been particularly pronounced in the years since the mid-nineteenth century, when the widespread use of fossil fuels encouraged the most significant transportation innovation of that century, the completion of a national system of railroads. Coal fueled the trains that transformed the American landscape, introduced a new scale of business activity, and created a much broader national market. Then oil fueled the cars and planes that extended that transformation. The burning of coal, oil, and natural gas generated the bulk of electricity and provided most of the fuel for modern industrialization. Taken together, these fossil fuels have provided far and away the most significant supplies of energy in the modern economy; their adaptation to industrial uses and transportation systems fundamentally altered the economy by broadening the scale of markets and the scope of business activities.

One result was the incorporation of many regional markets into national and international ones. As national markets grew, fossil fuels helped expand the size and economic reach of urban centers by fueling the transportation and communication systems that tied ever-larger sections of the nation's hinterlands into urban-centered economies. Such urban growth pushed the effective boundaries of cities out into surrounding areas, creating the "standard metropolitan areas" that now define urban life. Within the resulting regions, coal and electricity shaped the rise of a distinctive pattern of urban growth in the northeastern and midwestern United States in the years after the Civil War; the automobile spawned a somewhat different Sun Belt model of city development in the twentieth century.

Fossil fuels helped transform the modern city, altering the physical environment in new and significant ways. The most obvious impact was a fundamental change in land-use patterns in and around cities, which reached out and absorbed once-rural land surrounding the sprawling urban centers. The concentrated usage of energy in the production of goods brought a new scale of industrial pollution to cities. Growing energy use for transporting people and goods added another layer of pollution to the mix, particularly in cities that grew rapidly only after the advent of the automobile. In these and many other ways, as the lure of jobs and better opportunities from urban industrial growth attracted larger populations, the environmental impacts of increasing energy use also grew dramatically.

Although the strong and complex connections between energy use, urban growth, and environmental issues are intuitively obvious, they have been slighted by historians. Perhaps the connections are simply too deeply embedded to be easily analyzed. Also, the study of energy history has not yet developed as fully as the vibrant fields of urban and environmental histories. One way to begin to examine more thoroughly these related issues is to focus on extreme cases, which show most dramatically the relationship between energy, environment, and urbanization.

To more fully examine the intersection of energy, environment, and urbanization, this collection of essays offers the example of Houston. The Bayou City is currently the nation's fourth-largest city, and it sits at the center of the seventh-largest metropolitan area in the United States. It is the home to over 100 racial and ethnic groups, making it one of the most diverse cities in the nation.[2]

As both a consumer and producer of energy, especially oil and natural gas, the city has few rivals around the world. Other parts of the nation can boast of significant oil-producing and -refining areas, most notably the original oil fields in Pennsylvania; the substantial refining complexes in Cleveland and New Jersey; southern California's concentrations of refineries, oil fields, and transportation hubs; and the refining region of southern Louisiana. None of these areas, however, matches the Houston region in its concentration of oil refining, petrochemical production, oil and natural gas transportation, and oceangoing tankers. For better and for worse, the Houston area has developed around oil and oil refining, and it provides a unique case of an energy-intensive metropolis more than other Sun Belt cities such as Atlanta, Dallas–Fort Worth, Oklahoma City, Phoenix, or Los Angeles.

Oil shaped Houston's modern economic and environmental history. In every industrial region, leading industries produce particular patterns of pollution. In the case of Houston, the production, processing, and shipment of oil and natural gas gave the city a distinctive identity within the national economy while also creating distinctive levels and forms of air, water, and ground pollution.

In every large city, the predominant fuel used in regional transportation and industry emits significant pollution. In twentieth-century Houston, oil and natural gas supplied the bulk of the fuel used to transport people and to produce and transport goods. Because petroleum was both the major industry and the major fuel for modern Houston, this self-proclaimed "energy capital of America" has also been the de facto "oil pollution capital of America."[3]

Houston's emergence as a major metropolis was shaped by fundamental changes in the national and international energy industries. Since the late nineteenth century, fossil fuels have steadily replaced wind, water, and animal power; in the process, local sources of energy have been replaced by sources supplied by an increasingly specialized and concentrated energy economy.[4] Oil and natural gas surged forward during the twentieth century to become the dominant energy source in the industrial world, and Houston, more than any other city, benefited economically from this development. Houston became synonymous with oil much as Pittsburgh was with steel or Detroit was with automobiles or the Silicon Valley became with microprocessing. It was a center for specialized activities needed by the national and international petroleum industry, one of the most dynamic industries of the twentieth century.

Nature endowed the Houston region with the abundant natural resources, innumerable sunny days, the lack of harsh winters, and the geographical location that allowed it to prosper as a center of oil production and refining. But good luck and good timing also help explain its dominant role in the development of oil and natural gas. With the discovery in 1901 of the epoch-defining Spindletop oil field in nearby Beaumont, Texas, the region's oil-related economy sprang to life. Houston and surrounding areas on the Gulf Coast quickly became the focal point for the expansion of the oil and gas industries in the southwestern United States, an area that produced the bulk of the world's oil from the turn of the twentieth century through the 1960s. Just as the nation entered the automobile age, with its surging demand for refined oil products, Houston found itself perfectly positioned to become the center for the regional and then the national petroleum industry. Even the subsequent movement of the locus of world oil production to the Middle East could not easily displace the economic advantages embedded in the region's infrastructure during the long dominance of southwestern oil. Adding strength to the regional economy after the 1930s was the spectacular growth of the natural gas industry in the United States, which was centered in Houston.

Over the course of the twentieth century, the regional economy diversified steadily to include more and more oil-related activities. The region's sturdy oil-related industrial core evolved to include the production, refining, and shipment of oil; the production and shipment of natural gas; the production and shipment of petrochemicals; management and research in oil, natural gas, and petrochemicals; specialized construction for these industries; the manufacture of tools and supplies needed by the petroleum complex; specialized technical and management services; and highway and residential construction undertaken in part to meet the needs of the workforce of the oil-related complex. From these activities came both the economic growth and the severe industrial pollution that characterized the modern history of the Gulf Coast region surrounding Houston.

The industrial heart of this oil-related complex was refining. The giant petroleum refineries and petrochemical plants that processed crude oil and natural gas into a variety of products also created tens of thousands of industrial jobs that attracted generations of workers to the region. The center of the massive refining complex that stretched from Corpus Christi, Texas, to New Orleans was the Houston Ship Channel, which reached southeastward from Houston forty miles to the Gulf of Mexico. In the mid-twentieth century, this region along the Gulf Coast produced as much as a third of the nation's refined goods and half of its petrochemicals. Into and out of this highly specialized industrial complex flowed millions of barrels of crude oil and refined oil and tons of petrochemical products per day, and the processing and shipping of these products gave the region its identity in the national economy. Even in the late twentieth century, after automation sharply reduced the number of employees in the refineries and the Houston economy successfully diversified into other economic activities, refining remained a major contributor to regional prosperity.

The manufacturing jobs in these and other oil-related factories gave the region a distinctive industrial working-class tone to a greater extent than Dallas, Fort Worth, Austin, or San Antonio. This was a place where people from the rural hinterland of Texas and Louisiana came in search of greater opportunities for themselves and their children. Bringing with them the racial attitudes inherited from the strict and violent segregation practiced in the small-town and rural South, they helped make Houston the largest Jim Crow city in the nation by 1950. But despite segregated housing and public services and segregated workforces in the area's refineries and factories, these workers—black and white—also brought the migrants' faith that those who worked hard could create a better future for their families in Houston. This fundamental optimism that the city held opportunities for those willing to work hard floated in the air of Houston along with the fumes from the refineries. Jobs, not air pollution, remained the primary concern of millions of people who migrated to the region in the twentieth century.

Sustained growth, of course, brought a new set of challenges. Where would those who flocked to the city live? How would the urban services needed by a growing population be provided? Transportation was one key issue facing the booming region. A prototypical Sun Belt city, Houston grew up with the automobile. With no widespread investment in public transit before the coming of oil in 1901 and the opening of the Houston Ship Channel in 1914, the city expanded rapidly just as cars came into general use. It is symbolically fitting that the decade of the city's fastest growth in the twentieth century was the 1920s, when auto use took off in the region and around the nation and the ship channel refining complex boomed. In that decade, the first substantial suburbs connected to the central city by jobs and roads were the refinery towns east of the city throughout the industrial corridor along the ship channel—Pasadena, Galena Park, Baytown, Deer Park, and Texas City.

In subsequent decades, the city expanded in every other direction. Favorable state laws, such as the Municipal Annexation Act (1963), also allowed the city to aggressively annex adjoining areas and thus further enlarge its territory. By 1999, this generous annexation policy had allowed Houston to reserve approximately 1,289 square miles (excluding the areas of the city within it) for future annexation.[5] Land surrounding the city provided living space for the millions of migrants to the region; inexpensive gasoline provided the fuel needed to commute longer and longer distances; and inexpensive electricity produced primarily by abundant and low-cost natural gas allowed for the air-conditioning that made the city livable during its long, harsh summers. Local, state, and federal governments responded to citizens' demands to build more and more roads reaching farther and farther out from the city. Individuals responded by hustling up the resources to acquire cars and, if possible, homes in the suburbs. The region as a whole moved easily and with little public debate toward a "mass transit" strategy of more highways filled by more cars, often with one driver per car.

Over decades of sustained outward urban growth, both energy costs and environmental costs became embedded in regional transportation systems. Once infrastructure for the transportation of goods and people had been built, change was most difficult, in spite of shifting political calculations of economic and environmental costs. By the end of the twentieth century, most people living in a thirty-mile radius of the central city had become "Houstonians" who were tied into the economic and cultural life of the city primarily through a sprawling system of roads and freeways. The specialized transportation system of pipelines, tanker trucks, railroads, barges, and oceangoing tankers used by the region's petroleum-related core of industries bound together the region's economy and tied it into national and global markets.

The Houston region grew from a frontier town in the early nineteenth century, to a small city of about 45,000 in 1900, and finally to an expansive metropolis of more than five million in 2006. Urban growth has been a core objective of the city from its modest start, spearheaded by John Kirby Allen and Augustus Chapman Allen in 1836, until this very day. Shipping, rail, and especially automotive and truck transportation helped push Houston's borders and Houston's influence beyond its initial location along Buffalo Bayou (see figure I.1).

Although the region expanded geographically before oil, the booming new industry and the growing use of the automobile greatly accelerated the city's expansion out into the surrounding countryside in the twentieth century. By the turn of the twenty-first century, "Houston"—as defined by economic and commuting ties, not by political boundaries—had become one of the world's largest cities in area, spanning perhaps 2,000 square miles that stretched thirty or forty miles from downtown Houston in every direction over a broad area of the Texas Gulf Coast (see table I.2). Much of this land had been farmland in the early twentieth century, but by the year 2000, from Katy to Conroe to Baytown to Galveston to Sugar Land, each exit looked much the same, the homes in each subdivision merged into several generic floor plans, and all roads led into

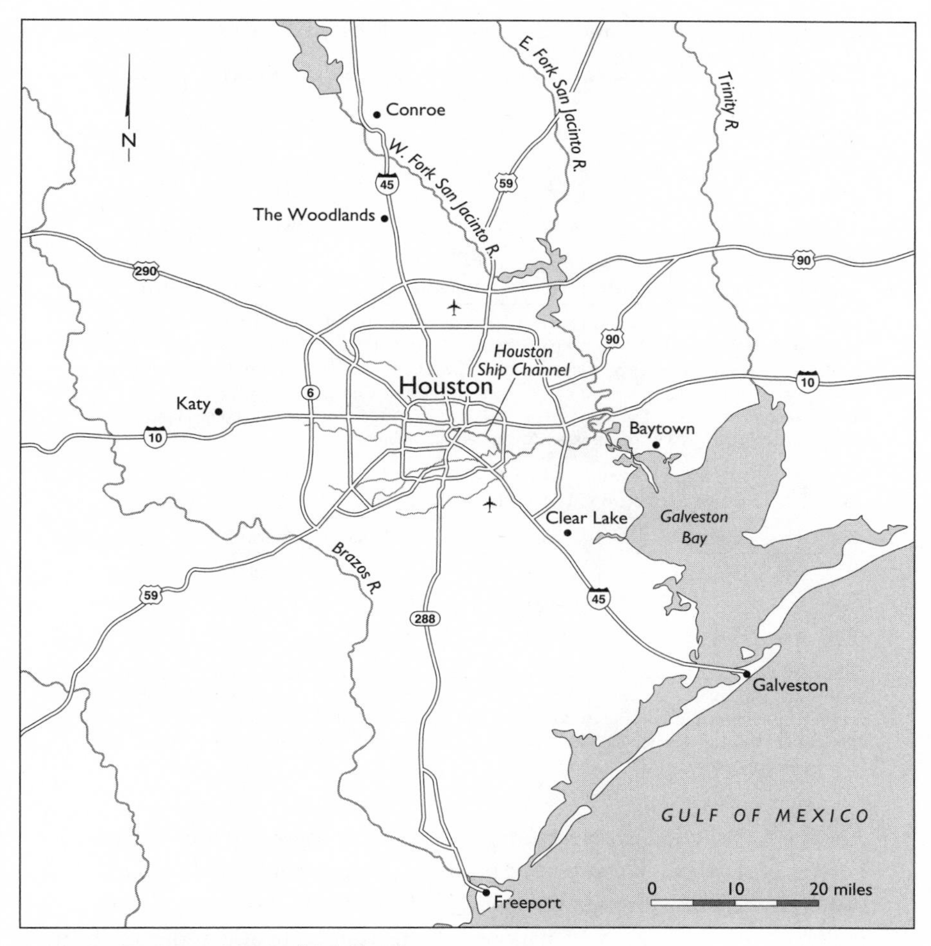

Figure I.1. Houston. Map by Bill Nelson.

and out of Houston. The city exhibited the worst kind of urban sprawl—patternless, unplanned (Houston is the largest city in the United States without zoning), and highly decentralized. In the post–World War II years, Houston had half the population density of Los Angeles and reached into ten counties.

Of course, cars commuting around the sprawling city combined with the refineries and petrochemical plants to produce serious air and water pollution. Adding to the city's environmental woes is its susceptibility to severe weather conditions such as tornadoes and hurricanes, and a propensity to flood often and intensely. Volumes of water exacerbate nonpoint pollution problems, as all kinds of toxic materials—from lawn fertilizers to heavy metals—run into the city's extensive network of bayous and ultimately spill into the Gulf of Mex-

Table I.1. Population, Houston Metropolitan Statistical Area (Houston–Baytown–Sugar Land, TX MSA), 1837–2005

Year	*Population*	*Year*	*Population*	*Year*	*Population*
1837	1,500	1900	122,785	1960	1,364,569
1850	14,773	1910	176,589	1970	1,903,192
1860	29,801	1920	256,023	1980	2,753,155
1870	42,962	1930	439,226	1990	3,342,247
1880	63,729	1940	627,311	2000	4,715,407
1890	76,959	1950	908,822	2005	5,280,077

Sources: Proximity: Resources to Create and Apply Insight, http://www.proximityone.com/metros.htm#top10; Houston Facts, 2000, prepared by the Greater Houston Partnership Research Department, http://www.houston.org/tophoustonfacts/houstonfacts2000; "Houston,Tex.: Metropolitan (MSA) Population and Components of Change," Real Estate Center, Texas A&M University, http://recenter.tamu.edu/Data/popm/pm3360.htm; David McComb, "Houston," Handbook of Texas Online, http://www.tsha.utexas.edu/handbook/online.

Table I.2. Area of Houston City Limits, 1910–2000

Year	*Square Miles*	*Year*	*Square Miles*	*Year*	*Square Miles*
1910	17.4	1950	160	1980	556.4
1920	36.5	1960	328.1	1990	539.9
1930	71.8	1970	433.9	2000	617
1940	72.8				

Sources: U.S. Bureau of the Census, "Population of the 100 Largest Urban Places," 1910–90, http://www.census.gov/population/www.documentation/twps0026.html; Houston Facts 2000, prepared by the Greater Houston Partnership Research Department, http://www.houston.org/tophoustonfacts/houstonfacts2000.

ico. Severe pollution problems arose from the transportation of the millions of barrels of crude oil and refined products per day that flowed through these plants on their way through the global oil economy. In this sense, the Houston region served not only as the nation's refining center but also as one of the national economy's primary dumping grounds for oil-related and other forms of pollution.

Just as Houston earned great economic benefit from its specialized role as a center of oil-related activities, it also paid a high environmental price. Petrochemicals presented their own special set of problems from air and water emissions, along with the disposal of a variety of solid and liquid wastes. Altogether, the region suffered the triple dilemma of dealing simultaneously with mounting oil-related pollution from the exhausts of gasoline-powered automobiles, the production of oil and chemical products from local plants, and myriad urban pollutants.

Efforts to find an acceptable balance between the costs and benefits of petroleum processing and petroleum pollution proved difficult. For most of the region's history, a broadly shared societal consensus that included a majority of the population, rich and poor, favored oil development largely unrestrained by pollution controls. "Opportunity" and "economic growth" were the twin tenets of the local religion of boosterism, and the church did not have much tolerance for doubters who voiced concerns about the quality of life.

Those who called for stricter controls of pollution had to overcome more than regional attitudes favoring growth. Local politics, as well as civic leadership, were dominated by business leaders whose idea of a "healthy business climate" included low taxes, weak unions, and very limited regulation. Several scholars have provided the useful label "free enterprise city" to describe the dominance of Houston's political and civic cultures by conservative businessmen.[6] Granted such business power is hardly unique among cities in capitalist America, and further granted that much of the power and the behavior of the local elite can be explained with reference to their commitment to segregation as easily as to their commitment to free enterprise, the fact remains that business and civic leaders in Houston historically represented a strong and consistent barrier to the passage and enforcement of effective pollution controls.

Political realities also included the entrenched power in the political process at the state and federal levels of the well-organized interests of the major industries that produced the bulk of the region's industrial pollution. The basic decisions about Houston's oil-related development were made by private corporations in the global energy economy. In Houston, as around the world, price dictated the key decisions on energy use and, to an extent, the approach to pollution control. But the political process played a pivotal role in channeling government promotion and blocking government regulation. The economic importance of oil in the state of Texas, in general, and in the city of Houston, in particular, skewed political decisions toward policies that promoted the oil industry and away from policies that constrained the industry, at least until the federal government preempted much of the traditional authority of states over pollution control after the 1960s. The state's one-party political system through the 1960s also proved to be a barrier to change, as those who advocated states' rights in defense of Jim Crow had ample reason to support the states' rights arguments of those who fought against federal government involvement in pollution control.[7]

An important part of the economic/environmental history of the region has been the ongoing efforts to create more effective pollution controls while also encouraging continued economic growth. Only in the recent past have segments of the American public recognized that more efficient energy use in response to higher prices has broken the traditional coupling of growth in energy use and economic growth. Only recently have segments of the American public—including some civic leaders and grassroots organizations in Houston—recognized that high levels of pollution are not only a threat to public health and the quality of life but also can become a barrier to economic growth and the creation of jobs in the region and the nation. For most of Houston's history—in politics and in practice—there was little effort to "balance" energy and environmental needs, since pollution control was treated as distinctly secondary to economic growth. Yet there are at least tentative signs that a new balance, with greater concern for environmental quality, is politically possible.

Unlike many other oil-producing regions around the world that have been seemingly "cursed" by the problems of oil-led development, the Houston area has grown spectacularly since the discovery of oil in the region in 1901.[8] Since that time, the region has successfully absorbed the dynamic oil industry, using it as the engine of growth that transformed the city. Historically, the timing of the discovery and development of oil was fortuitous for Houston, which got in on the ground floor at the birth of the modern petroleum industry. The giant refineries built in the region in the early twentieth century could not be easily moved as oil production itself subsequently moved away from the region.

The area around Houston was a thriving regional center before the discovery of oil. The rise of shipping and commercial development along Buffalo Bayou in the nineteenth century was an essential precedent for the construction of the Houston Ship Channel and the emergence of Houston as an international focal point for oil production, refining, and petrochemicals. The business community proved well equipped to take on the new economic activities spawned by the industry. It had an established legal system capable of managing the demands of the new industry, and well-developed local corporate law firms. Most important, oil development went forward under the direction of transplanted Texans and native Texans who reinvested much of the profits from the giant new oil fields in regional developments such as giant refineries and pipelines connecting these refineries to other major oil fields outside the region.

The long-term prosperity generated by the regional economy also reflected its historical capacity to diversify away from oil production—from a regional perspective, the most difficult to retain branch of the oil industry. In the mid-twentieth century, such diversification occurred within the expanding oil-related core, where first refining and oil-tool manufacturing, then petrochemicals, and finally the production and transportation of natural gas provided dynamic areas of economic expansion. Then in the 1960s and accelerating in the mid-1980s, after the dramatic drop in oil prices led to a devastating regional depression, diversification outside of oil bolstered the region's growth.

Yet even much of this "new, nonoil economy" in Houston had indirect ties to the oil industry. The coming of the Lyndon B. Johnson Space Center to Clear Lake, south of Houston, was orchestrated by an alliance of George Brown of Brown & Root (a construction company with many strong ties to oil industry markets), the president of Humble Oil, and Vice President Lyndon Johnson. The growth of the Texas Medical Center, which became a major employer in Houston by the 1980s, was fostered by the philanthropic support of many Houstonians who had made their original fortunes in oil, as was the growth of Rice University and the University of Houston. Even the real estate development industry that built the Houston suburbs had strong ties to the oil industry; for example, Friendswood Development, a major regional developer that built much of Clear Lake and Kingwood, originally was affiliated with Exxon and named after one of its important regional oil fields, and the Woodlands north of the city was inspired by the efforts of oilman George Mitchell.

Yet despite the Houston area's capacity to "absorb" the oil industry, using its oil wealth to sustain long-term economic growth, one significant part of the "curse of oil-led development" around the world proved most difficult: the management of the environmental impacts of the production and use of petroleum. As we continue to grapple with these issues in the early twenty-first century, historical perspective can be useful in analyzing what we have and have not done in the past. This collection of essays is a contribution to historical perspective on Houston and its environment. It seeks to provide context for a larger set of questions: what can and cannot be done in the future?

The Physical Setting

As an energy-intensive metropolis, Houston has been shaped by natural and human forces. This volume focuses primarily on the latter. However, to fully understand the environmental history of this urban area—including Galveston—requires attention to its physical realities, especially its location within the Gulf Coast region of southeastern Texas. The circumstance of where the metropolis is situated speaks volumes about its dynamic history.[9]

Houston is foremost a product of the Texas Coastal Zone, an area of approximately 20,000 square miles consisting of about 2,100 square miles of bays and estuaries, 375 miles of coastline, and 1,425 miles of bay, estuary, and lagoon shoreline. The shoreline itself is composed of interconnected natural waterways, restricted bays, lagoons, and estuaries with modest freshwater inflow, elongated barrier islands, and a very low astronomical tidal range.[10]

Texas actually has two shorelines—one running along the Gulf of Mexico and another along the bays. Bolivar Peninsula, Galveston Island, and Follets Island are grass-covered barrier flats and sandy beaches of one to three miles in width that separate the bay areas from the Gulf of Mexico. Galveston Island is the best known of the barrier islands, approximately thirty miles long and two-and-a-half miles wide, with no underlying bedrock and consisting of mud flats on the side facing the bay. As David McComb has noted, "Lying parallel to the coast two miles away, Galveston stands as a guardian protecting the land and the bay from the Gulf."[11]

Behind the barrier islands, Galveston Bay (composed of four major subbays: Galveston Bay, Trinity Bay, East Bay, and West Bay) and some smaller bays comprise almost 600 square miles of surface area. West Bay is a lagoon separating Galveston from the mainland. Galveston Bay and its continuation as Trinity Bay constitute an estuary. The entire Galveston Bay watershed, or drainage basin, covers 33,000 square miles of land and water from the Dallas–Fort Worth Metroplex to the Texas coast—a substantially larger area than the water that the bay encompasses. Extensive marshy areas less than five feet above sea level extend along the landward side of West and East bays. Bolivar Roads and San Luis Pass are natural passes that connect the bays to the Gulf. Rollover Pass, extending through the Bolivar Peninsula, connects the Gulf of Mexico and East Bay.[12]

Above the bays, two major river valleys—the Trinity and the San Jacinto—and several minor valleys of headward-eroding streams—Cedar Bayou, Buffalo Bayou, Clear Creek, Dickinson Bayou, Halls Bayou, Chocolate Bayou, and Bastrop Bayou—cut into the coastal plain. The Brazos River and Oyster Creek flow through the western portion of the Texas Coastal Zone, but not within deep valleys.[13]

The Houston-Galveston area consists of approximately 2,268 square miles of land, which is a broad region of flat coastal plain situated between the coastal marshes and the areas of pine and hardwood forests along either side of the Trinity River and north of downtown Houston. The coastal plain itself inclines from the Gulf at two to five feet per mile. The maximum elevation of the coastal plain is about ninety feet above sea level in the northwestern part of the coastal zone.[14]

Situated approximately forty-nine feet above sea level on prairie some fifty miles from the Gulf of Mexico, the city of Houston is linked geographically, geologically, and climatically to the Texas coast. The coastal plain of which Houston is a part comprises gently dipping layers of sand and clay. The slope of the impermeable layers of clay, shale, and gumbo interbedded with permeable, water-bearing sands and gravel is greater than the slope of the land surface. These conditions are favorable for artesian water, and the city historically has drawn water from the Chicot and Evangeline aquifers running southeast to northwest from the Gulf Coast through the city and to its north. Extraction of the groundwater, however, can lead, and has led, to land subsidence and saltwater intrusion if water is drawn out too aggressively.[15]

On the surface, water from Houston drains into the Gulf of Mexico via an elaborate network of bayous. With an average yearly rainfall of forty-two to forty-six inches from often torrential downpours, the area is subjected to frequent flooding. Since urbanization removes much of the filtering capacity of the soil, runoff has exaggerated Houston's tendency to flood as the city continued to expand. Since the city also is susceptible to hurricanes and tornadoes, water and the pollutants it often carries have been the greatest natural threat to Houstonians and its neighbors.[16]

Along with its extensive waterways, Houston also is a heavily vegetated city. Large portions of the region are forested, with substantial tree growth along the bayous. Loblolly pine is the tree species that dominates the region. Much of the west side of the area, which contains acres of native prairie, has been converted to developed private property where trees have been planted. On the north and northeast sides, the natural land cover has densely vegetated canopies over approximately 50 percent of the area. To the south, the area is covered with a combination of prairie, marsh, forest, and abandoned agricultural lands. At the highly developed city center, however, the ground and canopy cover has diminished markedly. A "heat island" effect is most pronounced in this area. A study by the National Aeronautics and Space Administration (NASA) in August 2000 found "hot spots" of approximately 149 degrees Fahrenheit in the warmest locations and 77 degrees Fahrenheit in cooler areas.[17]

In many respects, Houston's climate is one of its most identifiable features. An ill-fated publicity campaign once used the phrase "Houston is Hot!" to promote the city. Indeed, Houston's climate is subtropical and humid, with prevailing winds bringing heat from the deserts of Mexico and moisture from the Gulf. The sun shines for much of the year, with an annual growing season of almost 300 days. The average low temperature is 72 degrees Fahrenheit in the summer and 40 degrees Fahrenheit in the winter; the average high is 93 degrees Fahrenheit in the summer and 61 degrees Fahrenheit in the winter. Humidity in June is typically about 63 percent.[18]

As one geologic study noted, "The attributes that make the Texas Coastal Zone attractive for industrialization and development also make it particularly susceptible to a variety of environmental problems."[19] The deepwater ports, intercoastal waterways, good water supplies, large tracts of arable land, and relatively mild climate have been valuable assets. In addition, the region has a variety of other exploitable resources, including timber, sulfur and salt, sand and gravel, shells for lime, abundant wildlife, shellfish and finfish, and petroleum reserves.[20] The agricultural, commercial, industrial, and recreational possibilities for the region, however, have come with a price due to natural and human impacts.

Probably the most dramatic natural events have been tropical hurricanes and tornadoes. As far back as 1776, a hurricane of unknown intensity destroyed a mission in the Galveston area. In the nineteenth century, major hurricanes struck the upper coast in 1854, 1867, and 1886.[21] The "Great Storm" that hit

Galveston on September 8, 1900, is the best known of the South Texas hurricanes, especially since it killed more people than any other natural disaster in the history of the United States—6,000 people in Galveston alone and probably 10,000 to 12,000 total—and devastated the city itself.[22] Although no storm since that time recorded loss of life even close to the one in 1900 (the death toll from Hurricane Katrina in 2005 exceeded 1,800 people), hurricanes and tropical storms have been regular visitors to the upper Texas coast. In recent years, powerful storms such as Hurricane Carla (1961), Tropical Storm Claudette (1979), Hurricane Alicia (1983), Tropical Storm Allison (2001), and Hurricane Rita (2005) have wracked southeast Texas.[23] Along with loss of life, property damage, and severe flooding, the storms also have contributed to substantial shoreline retreat.[24] However, human action also encourages shoreline recession. For example, until a 1970 bill was passed by the Texas legislature, beach sediments were removed for road building. Building of jetties also disrupts the normal transportation of sediment and produces recession, as in the case of the Sabine Pass jetties and jetties along Bolivar Peninsula.[25]

Human impacts on the Texas Coastal Zone are wide ranging. Extensive dredging of channels and passes has resulted in the discharge of sediment into bays, ultimately modifying natural bay circulation patterns, and affected water quality and estuarine plants and animals. Because of increased cultivation, the construction of irrigation and drainage canals and urban paving result in many streams accelerating the transport of sediment into bays as well as increasing nonpoint pollutants, including pesticides and herbicides from runoff. Straightening and lining streams with concrete—as in the case of several bayous in Houston—encourage flash flooding. Thermal effluents from various manufacturing processes and power generation can be lethal to fish. Aggressive withdrawal of groundwater causes land subsidence and saltwater intrusion, and activates faults. Discharge of organic materials, trace metals, and other materials too numerous to mention from a variety of sources—including oil production, pipelines, spills, and chemical production—adds significantly to the pollution load of all water courses.[26]

Human actions also have had significant impacts on the major habitats of the bays and environs, including oyster reefs, submerged aquatic vegetation, intertidal marsh vegetation and animal life, and freshwater wetlands. For example, between 1950 and 1989, approximately 54 percent of the freshwater marshes in the Galveston Bay watershed were lost because of draining wetlands and conversion to upland areas.[27]

Houston and other upper Texas coast cities are subject, first and foremost, to the geologic, geographic, and climatic features that help to define them in physical terms. In Martin Melosi's chapter on Houston's sanitary services, one observer is quoted as saying, "Harris County doesn't have earthquakes . . . doesn't have blizzards . . . doesn't have avalanches. We have flooding." The unique physical characteristics of southeast Texas go hand in hand with the human modifications to the region that define its environmental history—its potential, its shortcomings, and its evolution. Part of what makes Houston an energy-intensive metropolis existed before people put their stamp on this part of the world—forces that resist as well as accommodate urban development.

PART I — ENERGY AND ENVIRONMENT

Gigantic oil strikes in the Southwest and California in 1900 and 1901 changed the nation west of the Mississippi. Beginning as a regional phenomenon, the discovery of vast oil fields in Texas, Oklahoma, and California created a new form of wealth in what would become the American Sun Belt, accelerating urbanization in its wake. In explaining Houston's rise from a small town into a metropolis, it has been common to identify World War II as the great dividing line. After all, three-fourths of the physical city was built in the postwar period, having received a tremendous boost from the influx of wartime industries, the great demand for oil, and the inflow of thousands of new workers. Without question, both World War I and World War II played pivotal roles in the city's growth, but it is difficult to imagine modern Houston without the Spindletop strike and the emergence of the city at the heart of the refining region along the upper Gulf of Mexico.

Drawing on the writings of wildlife biologist Aldo Leopold and the scholarship of Carolyn Merchant, Brian Black explained the history of the first oil strike in Titusville, Pennsylvania, in 1859 and its impact on the Oil Creek valley as a "biotic reaction" when "the naturally occurring pockets of fossil fuels met the human enterprise and dogged pursuit of wealth that created the period of American industrialization." America's first oil boom was, according to Black, emblematic of one of the earliest examples of "the culture of massive disturbance—the culture that remains a mainstay of American economic development."[1] Black's work blended human determination to exploit resources for monetary gain with the resultant impact of that exploitation.

The rise of the oil and refining industries in and near Houston created a "biotic reaction" of its own. It might be overly dramatic to characterize the rapid and sustained growth of Houston as simply "a massive disturbance," but there is little doubt that to slight the role of oil in stimulating the growth of Houston and its environs is to lose sight of one of its unique characteristics as an urban place.

The first two essays focus on the production end of energy and industrialization, but also link economic activity to Houston's growth. In his lead piece, Joseph Pratt provides an introduction to the growth of oil production and refining that so powerfully shaped Houston. He suggests that Houston prospered before the coming of oil, but since Spindletop, the city has depended on "the oil-related core of industries" that helped shape it. In the next essay, Hugh Gorman discusses in some detail the importance of the Houston Ship Channel—a vital link to the Gulf and a critical economic engine for the city and region—and its relationship to industrial pollution along the city's eastern border. The piece makes clear that environmental repercussions of industrial growth were severe, and initially the abiding ethic for confronting industrial pollution "was an ethic based on increasing efficiency and eliminating waste, not managing environmental quality." But Gorman also connects the emerging industrial landscape along the ship channel with the overall growth of the city of Houston. As he reminds us, the ship channel is "as much a part of the city's landscape as its skyline, urban core, or sprawling network of commercial and residential developments."

The final two essays in this section look primarily at the consumption side of energy development. Both examine the more recent history of the city, a period that fits well into David Nye's conception of a "high-energy society," dependent on systems and technologies that provide all the modern conveniences.[2] Robert Fisher discusses a central component in Houston's serious air pollution dilemma—ozone—derived from power generators, refineries, and other industrial sources, and from the city's staggering automobile dependence. Fisher's story is one of failure for not yet effectively confronting one of the most pernicious results of urban/industrial growth. Robert Thompson explores the rise of air-conditioning in the metropolis. Energy, this essay clearly demonstrates, is not simply a commodity, but the means to an end. Not only is Houston an unthinkable place without air-conditioning, but, as Thompson argues, "Houstonians used climate control to transform the cityscape, as the technology required the construction of environments in which weather conditions exercised few constraints on human comfort or economic efficiency."

Unfortunately, in the "most air-conditioned city in the world," a gap remained between the primary benefactors of the new technology and those who remain without. The aggregate use of electricity for air-conditioning has been monumental, but the distribution mirrors other resource use between haves and have-nots.

Taken as a group, the essays in this section demonstrate an intimacy among energy development and use, the growth of the city, the changing landscape influenced by energy production and consumption, and problems of pollution.

"It's a mixed blessing."

"It's a mixed blessing." Cartoon by Douglas Florian. Reproduced by permission from the *New Yorker*, November 21, 1977.

CHAPTER 1

A Mixed Blessing

Energy, Economic Growth, and Houston's Environment

JOSEPH A. PRATT

A cartoon in the *New Yorker* magazine shows two well-dressed men sitting and talking in a comfortable home. Yet in the midst of this picture of prosperity and comfortable satisfaction sits a miniature oil derrick spewing oil all over the room. The caption reads simply: "It's a mixed blessing." This cartoon captures the dilemma of Houston's oil-led development. Oil and related industries transformed the city into a major metropolis, but the costs included profound and long-lasting impacts on the regional environment.[1] The region benefited economically from the heavy concentration of many of the nation's oil-related industries, but it also suffered from the concentration of many of the nation's oil-related environmental problems. Much of what is unique about Houston environmental history reflects the mixed blessing of its historical role as the nation's oil capital. Oil helped make Houston a symbol of opportunity in a poor region of the nation. But as the region's oil-related industries and its population boomed, air and water pollution from oil-refining and petrochemical production combined with the exhaust fumes from the gasoline-powered cars that clogged its freeways to create serious problems that grew to threaten its future growth.[2]

The Region before Oil (1836–1901)

Houston prospered before the coming of oil. In the sixty-five years between its founding in 1836 and the Spindletop discovery in 1901, the city grew steadily into a significant regional center of transport, trade, finance, and legal services. What began as a relatively isolated and undeveloped area gradually forged stronger ties with the industrialized Northeast. Using energy sources other than oil and natural gas, the city built a healthy economy that benefited from the trade of cotton and timber from interior regions. As Houston's economy grew, the city confronted environmental problems that seem minor only in retrospect. But more significant for modern Houston than the specific problems faced were the attitudes and institutions that developed in the years before oil and continued to shape regional approaches to pollution control far into the twentieth century.

Other than the availability of ample water and natural resources, the upper Texas Gulf Coast was not a particularly promising location for a thriving city. The terrain was flat for as far as the eye could see in all directions. Heavy rainfall often made large sections of the region impassable marshland, and frequent deluges regularly caused serious flooding along regional rivers, streams, and bayous. The threat of hurricanes loomed for six months a year, from June to November, and the same months often brought oppressive, debilitating heat and humidity. Flat, ugly, wet, and hot—no wonder Frederick Law Olmstead, who traveled through the region from Houston to New Orleans in the 1850s, remarked: "this is not a spot in which I should prefer to come to light, burn, and expire; in fact, if the nether region . . . be a boggy country, the Avernal entrance might, I should think, with good possibilities, be looked for in this region."[3]

Many of those who migrated to this boggy country often made a two-part journey. The first was a move from such states as Tennessee and Georgia to the agricultural areas in eastern Texas and western Louisiana in search of land and opportunities. The second often came a generation or two later, as the children and grandchildren of the early migrants moved to nearby cities to escape the hard lot of the small farmer. To such migrants, Houston was an island of urban opportunities in a sea of rural poverty. Those who found their way to the Houston area were a part of the general westward movement that shaped the nation's expansion in the nineteenth century. At least until the Civil War, they also were often a part of a "southern movement," as plantation owners, their enslaved labor force, and those who supplied goods and services to the plantations pushed westward from the Deep South in search of new land that could be adapted to a plantation system based on slave labor.

Early in its history, Houston benefited from its proximity to the rich bottomlands of the Brazos River, where cotton plantations flourished in the decades before the 1860s. These large plantations became a focus of economic activity in southeastern Texas, and merchants in Houston and its rival city to the south, Galveston, hustled to create profitable ties with the cotton-growing regions in the interior. This was a difficult challenge given the rudimentary transportation system and the lack of dependable supplies of energy then available to connect the interior and the coast, and Houston's early growth resulted from its successful responses to this challenge.

One key problem was to develop efficient transportation and trade ties in an area not blessed with easily navigable rivers. Of all of the region's rivers, the Brazos came closest to being navigable from the coast to the interior, but difficulties with sandbars near its mouth hampered its use. Despite an abundance of readily available wood for wood-burning steamships, the region's waterways did not provide dependable, efficient transportation to carry cotton crops from the Brazos bottom to market. The use of wind, water, wood, and muscle power limited development in a sparsely settled region on a broad expanse of marshland and prairie.[4]

From its founding in 1836, Houston's civic and business leaders set about to find transportation improvements that could quicken the pace of development. Their top priority was the improvement of waterways, including Buffalo Bayou, the meandering stream on whose banks they had founded their new city. But they also worked to build other means of transport. The creation of better roads to the interior became a pressing concern, since travel by animal -drawn carts was almost impossible after heavy rains. Here the region went through a cycle of road building similar to that described in the northeastern United States in George Rogers Taylor's classic history of the transportation revolution.[5] One difference between the South and the Northeast was the availability of enslaved labor to work on these roads in the South, but little could be done with existing technology to protect graded dirt roads from becoming steams of mud during the frequent heavy rains that plagued the region. An effort to build and maintain a plank road between Houston and the interior failed because of the prohibitive cost and the lack of durability of planks in the harsh environment of the area.

It was thus not surprising when merchants in Houston and Galveston became strong supporters of the construction of railroads from the coast to the cotton-growing regions. Boosters in the two towns battled to find the capital and the political support needed to construct railroads that could bring cotton to their towns while shipping goods back to the plantations. By the 1850s, the first relatively short lines for wood-burning railroads had been completed. By 1860, Houston was the starting point or terminus for some 80 percent of the approximately 500 miles of track that had been laid in Texas, and its rails reached out up to 90 miles into the hinterland in several directions. In the process of building these early railroads, Houston leaders also clearly established what became a defining characteristic of the city's civic elite—the tradition of boosterism.[6]

"Opportunity" and "expansion" early on became the central tenets of Houston's religion of boosterism. Whatever it took to foster economic growth in the region would be done. Generation after generation of Houstonians understood that they could prosper individually only if their city on its unlikely location on the flatlands of the Gulf Coast also prospered. Many of the most fervent boosters were individuals who had migrated to Houston from the small towns in the interior of Texas in search of a larger stage on which to pursue their ambitions. Such adopted sons and daughters established a tradition of voluntarism in civic affairs that became an identifying characteristic of their city. In a city whose leaders trumpeted the positive and ignored the negative, pollution was not likely to be high on the civic agenda.

The coming of the Civil War changed everyone's agenda, temporarily stopping the expansion of regional rail lines and most other economic activity. At war's end, the city had to come to grips with the new world brought by the dislocation and chaos wrought by the South's defeat—and the end of slavery. Many black and white sharecroppers and subsistence farmers who before the war had fought the soil for sustenance in the Texas interior looked to Houston and other cities on the coast for a change in circumstances. Segregation and discrimination remained overpowering burdens on blacks throughout southeast Texas, but Houston afforded the freemen and women somewhat better job and educational opportunities, and even a measure of increased personal freedom in the large black communities that grew in the city.[7] Largely undamaged by the war, Houston rebounded from the defeat of the South more quickly than most other Southern cities.

The most significant change in the regional economy in the postwar years was the extension of national railroad systems through the region. In a flurry of activity in the 1880s and 1890s, national railroad systems such as the Southern Pacific completed connections into the region. While fundamentally altering the region's transportation system, the railroads also quickly became the largest employers and the most important financial entities in the region. Such railroads have been called "the nation's first big businesses" because they brought a new scale of activity and innovations in management and finance. They also assured Houston's future growth by cementing its status as the transportation hub that would connect the upper Texas Gulf Coast to the Midwest, while serving as a logical stopping point in east-west trade and travel between Florida and California. By the turn of the century, the railroads had tied the region into the national economy, allowing for the shipments of cotton, timber, and other natural resources to national and even international markets.[8]

These railroads also encouraged an important transition in energy use, the move from wind, wood, water, and muscle power to coal in the region's major industries.[9] The railroads provided an efficient way to transport coal from the ports on the Gulf Coast into the surrounding areas in the interior. In the late nineteenth century, the railroads and other leading regional industries burned coal shipped down the Mississippi River from Pennsylvania and Alabama or brought by the railroads themselves from the same areas. Major fuel-burning industries quickly made the transition from wood to coal, a new and better-burning fuel. Railroads, steamships, breweries, sugar refineries, timber processing, and, later, production of electricity and manufactured gas led the way in the consumption of coal. A coal-led industrial revolution began to transform the region in the 1870s, and the railroads were the key agents of this change. In this sense, they were more than the first big business in the region; they also were the first business to transport large volumes of coal to markets throughout the area and the first major industry to burn the fossil fuels that ultimately transformed the area's economy and its environment.[10]

Economic opportunities spurred by the railroads attracted a growing population, which reached about 45,000 by 1900, marking Houston as a substantial regional trading center. In some ways the city's expansion before oil mirrored its subsequent growth. Both before and after oil, the transportation and processing of raw materials for export constituted the engine of growth. The strong banks, law firms, and trading capacities developed before 1901 served the oil industry well after that date.

The environmental ethic of the preoil era reflected the attitudes of late-nineteenth-century American capitalism, with its emphasis on harvesting resources as rapidly as possible with no concern for the long-term depletion of these resources. Within a large region with relatively few people, few questioned this process. Cotton growers depleted the soil and then moved on. The rapid clear-cutting of the pines and hardwoods of East Texas for short-term profit went forward with little thought about the future. No one stopped to envision a time when companies could not just move on to previously uncut stands of timber in these giant forests. To most people in the region, the smoke produced by households burning wood and by the railroads and industries burning coal was a symbol of progress, not a warning of potential environmental problems. The attitudes of these formative years in regional development, when a vast area with a good natural resource base seemed capable of sustaining rapid growth far into the future, carried forward into the first decades of oil development.

Had oil not been discovered near Houston, the continuation of this coal-based economy would have produced a much different economic and environmental future for the region. The costs of coal imports would have posed a barrier to rapid, sustained economic growth. Large investments would have been required for manufacturing plants to convert coal to gas for a variety of uses, a movement that had begun in Houston by the 1890s. As in many coal-burning cities, serious localized air pollution undoubtedly would have emerged as the city grew.

Beyond such general observations, the magnitude of the differences between Houston as a coal-burning regional center of trade and Houston as a national center of oil and natural gas production is strictly conjecture, although several easy generalizations are possible. First, since oil and later natural gas were superior and cleaner-burning fuels than coal, the transition to oil undoubtedly reduced the most visible air pollution from the region's railroads, its industries, and its power-generating plants. But, of course, much of the use of oil for fuel came in new markets created by the economic boom in the region after the discovery of oil. Although it is impossible to calculate the overall environmental impact of the transition away from coal, it is clear that the coming of oil pushed the region onto a new economic and environmental trajectory.[11]

Era of the Gusher (1901–1930s)

Oil came to the Gulf Coast with a roar on January 10, 1901, with the discovery of the Spindletop gusher near Beaumont, Texas, some ninety miles east of the city of Houston. The inscription on a memorial later erected near the site of the discovery immodestly claimed, "On this spot, on the tenth day of the twentieth century, a new era in civilization began." This inscription correctly describes the impact of Spindletop on the Texas Gulf Coast, where unrestricted flush production from this field and others subsequently found in the region began a fundamental economic and environmental transformation. Indeed, the coming of an oil-led economy greatly accelerated the pace of development, while redefining the boundaries of the region itself. The coastal region from southwest of Houston to south of Lake Charles, Louisiana, quickly attracted the capital, expertise, and workforce needed to develop this newfound resource, whose importance in national and international markets made the region a magnet for outside investment. In the process, control of much of the decision making about the region's future flowed into the major oil companies that came to dominate oil production and refining in the region.[12]

Amid the excitement caused by the Spindletop discovery, the upper Texas Gulf Coast followed the pattern of boomtown development common in areas near mineral deposits throughout the western United States. An orgy of oil exploration and production followed Spindletop, with oil specialists from around the nation drawn to the Gulf Coast by the promise of new fields that rivaled the unprecedented size of the Spindletop field, which was far and away the largest yet discovered in the United States. After 1901, the locus of oil production in the United States shifted sharply from the Northeast to the Southwest, heralding a new era in the scale of the oil industry and in the use of oil for fuel. As explorers fanned out in all directions from Beaumont, they quickly found new fields throughout the surrounding sections of the Gulf Coast. Several of these fields moved the center of production closer to Houston, notably the opening in 1905 of the giant Humble field twenty miles north of the city and the Goose Creek field about the same distance to the east, which was discovered in 1908 and developed a decade later.[13]

The rush to develop the area's oil wealth created jobs and economic prosperity, but it also produced an oily mess of pollution. Brian Black's excellent study of the Pennsylvania oil boom of the late nineteenth century provides a useful backdrop for understanding the social and environmental consequences of rapid oil development.[14] But the Gulf Coast oil fields were much larger than those in Pennsylvania, and the rush to pump oil from the ground had even more extreme impacts in Texas than in Pennsylvania. Texas experienced unfettered capitalism squared, as the "normal" cutthroat competition of the era was heightened by conditions peculiar to the oil industry and to the region. Under the rule of capture, the legal doctrine governing oil production in the United States until the 1930s, any leaseholder above any underground oil reservoir had the legal right to pump oil from the field. Thus once oil was discovered, the race was on to acquire a lease over the field and pump as much oil as rapidly as possible. Extreme waste from gross overproduction, considerable evaporation from makeshift storage, dangerous fires, and abundant runoff of oil into surrounding waterways resulted.

Early oil development around Beaumont provided the most extreme portrait of production under the rule of capture, but the development of smaller fields in the region later produced similar waste. Before it could be controlled, the Spindletop gusher shot a spray of oil estimated at 100,000 barrels a day 100 feet above the derrick for nine days. Then hundreds of other wells from surrounding leaseholders quickly tapped the same reservoir. The rush to market left oil covering everything in sight. In the absence of an existing local industry capable of building the wooden or metal storage tanks used in the eastern oil fields, mule teams dug giant earthen pits that held as many as three million barrels of oil. The Texas heat encouraged the evaporation of oil fumes from these lakes of oil into the atmosphere. Massive, deadly fires plagued the southwestern oil fields; several fires destroyed much of Spindletop itself in the first years of its existence. Such extreme conditions were not, however, a matter of pressing public concern. Indeed, the gusher became a symbol of pride and of a bright oil-led future for the region. Operators at times treated excursions of tourists and potential investors to the spectacle of minigushers created by opening valves and allowing oil to spurt freely into the air.[15]

In this era of frenzied production, severe pollution was not limited to the immediate area surrounding the oil fields. Runoff from the fields into surround-

Oil workers at Goose Creek Field, 1910s. Courtesy Houston Metropolitan Research Center, Houston Public Library.

ing streams carried a wave of crude oil into the region's rivers, bays, lakes, and harbors. In this relatively undeveloped region, even serious stream or river pollution by oil went largely unchecked, at least until regional growth brought new claims on the use of these waterways by people and industries outside the oil industry.

Most of the oil produced in the region flowed through pipelines to large refineries built near the new oil fields to process crude oil and ship crude and refined products. Immediately after Spindletop, the Beaumont–Port Arthur area became the home of three major refineries of prominent oil companies: Gulf Oil, Standard Oil of New York (Mobil), and the Texas Company. These plants grew to rank among the largest in the nation throughout the twentieth century. The Houston area began to attract refinery construction in the decades after the opening of the Houston Ship Channel in 1914. With large tracks of inexpensive land, ample supplies of fresh water, abundant natural gas available for fuel, and access to the sea, the land on both sides of the ship channel remained a favored spot for refinery construction and expansion from the 1920s into the 1960s. The refining region between the Houston Ship Channel and Sabine Lake (near Port Arthur) became the location of the largest domestic refinery of most of the major American oil companies. The 1929 *Census of Manufactures* estimated that 19,000 refinery workers resided in Harris County (Houston) and Jefferson County (Port Arthur and Beaumont). By the end of the 1930s, this region alone accounted for about one-third of the nation's refining capacity. The giant industrial plants that refined crude oil became a primary source of industrial jobs and a powerful engine of growth for the upper Texas and Louisiana coasts, with the Houston Ship Channel gradually emerging as the center of this vast refining region.[16]

In the era of the gusher, however, these plants were dirty and dangerous. Oil being heated under pressure coursed through pipes and into stills, with significant leakages throughout these early refineries. The rapid expansion of these plants, the frequent introduction of new technical processes, and the lack of good training for many of the country boys who flocked to work in them set the stage for frequent explosions and fires. The general pattern was to "learn by doing." Thus when large quantities of oil with heavy sulfur content came to the region's refineries from Mexico in the early 1920s, sulfur fumes caused serious problems for workers and even for those living near the plants until modifications in refinery processes limited the escape of emissions high in sulfur content.[17]

Fumes from all phases of refining escaped into the air. Lacking knowledge of such emissions' potential implications for health, refinery owners had little reason to worry about closing up the refining system to minimize discharges. The same was true of oil leaks that contaminated the ground and water. Heavy rains in the region washed much of this leakage into the ground and surrounding waterways. The millions of gallons of fresh water used for cooling did not escape contamination before being pumped back into the ship channel or rivers and lakes near other plants. All in all, these early regional refineries were like old cars that burned and leaked oil, producing billows of smoke from their tailpipes while continuing to move powerfully along the road.

The shipment of crude oil into the refineries and the shipment of crude and refined products out of them remained a messy business in this era. Once the large investment needed to build giant refineries had been made, economic logic dictated the construction of expensive pipelines reaching out hundreds of miles to the major oil fields in the southwestern United States. While transporting hundreds of thousands of barrels of oil each day to the refineries, these pipelines experienced substantial leakage, which was excused at the time as the best that could be done with existing technology. The tankers that moved crude oil and refined goods from the refineries to global markets also spilled very visible pollution into the major waterways and into the Gulf of Mexico. Points of transshipment—from pipeline to storage tank, from tank to refinery, from refinery to loading docks, from docks to tankers—proved especially troublesome.

By the 1920s, oil spillages attracted the attention of other industries that relied on water transport and finally of the federal government.

One result was the first national survey of oil pollution near coastal cities, and this survey noted special problems in the Houston–Port Arthur area. Around the nation, coastal cities faced rising oil pollution due to an eightfold increase in the tonnage of oil shipped by tankers from 1914 to 1922 and the rapid substitution of oil for coal in fueling ships of all sorts. The opening of the Houston Ship Channel in 1914 and the building of refineries on both sides of the channel in the 1920s heightened pollution in the region.[18] In Port Arthur, the growing shipment of products from the giant plants of Texaco and Gulf intensified the pollution of the ship channel through Sabine Lake out into the Gulf of Mexico. The survey of conditions in 1922 and 1923 conducted by the U.S. Bureau of Mines in cooperation with the American Petroleum Institute (API), the primary trade association for the nation's oil industry, found the oil pollution in the coastal waters near Port Arthur and Houston to be the worst in the nation.[19]

By this time, a steady stream of oil-burning tankers visited oil ports along the Houston Ship Channel and in Galveston, Port Arthur, Beaumont, Baton Rouge, and Lake Charles. Galveston suffered severe pollution from oil-contaminated ballast water dumped near shore as tankers and other oil-burning ships prepared to enter the port. The beach at Galveston also suffered regular oil contamination as heavy rains frequently washed oil into the Brazos River from oil fields along its banks. Although the Brazos entered the Gulf some fifty miles southwest of Galveston, the surveyors found that the prevailing current in the Gulf often brought this pollution up to Galveston's beaches within a couple of days. The survey reported significant financial losses to Galveston merchants from closures of the beach due to oil pollution.

Conditions were just as bad along the Houston Ship Channel and the bayous and streams that emptied into it. The survey found that most manufacturers in this area were careless in the handling of oil for fuel. The surveyors noted that the "comparative cheapness of fuel oil and crude renders it less imperative from a financial standpoint to recover wastes here than in the north." This made Houston "among the worst polluted places we visited."[20]

The survey said the same about the Port Arthur area, where the pace of pollution control obviously had not kept up with the increased shipments of oil needed to supply an expanding demand for crude and refined products. So bad had the oil pollution become in Sabine Lake that oil companies had begun to use crews to skim oil off the surface of the lake and return it to the refineries. Oil pollution had forced the once-thriving lumber industry in the area to move up the Sabine River to Orange, Texas.

The report gave numerous examples of the "very elementary stage" of oil pollution control in the Houston region. Any system of pollution control that could not adequately deal with heavy rains was elementary, indeed, in a region noted for its frequent floods. Similarly, in many of the refineries the expansion of separators for removing oil from cooling water and for handling runoff had not kept pace with the growth in refining capacity, another clear sign that pollution control was not taken seriously. Often large spills at transshipment points could have been reduced dramatically simply by taking more care in the transfer of oil. In several places, the report made clear that facilities in other regions tended to be cleaner than those around Houston. After interviewing the public officials charged with maintaining the recently opened Houston Ship Channel, those making the national survey noted the absence of "the spirit of cooperation" needed to clean up oil pollution. After the publication of the Bureau of Mines–API report, oilmen with a strong interest in unfettered development along the ship channel forced the removal of local officials who had pushed for tighter control of pollution in the ship channel.[21]

The politics of pollution control were easy to understand, but difficult to change. The region had tied its economic fate to the oil industry, and those who owned the major companies—as well as the thousands of new workers who found employment in the industry—favored jobs and growth over all else. Oilmen had a strong aversion to government regulation of any kind, but they were hardly alone in their opposition to pollution controls and other regulations. Many of the region's refinery workers had migrated to their new jobs from dirt-poor conditions in nearby farming regions, and a steady paycheck more than compensated for polluted conditions. Local politicians and civic leaders sang the song of boosterism common to the emerging Sun Belt cities, and they repeated the refrain again and again: unrestrained economic expansion would put their city on the national map.

State-level government agencies had some incentives to sound the alarm over oil pollution, but very limited resources with which to work. The state attorney general's office, the Texas Fish and Oyster Commission, and the State Board of Health each had a partial mandate to address aspects of the problems caused by oil pollution. But their budgets were so small and their multiple responsibilities so great that there was little hope of a successful push to clean up the region's water. In this sense, oil pollution in the era of the gusher can be seen as an externality in search of an institution. Until at least the 1960s, no local, state, or federal agency had a clear mandate to make pollution control its top priority. When forced to allocate quite limited funds and personnel to specific issues, all of the existing agencies with limited powers over some part of the pollution problem chose to focus on issues that had a higher priority than pollution control. The State Board of Health, for example, had many more pressing demands on its resources than diseases related to oil pollution.

The politics of pollution control played out somewhat differently in the more developed regions of the nation. The survey of conditions in the early 1920s suggested that in the more-developed Northeast, oil pollution did substantial harm to fishing, to beaches, and to vacation homes on the shore. Those with economic interests threatened by the spread of oil pollution—and their insurance companies—fought back. Such interests led the call for action by the U.S. Congress on new pollution control laws, which in turn led to the national survey of oil pollution in the nation's coastal waters. But antipollution groups lacked the political clout to successfully challenge the oil industry.

By the 1920s, the oil industry had strong incentives to take control of the problem. The price of oil made some measures to stop leakages economically logical, but the drive for greater efficiency also had an engineering component that could go beyond short-term profits. The waste of the era of the gusher went against the grain of the training and experience of good engineers, who gradually moved to clean up the worst pollution in the interest of engineering efficiency. But perhaps the strongest incentive for improved self-regulation in this era was practical politics. Oilmen feared that if they did not clean up the most visible pollution, they would face government regulation. The threat of "outside" regulation was a powerful motivator for oil executives who desired to keep autonomy over their businesses.[22]

This was clear to see in the industry's response to the growing national political debate over pollution in the 1920s. After vigorous debate, in 1924 Congress passed a relatively weak law that regulated only oil pollution in coastal waters. The law contained a stipulation that Congress would reexamine this issue in 1926, with stricter regulatory measures to come in the absence of substantial progress in controlling such pollution. In response to this law, the oil industry kicked into high gear in creating an industry-wide initiative to reduce water pollution from its operations. The API led the way with a wave of studies and new recommended practices for reducing oily discharges into the nation's navigable waters.[23]

The oil industry's campaign for self-regulation proved to be a successful part of a broader campaign to block government regulation of oil pollution. By cleaning up the most visible pollution, the industry reduced the pressure for stricter government regulations. Industry representatives also proved very adept at political lobbying, especially in the southwestern states, where oil had become so economically important. One effective strategy was to appeal to southern legislators to fight federal pollution regulation on the grounds of states' rights, an argument guaranteed to get the attention of influential southern senators strongly committed to protecting their segregated social systems from federal challenges. As a well-organized interest group with clear goals, long-term time horizons, control of the data about their critical industry, and good lawyers and lobbyists, the oil industry proved quite successful in defining the terms of the political debate about oil-related pollution throughout the mid-twentieth century.[24]

Within the Houston region, few had to be lobbied by industry representatives on the issue of oil pollution controls. Oil had been very good to the region in the first decades of the twentieth century. Oil production had brought great personal wealth to "independent" oilmen who now made the city their home and headquarters. It had also attracted to the city the headquarters of several major oil companies. Oil refining was the sturdy backbone of industrial growth, providing the sort of factory jobs that attracted thousands of new workers and their families from the growing rural poverty of the hinterlands. Houston's population increased sevenfold from 1900 to 1930, when almost 300,000 people lived in the city. Times were good, and oil promised that the city would continue to prosper.[25] Amid this oil-induced boom, few Houstonians acknowledged that oil-related pollution might become a serious problem.

Mounting Pollution Under Self-Regulation (1920s–1960s)

From the 1920s through the 1960s, oil-related expansion drove the Houston economy forward, allowing the region to escape the worst of the Great Depression and fueling a sustained boom during and after World War II. Although the region became increasingly dependent on petroleum-based industries, it benefited from significant diversification within its oil-related core. As the giant refining complex continued to grow, it attracted a new generation of petrochemical plants often owned by the major oil companies and built adjacent to their existing refineries. The Houston-based natural gas industry also expanded dramatically with the coming of cross-country transmission from the Southwest to the Northeast. With the growth and the diversification of oil-related activities came more complex and diverse types of pollution.[26]

As the region prospered, Houston moved steadily up the ranks of the nation's largest cities. By 1970, the city's population was more than one million, with perhaps that many more inhabitants in a metropolitan area that spread in all directions along the flat coastal plain. An ever-expanding highway system held the city and its suburbs together, and automobile traffic steadily grew as more and more commuters drove more and more miles. By the 1960s, mounting air pollution from automobiles began to attract political attention as a serious and highly visible problem, but local officials had a most difficult time finding the political will and support to implement effective controls.

This was also true of other forms of oil-related pollution. Self-regulation by industry, put firmly in place in the 1920s, remained the norm until well into the 1960s. This generally meant that industry defined the primary approaches to pollution control, as well as the level of commitment. Various levels of government monitored this activity, cooperating with industry to assure more efficient handling of oil to reduce its discharge as pollution.[27]

This approach exhibited clear strengths in identifying waste and then reducing it with more efficient and less-polluting control systems and technology, but it also had inherent limits that became more apparent over time. One of the oil industry's strongest incentives to regulate itself was its desire to avoid coercive government regulations; when the threat of such regulation waned, so did the industry's commitment to stronger controls. A second weakness lay at the heart of the process of technological innovation in this era. The oil industry had very strong economic incentives to invest in new technologies of production, but few such incentives to invest in pollution controls. From the 1920s through the 1960s, techniques of pollution control failed to keep pace with techniques of production. Indeed, as practitioners of self-regulation in the industry made progress in cleaning up basic processes in the production, refining, and transportation of oil, new and more complex problems emerged as the scale and technological sophistication of oil-related activities increased.

In the refining of petroleum and the use of refined products, one telling example of this process was the introduction of tetraethyl lead (TEL) as the additive of choice for gasoline in the 1920s. The search for an additive capable of improving fuel economy reflected the perception for a brief time in the 1920s that the nation faced a shortage of crude oil. One way to address this potential shortage was to develop advanced refining technology that produced more gasoline from each barrel of crude oil; another was to develop additives to improve the burning of gasoline within cars. The industry took both paths with great short-term success. In the early 1920s, the Ethyl Corporation, a joint venture by Standard Oil of New Jersey, General Motors, and DuPont, announced a dramatic breakthrough in additive technology. An Ethyl Corporation representative hailed TEL as an "apparent gift of God" that "at small cost adds fifty percent to gasoline mileage." As the industry went forward with plans to produce and market the new additive, however, health specialists in government agencies and universities voiced fears of the possible environmental impact of an air-borne form of lead pollution that might be produced by its use.

Thus the mid-1920s witnessed an often fierce debate among health officials and executives of oil, chemical, and automobile companies about the possible regulation of TEL. Doctors who had long studied the illnesses associated with lead-based paint feared that long-term health problems might be caused by the widespread use of TEL. Those in industry countered by producing test results from studies of small samples of workers with limited exposure to TEL to argue that there were no observable health problems. When a highly publicized accident at an early TEL manufacturing plant resulted in the deaths of five workers in 1924, industry countered by instituting new safety procedures and closing up the manufacturing process to protect workers from direct exposure to TEL. The expert opinion of industry specialists gradually overcame the doubts and the resources of their opponents, and by the 1930s TEL became the most common gasoline additive.

By this time, the perceived oil shortage had been replaced by a glut of crude oil, and TEL was used to fuel heavier and more powerful cars, not to increase gas mileage. Despite a U.S. Public Health Service study in 1926 that called for continued monitoring for possible long-term health impacts of TEL, no such monitoring took place. Indeed, a Public Health Service study conducted in 1959 lamented, "It is regrettable that the investigations recommended by the Surgeon General's Committee in 1926 were not carried through by the Public Health Service. If data were now available on body lead burdens, with 1926 as a baseline, a more objective decision would have been possible."[28] It might not have mattered, because the tests used to study possible long-term health impacts of air-borne lead could not readily identify such impacts until the 1960s and 1970s, when breakthroughs in the measurement of lead in the human body revealed that high concentrations of air-borne lead from TEL could harm the health and even lower the IQ of people who lived near highly traveled urban highways. The mandatory use of catalytic converters in the United States in the mid-1970s led to the banning of leaded gasoline for new cars, since the lead destroyed the effectiveness of the converters. Then new and more thorough studies of the health implications of TEL led to the total ban of its use in the United States, but not until after it had been used in most cars for almost forty years.[29]

Such changes in refining technology greatly affected the Houston region, which remained the center of one of the largest refining regions in the world from the 1920s through the present. The broad area from New Orleans to the Houston area grew to supply more than 35 percent of the nation's refining capacity by 1970, with the Houston Ship Channel and Port Arthur–Beaumont accounting for about two-thirds of this figure.[30] The giant plants in this region applied the latest in new refining technology, which became increasingly sophisticated in the post–World War II era. Each new technique to improve the yield and the quality of refined products introduced new questions about the resulting emissions. Although the refineries became cleaner, more efficient, and more concerned with worker safety under the self-regulation of this era, the expansion of production and the application of new technology to production processes remained the focus of attention, with much less regard for innovations that contributed to better monitoring or greater reductions in emissions.

This also was true for the large plants that sprang up near the refineries along the Texas and Louisiana Gulf Coast to supply the bulk of the nation's rapidly expanding petrochemical production during and after World War II. Located on the Gulf Coast to take advantage of the feedstock from refineries, the abundance of natural gas, and the ready access to pipeline and water shipment, these petrochemical plants became the fastest-growing industrial facilities in the region. Combined with the refineries, they attracted tens of thousands of workers to the region while producing an array of new, petroleum-based products that found rapidly expanding markets in postwar America.

In retrospect, it is clear that this first generation of petrochemical plants held many new and poorly understood risks to human health. Those who worked in these early petrochemical plants before, during, and after World War II had secure, high-paying jobs in comparison to most other industrial jobs in the region. They also unknowingly served as human guinea pigs of sorts, with prolonged exposures over decades to largely unregulated discharges of many new, poorly understood, and potentially harmful emissions. The production of the butadiene needed to manufacture synthetic rubber in plants built in the region during and after World War II, for example, also produced emissions later linked to increased risks of leukemia. Benzene emissions, pollutants from vinyl chloride, and other potentially harmful emissions from the new plants were largely unregulated until the 1970s.[31]

Such emissions presented tricky problems for industry. Self-regulation that hoped to be effective over the long term would have required careful monitoring of the possible health impacts of pollutants from new technical processes used to produce new petrochemical products. But industry feared that the results of such studies might be used to justify stricter government regulations. Thus the safest approach for industry was that taken in the TEL case—quick studies with the promise of long-term monitoring. As long as government could not or would not assert an independent will to regulate the health impacts of such emissions, the logic of self-regulation as practiced in this era dictated a short-term, profit-oriented approach instead of a long-term, health-oriented one.

As early as 1922, British officials concerned about a high cancer rate among workers in the Scottish shale oil industry had written to the U.S. Bureau of Mines seeking any available information about the incidence of cancer among U.S. refinery workers. When the bureau forwarded the request to the medical director at Standard of New Jersey, which operated one refinery in Baytown near Houston and one in Baton Rouge, Louisiana, he responded that the cancer rate among refinery workers was probably lower than that among the general population because of the company's rigorous physical exams of new employees. He did, however, add that "this does not apply of course to the older men . . . and it is among this latter group that we have found the few cases that [we have] had under treatment." When the bureau wrote back asking for further clarification about these older employees, the company doctor abruptly ended the discussion, asserting, "I have known of no case of cancer that could be attributed, even remotely, to exposure to paraffin or petroleum products."[32] For the next half a century, the industry generally took this "don't ask, don't tell" approach toward potential health problems associated with emissions from refineries and petrochemical plants.

In this era, companies in the oil and petrochemical industries applied the same general approach to the dumping of toxic wastes. Unwanted by-products from production processes could often be disposed of by contactors at dump sites far removed from residential areas. But "far removed" changed over time, as Houston's suburbs stretched out into the coastal plain. Decades after the dump sites had been used for toxic wastes, the city might encompass them, raising problems for those who at times unknowingly moved into neighborhoods near abandoned toxic waste sites.

In several areas, self-regulation proved more successful. To its credit, the industry made good progress in addressing the major sources of pollution inherited from the era of the gusher. In the refineries, oil fields, and pipeline and shipping terminals, this meant "closing up" the system to prevent the escape of oil. The major companies also learned to take much greater care in preventing oil from refineries and shipping facilities from washing off during heavy rains. They made use of industry trade associations such as the API to define and promulgate best practices for the industry in limiting pollution from a variety of aspects of their operations. Although the API stopped short of agreeing to serve as an "industry policeman" with coercive authority to inspect facilities and force those who did not comply with best practices to do so, it did build a solid reputation within the industry as an organization that could help companies help themselves on the difficult problems presented by pollution.[33]

As the industry made progress in cleaning up its operations, it faced new challenges such as pollution from offshore drilling and production. The industry's initial move "offshore" involved drilling in lakes and rivers. In the Houston area, production in the years between 1910 and 1920 from the Goose Creek field east of the city required Humble Oil and other companies to learn the best methods for preventing oil from escaping into the water. The same conditions prevailed in the 1930s in drilling in Galveston Bay and off of a trestle into the Gulf of Mexico near High Island. After World War II, the industry moved aggressively out into the Gulf of Mexico.[34] Although most of the major finds were in Louisiana waters, sufficient production came from wells offshore the northeast coast of Texas to threaten area beaches with severe oil pollution. This was a demanding environment, complete with hurricanes and the realities of blowouts and accidents that could place workers in life-threatening situations while presenting new challenges for the control of pollution. In the early decades of offshore development, area residents came to accept clots of oil on local beaches as a necessary evil. As with the smell of the air in the refinery towns, such pollution of beaches was either shrugged off by most as inevitable or accepted as a symbol of oil-led prosperity. Oil pollution more directly threatened the prosperity of those who made their living fishing and shrimping and harvesting oysters in the Gulf and Galveston Bay, and they led largely unsuccessful protests against the harmful impact of oil pollution on their industries.

A variety of government agencies found themselves in the middle of the growing tensions between the industries that produced oil pollution and those directly harmed by it. The U.S. Coast Guard had responsibilities to regulate pollution from offshore oil production and shipment, and it entered into a generally cooperative partnership with offshore companies to monitor and control oil pollution. Several different state agencies in Texas and Louisiana that had jurisdiction over issues involving fish and game, pollution control, and health also had occasion to monitor oil pollution and to seek to mediate disputes between the oil industry and other industries. They could do little else, given their lack of resources and technical expertise.[35] Even when officials effectively addressed specific problems from pollution, other new and often more difficult problems were arising from the growth in scale of transportation and production of petrochemicals and the applications of new technologies in the production process.

One energy-related development that had a positive impact on regional pollution was the coming of natural gas as a primary fuel source in the region. Natural gas was found in abundance near Houston in the first decades of the twentieth century. Initially, it was seen as a nuisance and eliminated through flaring (burning off into the atmosphere), but in the 1920s the completion of pipelines into the city from surrounding gas fields allowed this gas to be used in and around the city. Natural gas quickly became the fuel of choice for the region's major industrial plants, including those plants that generated electricity. The use of natural gas to produce electricity made a significant contribution to cleaner air, particularly after the coming of air-conditioning greatly expanded the demand for electricity. Natural gas was cleaner burning than oil, which itself was cleaner burning than the coal it had displaced. In the postwar years, natural gas was not yet touted as a "green fuel." Its use grew because it was cheaper and easier to use than other fuels, not because it was cleaner burning. Yet acknowledged or not, easy access to abundant natural gas for fuel reduced pollution that would have been produced by coal or oil, the available competing sources of fuel.[36]

The region greatly needed any such reductions it could find, since rapid industrialization and suburbanization combined in midcentury to create mounting pollution. The most obvious result of suburban growth was the increased use of the automobile. The number of cars, the miles driven, and the network of paved roads grew steadily over time, as did the amount of pollution coming out of the tailpipes of cars. Little thought was given to auto-related pollution until the 1950s, when Houston officials followed the debate in the Los Angeles area over the sources of smog. Both regions had growing traffic and a large cluster of oil refineries. Both had air pollution that could often be smelled and at times seen. But Californians placed a greater emphasis on air pollution control than did Texans, and most of the early debate about the causes and effects of smog remained in California.[37]

The era of self-regulation by the oil industry ended earlier in California than in Texas for a variety of reasons, including the physical beauty of California, its more diversified economy, and its two-party political system. The state of Texas and the Houston region had what might be best described as "oil-friendly" politics. The oil industry was well organized at the local, state, and national levels, with expert lawyers and lobbyists and access to authoritative data on most industry-related issues. In the one-party South, the ruling Democratic Party held firmly to the concept of states' rights, which could be used to oppose new federal government powers over pollution as well as civil rights. Until the 1960s, few well-organized public interest groups or labor unions registered effective opposition to oil-related pollution. But the political clout of "big oil" in Texas went far beyond interest-group politics or campaign contributions. In this state—and particularly in and around Houston—the oil industry was the largest employer, the most significant taxpayer, and the fundamental engine of growth. In the politics of this era, jobs trumped cleaner air and water.

By cleaning up the most visible pollution of the 1920s and putting in place cleaner operating procedures, the industry contained serious challenges on the pollution front until the 1960s. The external political environment helped deflect attention from pollution. The Great Depression of the 1930s focused politics on economic recovery while concentrating on cleaning up municipal waste, which was a good target for public works programs. The progress made by several levels of government in treating municipal wastes helped clean the nation's waterways. According to one waste-reduction specialist in the oil industry, such progress made oil pollution in rivers and streams more visible, assuring that it would become a bigger target for those seeking stricter pollution controls.

World War II suspended concern for pollution control, and the postwar boom in the nation and in Houston encouraged returning veterans to move ahead in pursuit of their personal ambitions. Almost half a century passed from the World War I era until the strong environmental movement of the late 1960s, and in this long period the oil industry retained central control of the ability to define and address the issues raised by petroleum-related pollution. This not only reflected the industry's success in removing much visible pollution but also the combination of the industry's strong political clout and the political focus on other, more pressing issues. It also reflected the lack of well-organized groups capable of asserting the demand for stronger pollution controls in Texas and Houston.

Yet by the 1950s, there was a growing sense that oil pollution was reaching levels unacceptable to some outside the oil industry. A few government officials at the local level began to make at least an implicit argument that oil-related pollution was a potential barrier to regional growth. Walter A. Quebedeaux, who in 1954 became the head of the newly created air and water quality control section of the Harris County Health Department, was one of the first officials to tackle the problems head on. Quebedeaux had a good background for this job, having previously served as air and stream pollution director for the Champion Paper Company. In a series of initiatives, he directly and very publicly challenged the oil industry to clean up its pollution in the Houston Ship Channel. Despite his lack of sufficient staff and funding to effectively challenge the oil and petrochemical industries, he did not go down without a fight. By collecting data on oil and water pollution and publicizing the results in local media, he grabbed the attention of many in the Houston area. For a time in the 1960s, the major union for workers in the refineries and petrochemical plants, the Oil, Chemical, and Atomic Workers (OCAW), also conducted a campaign aimed at addressing issues of air emissions as a threat to the health and safety of workers in the plants and of those who lived near the plants.[38]

The era of relatively quiet acceptance of oil pollution was ending. Self-regulation by individual companies could not identify and address the cumulative impact of the expansion of the refining complex and the growing use of automobiles in the region. Nor did self-regulation tend to look aggressively for new problems associated with new types of emissions from technical advances in refining or from the new plants that produced a myriad of complex petrochemical products. W. B. Hart, one of the leading industry voices for self-regulation in the mid-twentieth century, acknowledged a final weakness of this approach. Self-regulation was a strategy to remove the prospect of government "compulsion"; yet the industry's success in using political lobbying to contain stricter government regulation removed some of its incentive to aggressively pursue efforts to monitor and control its emissions.[39] Not until mounting pollution in the 1950s and 1960s focused intense renewed public scrutiny on oil pollution did the industry have strong political incentive to respond with its own new wave of more effective regulations. By then it was too late for industry to retain control of this increasingly charged issue.

The need for change in the Houston area was symbolized by the return in the 1960s of the very visible water pollution of the 1920s and by the city's first serious air pollution "crisis" in the late 1960s. In 1967, federal investigators called the problem with pollution in the Houston Ship Channel "overwhelming," labeling conditions there as "by far the worst example of water pollution observed . . . in Texas." A survey of the region's air quality carried out by local authorities in 1965 concluded that "the most significant change [since the city's original survey in 1956–1957] . . . is the occurrence of the type of chemical reactions in the atmosphere which are characteristic of the well-known Los Angeles smog situation."[40]

Those living in the region did not need this report to note the coming of a smog problem to their city; they had only to look outside on an increasing number of days each year to see the fouled air. A series of very bad air days in the late 1960s and early 1970s brought indignant headlines in local newspapers, complete with photos of the smog-filled air hovering over the city. Television and newspaper coverage of pollution heightened awareness of the problem; well-organized environmental activist groups pointed to the need for change; and the growing wealth of the nation meant that more people could look beyond short-term economic considerations to issues involving the quality of life. But a fundamental reason for the coming of an era of stronger environmental regulations should not be ignored. Pollution was getting worse—especially highly visible air and water pollution.[41]

Yet even considering the mounting and increasingly visible oil pollution in the region, it is doubtful if significant new regulations would have been forthcoming without the national environmental movement. The federal government finally responded to the political pressures for stricter pollution controls with a wave of new laws in the 1960s and 1970s. Ultimately, the Gulf Coast refining region did not abandon its long-standing commitment to jobs over environmental quality; it was forced to assert stronger controls over pollution by the intervention of regulators from outside the region.[42]

Federal Regulation and Regional Pollution (1960s–2000)

The transition from industry self-regulation of pollution to command and control regulations, in which the federal government enforced strict new standards on the discharge of pollutants, moved forward quickly after the mid-1960s. The passage of strong new laws, the implementation of the requirement for environmental impact statements, and the creation of the federal Environmental Protection Agency (EPA) dramatically altered the legal and environmental framework of pollution control. Since the operations of oil refineries and petrochemical plants and automobile emissions attracted special scrutiny in the new generation of environmental regulations, the region witnessed considerable tensions between its traditional commitment to economic growth and the federal government's strengthened efforts to clean up air and water pollution.[43]

The post–World War II boom in the regional economy intensified after 1973, when the assertion of power by the Organization of Petroleum Exporting Countries (OPEC) over both the level of oil production and the price of oil led to a sudden quadrupling of oil prices. At the end of the decade, the hostage crisis in Iran then doubled prices. Although higher energy prices harmed much of the rest of the American economy, the nation's energy capital enjoyed an exhilarating period of prosperity from the early 1970s through the early 1980s. But as suddenly as the energy roller coaster had climbed up to new heights, it plunged down toward earth as real oil prices fell in the mid-1980s back to near pre-1973 levels. Houston's economy staggered, with conditions resembling those in the city in the early years of the Great Depression. As it slowly recovered, the region gradually moved toward greater diversification of its economy, with space, medicine, high-tech industries, and education taking up some of the slack created by the devastating blow to the oil industry. Even after oil-related activities rebounded to reclaim a central role in the economy, representatives of other industries also claimed seats at the table in the ongoing discussion about the direction the city should take, including the need to find more effective ways to control oil-related pollution.

From the mid-1960s forward, however, the table itself moved to Washington, D.C., as the federal government dramatically altered the process of environmental regulation. National politics shaped the process of change, driven by growing discontent with the results of industry-led initiatives to control pollution. The oil and automobile industries came in for harsh criticism in the face of mounting and highly visible smog, which became a symbol of the need to do something different. The oil industry also became the focus of political discontent in 1969, when the much-publicized Santa Barbara oil spill became a symbol of the industry's inability to control pollution. Such events and issues fueled the growth of increasingly active public interest groups that pushed for stronger controls and a new way to regulate pollution. Although the national environmental movement initially was not particularly active or influential in Texas, it ultimately had a significant impact on the region as it helped forced Congress to rewrite the nation's fundamental laws on pollution control.[44]

Stronger national legislation poured forth from the mid-1960s through the 1970s, with another surge of legislation after the Exxon Valdez oil spill in 1989.[45] In general, these laws embodied an approach aptly characterized as "command and control." Although some room remained for industry influence in the writing and in the implementation of these regulations, the new laws embodied attitudes and procedures markedly different than those of the long era of self-regulation. Indeed, this wave of environmental regulations contained an element of the affirmative action impulse that characterized the civil rights laws of the same era. The regulators who enforced the new laws had a difficult dual task: to clean up pollution inherited from the past while also implementing new standards and processes capable of reducing the contamination of the air, water, and land in the present and future.

Nowhere was the need for such affirmative action clearer than in dealing with potential deadly pollution from toxic waste sites. (For a case study of this process, see Kimberly Youngblood's essay in this volume.) Environmental regulators at all levels of government faced daunting challenges under the new Superfund laws. Toxic wastes presented particularly vexing problems in the Houston area, with its high concentration of refineries and petrochemical plants. Those who sought to clean up toxic waste disposal sites in the region had to grapple with the historical problems caused by the rapid spread of the city out into suburban areas previously used as dump sites while at the same time creating new procedures to regulate the ongoing disposal of wastes from regional plants. This process often pitted local, state, and federal regulators against each other while also calling forth waves of lawsuits aimed at deflecting blame for past abuses.

Although a measure of discretion for the states remained in some of the new laws, most relied heavily on national standards defined in federal legislation and interpreted and enforced by the EPA, a strong new federal regulatory agency created in 1970. Command and control generally involved strict new standards or requirements backed by harsh punishment for those who did not meet the standards. Thus when Congress originally decided to mandate dramatic reductions in automobile emissions in 1970, it passed a law with a tight timetable for phasing in cleaner-burning car engines. This timetable dictated the addition of catalytic converters to new automobiles. Carmakers faced prohibitive fines if they did not or could not meet the schedule to comply with these new standards. As with many of the command and control regulations, those charged with enforcing the law retreated from the original timetable for compliance by granting extensions, but they did not back away from the final goal of dramatic and rapid reductions in harmful emissions.

Air and water standards, regulations of toxic wastes, and requirements for safer oil tankers all contained elements of the command and control approach. The political message to the oil and petrochemical industries was clear: you had a chance to take care of these problems, and you failed to do so. Now angry and well-organized activists demanded immediate action, with little regard for the cost or the inconvenience to industry.

The oil-refining industry was particularly hard hit by these regulations. Several of the major regulatory initiatives issued by the federal government required substantial investments by refiners. Especially demanding were two laws aimed at reducing harmful auto emissions: the requirement for catalytic converters after 1975 and the requirement for reformulated gasoline after 1990.

Both of these government mandates forced petroleum refiners to make fundamental adjustments in the refining process. The use of catalytic converters required the phasing out of TEL from gasoline over a relatively short period. This meant that refiners had to find new ways to create the high octane levels in gasoline used by automobiles since the 1920s. Investment dollars previously available for other endeavors now went toward meeting the government-mandated challenge of lead removal on a short timetable.[46]

The process repeated itself in the 1990s, when amendments to the Clean Air Act in 1990 mandated cleaner-burning gasoline for regions that did not meet the EPA's ambient air standards. This meant that Houston and other major cities in America would require new types of gasoline "reformulated" to burn more completely, leaving fewer, cleaner emissions. Again, the law mandated the phasing in of such gasoline on a relatively short timetable, forcing refiners to make large investments in the fundamental technology of petroleum refining. Some companies based in the region also undertook to supply MTBE, the gasoline additive of choice for refiners seeking to comply with the new law. These companies faced additional difficulties after questions about the environmental impact of leaks of MTBE into the water table demanded further investments while entangling industry in litigation to sort out responsibilities for absorbing the variety of costs associated with the detour into MTBE's production and use in reformulated gasoline.[47]

As the refining industry adapted to such government mandates, it also had to meet stricter standards for air and water emissions. Best practices became much more demanding in the decades after the 1960s, as government standards forced the industry to move beyond what it had done previously under self-regulation. Meeting the new standards required additional investments while creating uncertainties for the refining industry. This encouraged much greater caution in building new plants, as did the difficulties of siting new plants under the requirement of an environmental impact statement open to public input.

New environmental constraints—combined with economic factors such as the chronic overcapacity in refining and relatively low profit margins in refining compared to those in other parts of the operations of vertically integrated oil companies—effectively stopped the construction of new refineries in the United States after the mid-1970s.[48] The refining region surrounding Houston thus faced a fundamental transition from boom years of growth in capacity through new plant construction to what might be called "expansion in place," which involved expanding and improving existing plants by installing new processes in old plants. During the late twentieth century, refineries, like most other large manufacturing plants, also underwent a revolution in computerization, with new applications of computing power to manage and monitor the refining process more efficiently, in the process allowing for better measurement and control of potential emissions. Such initiatives were expensive, as were the development and application of new refining technologies, and they competed within companies for funds required to meet the demands of new environmental laws.

As refiners struggled to remain competitive and innovative while meeting new environmental requirements, they also began to face growing questions about the long-term health impacts of their refineries' emissions. Such questions had been raised since at least the 1920s, and the basic answer of the refining and petrochemical industries remained roughly the same throughout the century: No medical research establishes definitive links between our emissions and significant health problems for our workers or those living near our plants. Until such research can sort through the complexities of the varied causes, including smoking, of the many forms of cancer and of severe respiratory diseases such as asthma and show direct links between the operations of our plants and public health, we will continue to try to be good neighbors and employers by monitoring our emissions and the health of our employees. In the late twentieth century, union representatives, public interest groups, newspapers, and even local officials in Harris County (Houston) and Jefferson County (Port Arthur and Beaumont) periodically demanded actions to address concerns that the emissions of benzene, butadiene, and other toxics produced along with refined goods and petrochemicals might increase the incidence of some forms of cancer. But government at all levels proved reluctant to venture into the gray area of the causes of variations in the cancer rates. By default, such issues seem likely to be addressed in future class-action court cases.[49]

More immediately troubling to the industry and the government during the years after the energy crisis of the 1970s, however, was the growing shipment of oil around the world, as the industrialized nations became more dependent on ever-larger shipments of oil in tankers. The extraordinary growth in tanker shipments placed a premium on preventing giant oil spills, and the political fallout from the *Exxon Valdez* oil spill in Prince William Sound, Alaska, in April 1989 forced all involved to place a greater emphasis on spill prevention and containment. The Oil Pollution Act of 1990 responded to the *Exxon Valdez* spill with new mandates phasing in double-hulled tankers for all oil shipments to U.S. ports and establishing strong incentives for industry groups to organize and operate emergency response teams capable of reacting quickly to regional oil spills. Although such measures have not and cannot eliminate major oil spills from the region's waters, they have encouraged increased safeguards against spills and better industry preparation and training to contain such spills. As with many other environmental laws passed since the 1960s, this new law, which greatly affected a region with large tanker shipments in and out of its ports, came in response to events outside the region and was put in place at the federal level.[50]

The greatest tension between the federal government and state and local officials in this era came over the controversial issues of smog and ozone depletion. In response to problems within the Houston area in attaining air pollution standards set forth by the federal government, federal regulators urged serious restrictions that fundamentally challenged existing attitudes and institutions in the region. The effort to contain air pollution brought federal, state, and local governments into conflict while also raising questions about the basic patterns of gasoline use in sprawling postwar Houston.

The number of cars in the metropolitan area grew spectacularly from the 1920s forward, with individual cars often carrying one passenger and fueled by inexpensive gasoline serving as the region's de facto "mass transit" system. Throughout its existence, the city steadily reached out and encompassed much of the surrounding countryside, and by the 1970s urban sprawl had produced a giant metropolitan area that relied on an ever-expanding system of highways, whose construction had become a central part of the regional economy. By the late twentieth century, the practical definition of "Houston" had come to mean the very broad system of city and suburbs that stretched out for thirty to forty miles in every direction. Such sprawl had played an important role in Houston's sustained growth. The availability of jobs attracted new immigrants, while relatively inexpensive housing made possible by ever-expanding suburbs kept housing costs low. Inexpensive gasoline and an excellent system of roads completed the equation for the region's expansion into a major city.[51]

When the implications of this pattern of home building and commuting for the region's air became impossible to ignore, federal authorities proposed drastic measures that challenged the historical pattern of regional commuting. In the early 1970s, after it became clear that Houston and other cities could not meet recently promulgated air standards, federal regulators called for a variety of measures to reduce gasoline consumption. Their seven-point plan in 1973 included the creation of carpool lanes on major highways and streets, a ban on construction of new parking facilities, possible gas rationing, and plans to force a 10 percent reduction in miles driven. Even the introduction of such measures for discussion called forth a wave of angry responses from political, civic, and industrial leaders in the region, who claimed that such restrictive policies would have a devastating impact on the area economy. In the face of such strong criticisms, which seemed to reflect the attitudes of many Houstonians, the EPA backed away from its proposals in 1975. A headline in the *Houston Chronicle* on May 29, 1975, captured the sense of a battle that was far from over: "EPA Apparently Abandoning Its Auto Cut Plans Here."[52]

The pattern developed in this initial skirmish repeated itself in several subsequent battles over the proper response to recurring problems with the region's air pollution. In the late 1970s and the early 1990s, the EPA once again proposed various restrictions on driving after the region failed to meet existing air quality standards. Each time, many of the region's political and civic leaders manned the barricades against the EPA, making politically effective arguments that its policies would be ineffective in Houston. The fundamental point was simple: the region's well-being should not be sacrificed to air pollution controls so strict that they would debilitate the regional economy. As one concession to the new realities of severe ozone problems, high-occupancy vehicle lanes began to be constructed in the center of the region's freeways, while existing freeways continued to be widened and new ones extended into the suburbs. The best indication of the attitudes of Houston's commuters toward these EPA initiatives was their continued movement out into the suburbs, despite air pollution and growing gasoline prices.

At the same time, civic and industrial leaders, two categories that consistently overlapped, made strong and repeated arguments about the need to avoid new regulations that imposed unnecessary burdens on the region's industrial core. Tensions between industry and government regulators remained high throughout much of the United States in the years after the 1960s, but they were particularly pronounced in a region containing a large share of the nation's petroleum refining and petrochemical production. Command and control regulations placed demanding and expensive new burdens on these industries. Their leaders perceived themselves as under siege, and they fought back with all the considerable weapons at their disposal. When political lobbying could not temper federal regulations, they lobbied more successfully at the state and local levels to moderate the implementation of standards. Political influence, good lawyers, control of technology and basic data about their industries, and a long-term time horizon all helped them establish a measure of elbow room with regulators. They lobbied for relief from new environmental laws, arguing for looser enforcement of environmental standards in the interest of job creation and energy independence. They made some headway, especially during periods of intense political concern about energy prices and supplies. All in all, however, they had only limited success in moderating the early wave of regulations or preventing the passage of new, stricter regulations of pollution.

Such regulatory give-and-take between energy industry groups and environmental groups remained a source of often intense political dispute. From the point of view of those who favored stronger pollution controls, industry seemed to own the Texas state government, dictating a series of grandfather clauses and exceptions that seemed to gut effective enforcement of much-needed restrictions on emissions. Houston's civic leaders, on the other hand, usually joined the effort to block stronger regulations. These leaders, including many representatives of the oil and petrochemical industries, believed that vital regional industries were too often made whipping boys for broader problems, including the difficulty of enforcing regulations that limited "lifestyle choices" such as commuting distances. Industry leaders knew firsthand that meeting environmental regulations was very costly, and they voiced fears that the high price of compliance was making them less competitive in global markets. When they spoke to each other and not to a broader audience, a common complaint was that people who did not understand the inner workings of markets and the complex technology of their industries too often made unrealistic demands that unnecessarily restricted their capacity to find cheaper, more efficient solutions.

Such conflicting interpretations of reality proved particularly sharp in one of the most controversial areas of pollution control, the effort to understand the long-term health costs of pollution and to fashion public policies that could balance these costs against other competing social and economic demands. As early as the 1970s, pioneering health studies suggested that the region had an unusually high concentration of certain cancers related to environmental causes. During and after the 1990s, additional studies began to receive greater publicity in the region's media, leading to heightened fears by those who worked in or lived around the industrial complex that remained an important regional employer. Those who suffered from respiratory diseases such as asthma did not have to be warned of the dangers of air pollution; they felt its sting when they exercised outside or even worked in their gardens. Those who contracted various forms of cancer looked with suspicion at a variety of chemicals spewing forth from local industrial plants. The debate over environmental causes of cancer could not produce definitive answers, since the existence of many possible causes of "excess" cancer rates made it impossible to pinpoint the impact of industrial pollution. Individual critics of industry could only fall back on personal observations of people all around them who had certain types of cancer. Spokespeople for the industries could not cite studies that "proved" the lack of links between their activities and the region's high cancer rates, but they could argue that no studies conclusively proved the counterargument.[53]

Amid such uncertainties, the companies that operated the refineries and petrochemical plants found several ways to limit the impact of such debates on their current and future operations. One approach was to buy out residential neighborhoods near their plants, thus removing some of the worst complaints by relocating the people who made them. This could be expensive, but many companies found the resulting gains in public relations worth the price.

Broader efforts by some of the region's major companies to take a more proactive approach to pollution control came to be labeled "green oil." All of these companies did not, of course, take the same stance on these issues, but several of the major international oil and chemical companies with large plants in the region led the way in the search for changes in behavior that might gain for them a better reputation on environmental issues, thereby perhaps giving them a louder voice in the debates over pollution control. Green policies became especially apparent in companies such as Amoco (absorbed into BP in 1998), which operated a large refinery and several large petrochemical plants in the region. In chemical production, Amoco joined a more general "green crusade," Responsible Care, which sought to find solutions to safety and health problems associated with its industry before being forced to do so by government regulations. Included were initiatives to identify and correct problems and to become better, more communicative neighbors in areas around their plants. This process became more pressing within the industry after the 1980s, when the federal government began to publish reports on the quantities of potentially dangerous chemicals released into the air by manufacturing plants. Companies such as Amoco responded by making the reductions of these numbers a matter of corporate strategy; environmental activists used the same numbers to rally political sentiment for stronger government enforcement of regulations.[54]

In historical perspective, the movement toward greener policies by some regional companies can be seen as an effort to reassert a measure of self-regulation by taking greater responsibility in improving the process of pollution control in their own operations. Unlike the process of self-regulation before the 1960s, however, such recent efforts go forward under the umbrella of strong government regulations and under the gaze of a highly skeptical public. The green initiatives by individual companies and by industry groups are a part of a broader change in business opinion that has emerged in Houston only in the very recent past.

In the early twenty-first century, the primary organization for the region's business community, the Greater Houston Partnership, has begun to voice greater concern for improved pollution control as an important part of an overall strategy for making Houston more attractive to new businesses. As discussed in Robert Fisher's essay in this volume, this new stance can be interpreted in several different ways, but it should not be ignored. The partnership stands at the end of a long line of Houston civic and business leaders who have pushed hard to create what they consider a "healthy business climate" that encourages regional economic growth. These individuals are the descendents of the people who first brought railroads to Houston, who helped finance the Houston Ship Channel, and who found ways to attract the Lyndon B. Johnson Space Center to the Houston area. It is good news when such an organization becomes more aggressive in searching for ways to improve the quality of life in Houston by advocating the reduction of pollution. This does not mean, of course, that other Houstonians will necessarily agree with the pace and timing of changes advocated by the Greater Houston Partnership, but it perhaps heralds a shift in the focus of debate over air pollution controls.

This change in business leadership's traditional view that stricter pollution controls might harm the area's economy rests on the simple proposition that severe pollution threatens the quality of life in the region, which might in turn block future business expansion. It is possible that a new consensus could emerge in the region that cleaner air and water further the long-term commitment to economic growth—that traditional boosterism might in the future begin to include the assertion that Houston is both the city of economic opportunity and a city committed to cleaning up pollution.

Conclusion

This restatement of the traditional argument about the possible impact of pollution on jobs has particular significance for the nation's oil capital. It reflects in part the changing reality that industries other than oil and petrochemicals now have a greater presence in the region. But it also reflects changes within the oil-related complex that has dominated the region's economy. A wave of mergers over the last twenty years has produced a greater dominance of the giant companies that historically have been most capable of absorbing the new costs of pollution control. These companies have almost forty years of experience in complying with command and control regulations, and they are increasingly eager to move beyond the old tensions on these divisive issues and concentrate more fully on the broader issues of global competition. The top priority of the general population in the region no doubt remains jobs and economic security, but a generation of improvements in pollution control has raised the bar on the level of pollution that is socially and politically acceptable.

Despite somewhat cleaner air and water, the struggle for cleaner air and water is far from over in Houston. Sustained growth in an ever-broader Houston metropolitan area means that pollution control in the region is a moving target. As progress is made in reducing pollution from individual automobiles or industrial plants, the addition of more cars and more industrial output creates more pollution in more diverse forms. Ongoing efforts to expand the region's freeway system illustrate this difficult dilemma. Once completed, more lanes of freeway inevitably carry more cars farther and farther out from the city, encouraging urban sprawl. Congestion ultimately returns, bringing with it additional pollution and renewed calls for stricter controls on auto emissions and for more freeways.

Although the regional economy continues to diversify, it remains dependent on the oil-related core of industries that historically have produced both good jobs and industrial pollution. The trend toward less reliance on petroleum refining and petrochemical production will continue, with significant long-term implications for the region's environment. But far into the future, the giant industrial complex that stretches from the Houston Ship Channel to Port Arthur will remain central to the region's economy. In terms of jobs and prosperity, the region could do much worse than remain the oil capital of the nation as it continues to diversify into other economic activities.

How the regional environment fares in the coming decades will depend on the choices of those who live in the region, those who have influence over business and civic decisions, and those who manage the vast structure of environmental regulations at the local, state, and federal levels. If the past is a guide to the future, it will be a daunting challenge to keep pace with the cumulative impact of a century of oil-related pollution. It will also be difficult and expensive to adapt the transportation system built during eras of low gasoline prices and little concern over air pollution to the demands of a new era. One key variable shaping future efforts to improve environmental quality will be the attitudes of the general population and their elected leaders. In a region built on the promise of economic opportunity, a stronger commitment to pollution control is likely only when poor air and water quality are widely viewed as significant threats to jobs and public health.

CHAPTER 2

The Houston Ship Channel and the Changing Landscape of Industrial Pollution

HUGH S. GORMAN

The Houston Ship Channel, which allows oceangoing vessels to reach Houston from the Gulf of Mexico, is as much a part of the city's landscape as its skyline, urban core, or sprawling network of commercial and residential developments. Indeed, in aerial photographs, the waterway stands out as a defining feature, snaking in from the east and penetrating toward the heart of the city. The first half of this fifty-mile channel cuts across the shallow waters of Galveston Bay and guides ships into the lower course of the San Jacinto River. Then the channel turns and follows Buffalo Bayou, allowing vessels to wind inland for another sixteen miles or so. Each year, nearly 200 million tons of cargo move along the channel, going to and coming from places such as Venezuela, Saudi Arabia, Algeria, and Brazil. In terms of foreign trade, the Port of Houston is the largest in the United States; it ranks second in terms of total tonnage.[1]

A trip up Buffalo Bayou makes it clear that the channel is more than a transportation route. Scores of facilities—including refineries, steel mills, and chemical plants—line its banks, creating a landscape of industrial muscle on a par with any in the world. Like other major industrial clusters in the United States, the one serviced by the Houston Ship Channel developed as American society began to extract, process, and consume material and energy on a scale never seen before. As the level of production and consumption increased, firms constructed larger and more efficient facilities but gave little thought as to how their processes and products interacted with the physical environment. Instead, decisions affecting the environment were often guided by a narrowly conceived ethic of efficient resource use, which placed the focus on increasing efficiency and eliminating waste, not managing environmental quality.

Since then, Americans have adopted an ethic of setting environmental objectives and implementing policies to achieve those objectives. Firms are now expected to manage their effluents, emissions, and solid wastes in a manner consistent with water and air quality objectives. Regardless of one's assessment as to how effective environmental regulations are, there is no doubt that significant changes in perception and practice have occurred.[2] If nothing else, most discharges of waste material into the air and water and on land are now measured and monitored as manageable transactions between firms and the environment.

Even Houstonians, long proud of their city's favorable business climate and correspondingly suspicious of the effect that environmental regulations might have on that climate, have come to see environmental quality as an important component of their city's economic stability.[3] For some, the turning point came as late as 1999, when ambient concentrations of highly reactive ozone exceeded federal air quality standards more often in Houston than in any other city in the United States.[4] A year later, a team of 300 scientists and technicians from over forty universities and government agencies descended on the city to study the problem, only to discover that even more violations would have been recorded if air monitors had been installed along the ship channel.[5] This new reputation of having the poorest air quality in the United States was not one that anyone in Houston appreciated. In addition to casting a pall over the city, this status of not being able to attain national ambient air quality goals also triggered special provisions in the federal Clean Air Act, forcing state regulators to address the issue in a more systematic fashion.

One of the problems in the Houston area is the sheer concentration of industry along the ship channel. For example, despite significant improvements in how facilities treat their effluents, the concentration of industry along Buffalo Bayou is so great that basic water quality goals cannot always be met. Even if individual facilities manage their effluents according to federal technology-based standards, the total discharge from all facilities and nonpoint sources can still overwhelm the waterway. To meet some water quality goals, state officials have had to move beyond the normal permit system and establish total maximum daily loads (TMDLs), a process that is tedious, time consuming, and expensive.[6]

Soil and groundwater contamination along the waterway are also ongoing concerns. Although the disposal of hazardous wastes is now systematically tracked and managed, industrial practices in the period before strict pollution control regulations inevitably contaminated much of the land. Such concerns make redeveloping pieces of property for new uses difficult, with few people wanting to purchase land that requires a significant amount of remediation. Anybody familiar with even one Superfund case cannot look at the aging facilities that lie along the ship channel without wondering about the level of potential contamination and the expense of cleaning it up.[7]

The challenges faced by Houston raise numerous questions about the city's past efforts to manage pollution. Was industrial pollution simply ignored in Harris County until federal legislation forced firms to manage their effluents, air emissions, and solid wastes? Did civic leaders have any influence over early industrial practices? How did firms along the channel manage their effluents, emissions, and wastes in the period before federal regulations? This essay addresses these and similar questions by tracing the development of industry along the channel and examining how efforts to manage pollution have changed over time. It also suggests that the industrial landscape associated with the Houston Ship Channel is an artifact of the twentieth century. Although such industrial clusters are likely to persist for years to come, several trends in the twenty-first century—of which the effort to manage environmental quality is only one—have changed the economic, political, and social context that gave rise to such clusters.

Establishing a Nineteenth-Century Transportation Hub

Little about the location of Houston ensured its development either as a deep-water port or major industrial center. In the nineteenth century, Buffalo Bayou was a shallow, sluggish stream, thick with overhanging branches and navigable only to vessels with little draft.[8] Although the ability to reach so far inland made the site attractive to land speculators, one can certainly imagine scenarios in which other locations, many of which had their own boosters, might have developed as the Gulf's commercial and transportation center. Geography, in this case, does not make a particularly strong case for the land along Buffalo Bayou emerging as a major industrial center.

Certainly the first major wave of settlers from the United States did not see anything special about Buffalo Bayou.[9] When they arrived in the 1820s, they encountered land not significantly altered by European practices. Though claimed by Mexico after its independence from Spain, the area was still occupied by the Karankawas, a fishing society that thrived along the Gulf Coast between present-day Galveston and Corpus Christi.[10] The remoteness of the region relative to population centers is testified to by Jean Laffite's choice of Galveston Island as a base camp in the years just before 1821, which is when a foreign power—the United States—evicted Laffite and his privateers, opening the door to more permanent settlements.[11]

By the time settlers from the United States declared themselves independent of Mexico in 1836, a handful of sawmills had been constructed along Buffalo Bayou. In the following year, a group of land speculators arranged for a steamship to push its way up the stream to the settlement they named, somewhat ambitiously, in honor of General Sam Houston. They also established an entity called the Port of Houston and secured funds from the Republic of Texas to make the bayou navigable to barges and shallow steamships.[12] The new population center, however, remained small. In 1850, according to the first census performed after Texas became part of the United States, the population in all of Harris County had yet to reach 5,000 people.[13]

The effort to make Houston an important commercial and transportation center gained momentum in the 1850s when boosters connected the Port of Houston to a handful of railroads, with lines fanning out to the Texas interior. Buffalo Bayou, though, remained a relatively modest shipping channel, with much of the land along its banks undeveloped. Steamboats and barges carrying cotton and grain traveled up and down the waterway, but those vessels were becoming more obsolete with every passing year. After all, by the 1880s Galveston was a deepwater port, and Houston was not. Rather than load cargo in Houston on barges with shallow drafts, only to unload that cargo again in Galveston, why not ship cargo to Galveston in railcars, where it could be loaded directly onto oceangoing vessels? Although the fate of Houston as a commercial hub might have been secure, the future of Buffalo Bayou as a major transportation channel was not.[14]

Pollution along the bayou was not a significant issue in the nineteenth century. One problem related to the disposal of industrial waste did emerge, but it had to do with navigation, not water quality. The concern was that sawmill operators often dumped their sawdust along the banks of the bayou. Heavy rains then washed the sawdust into the stream, clogging up portions of the waterway. Houston's city council addressed the problem by prohibiting the dumping of sawdust on the banks of the bayou. In general, mill owners saw this prohibition as legitimate and complied with the ordinance. It helped, of course, that they could burn their wastes or dump them elsewhere without generating other complaints. In short, mill owners did not have to significantly change their practices to comply with the ordinance.[15] And nobody, of course, had to define or enforce any effluent or ambient water quality standards. As long as barges could move up and down the waterway without getting stuck, people were satisfied.

Domestic sewage flowing out of the developed areas of Houston emerged as a greater concern. In the 1890s, city boosters lobbied the U.S. Congress for funds to dredge a deepwater channel all the way to docks in Houston, but Congress showed little interest in a project that would simply shift traffic from

Galveston to Houston. The project attracted serious attention only after a storm destroyed the Port of Galveston in 1900. Soon after, national legislators agreed to fund the project. Before starting work on the channel, however, the U.S. Army Corps of Engineers required that Houston collect and treat the sewage flowing into the waterway.[16] At a time when disease-causing bacteria posed a significant threat to urban populations and large northern cities were installing the infrastructure necessary to address this concern, civic leaders could hardly refuse.[17] When the city's new sewage collection and treatment system was completed in 1902, it satisfied the federal government's engineers.[18]

Although the U.S. Army Corps of Engineers articulated concerns about municipal wastes, they probably gave little thought to the discharge of pollution-causing wastes from industrial facilities. Most people still visualized Houston as a shipping hub, not an industrial center. And even those who did foresee a rise in industrial activity probably thought in terms of warehouses, storage facilities, and light manufacturing. No good reason existed for thinking that land along the channel would soon become home to an industrial cluster as large as any on the face of the earth. For one thing, no clusters of a similar magnitude existed in the nineteenth century. Even the late-nineteenth-century network of steel mills, glass works, and refineries that flourished in places such as Pittsburgh could not compare in scale with what lay ahead. Neither the Army Corps of Engineers nor the boosters of Houston had any idea of what was in store for the land along the proposed channel.

The Emergence of an Industrial Cluster

Survey work for the Houston Ship Channel began in 1904; eight years of dredging started in 1906.[19] By the time the deepwater channel reached its terminus, a turning basin several miles from the city center but within the boundaries of a still-expanding Houston, much had changed. In 1901, petroleum—at that time, used mainly in the production of kerosene for lamps—had been discovered about ninety miles to the east at Spindletop, the first major oil field in Texas. Although oil from that famous field flowed through Port Arthur, not Houston, it established the Gulf region as an important oil-production and oil-refining center. In addition, petroleum began flowing through Houston several years later when wildcatters discovered and producers developed another oil deposit, the Humble field, closer to the city. Discoveries of other nearby fields soon followed.[20]

Access to the Gulf of Mexico allowed several companies—the Texas Company (later Texaco), the Sun Company (refiner of Sunoco-brand gasoline), and Guffey Oil (later Gulf Oil)—to establish themselves in an industry over which Standard Oil previously held a stranglehold. They immediately built refineries near Beaumont and began shipping oil via oceangoing tankers. To keep the oil flowing, these companies also constructed long-distance pipelines to large oil fields in Oklahoma. In addition, they used the petroleum as an inexpensive source of energy for their tankers, establishing oil as an energy source in the Gulf region that could compete with coal.[21]

The most important change, of course, was a rapidly growing demand for fuel to power automobiles being mass-produced by northern companies. The resulting mixture of inexpensive petroleum, flat land with access to deep water, and a rising demand for gasoline proved explosive. By 1919, thirty-eight firms had constructed facilities along the channel; within seven years, that number had increased to seventy.[22] Two decades later, there were more than a hundred. Not surprising, three of the first large processing facilities to be constructed were refineries, with the Humble Oil and Refining Company locating a facility near the Goose Creek oil field, while the Sinclair Oil Company and the American Petroleum Company sited refineries closer to Houston, one south of the channel and one north.[23]

With the construction of refineries, industrial pollution became a concern. In 1920, the first director of the Port of Houston, Colonel Benjamin Allin, took action against several refiners for fouling the water by discharging oily effluent into the ship channel. An official with the Texas Fish and Oyster Commission also grew concerned and sent a letter to the U.S. House Committee on Rivers and Harbors, attributing a 90 percent decline in oyster production around Galveston to this pollution-causing oil and urging strong action against those responsible. But the refiners, who had no easy disposal alternatives, turned out to be less cooperative than owners of sawmills had been half a century earlier. One company called for Allin's dismissal, describing him as exhibiting a "persistent disposition to unnecessarily harass the management of various industries along the ship channel and to interfere with and hamper their legitimate activities."[24] In this case, the mayor of Houston, A. Earl Amerman, and Houston's city council both supported Allin's efforts. The mayor, in his response, noted that "I have personally inspected the outlets from the refineries into the Channel and have seen with my own eyes large quantities of oil flowing into the Channel. By this, I do not mean the milky discharge, but black oil in quantities sufficient to cover the water to the side of the outlet for a considerable distance."[25] Allin kept his job as port director, but backed off in his attempts to prevent such discharges.[26]

Houston was not alone in experiencing increased levels of oil pollution in the years after World War I. Similar complaints were being heard up and down the East Coast, with fire insurance companies, representatives of the fishing industry, resort owners, and sportsmen all demanding that something be done about the quantity of oil flowing into busy ports and covering the water with a layer of scum. The main problem was that the quantity of petroleum being distributed and consumed in major urban areas had increased dramatically, and a lot of oil was spilling, leaking, and being flushed into streams and harbors. Because many of those waterways crossed state boundaries, these groups pushed for federal legislation that would make it illegal to discharge oily water into the nation's waterways.[27]

A variety of different House and Senate committees held hearings over these proposed pieces of legislation, with some hearings quickly expanding into a debate over how to control all sorts of industrial pollution, including wastes from chemical plants and drainage of acidic water from mines. However, none of the legislation, not even the strongest pieces, would have expected firms to measure or monitor the disposal of their wastes. Instead, the bills called for a complete prohibition against the discharge of oil and other chemicals, discouraging any serious discussion of effluent standards.[28]

The hearings almost moved in the direction of effluent standards when oil company representatives opposed the prohibition. They pointed out that the effluents of a refinery would always contain some oil. Was one drop in a hundred gallons of water enough, they asked, to represent a violation? Where should one draw the line? For example, a spokesperson for Jersey Standard estimated that its main refinery discharged two barrels of oil in every twenty million gallons of water, representing an oil concentration of four parts per million.[29] Was that, he asked, good enough? Answering that question and seriously addressing the question of effluent quality would have been a significant first step toward an effective regulatory system, but the debate in Congress did not go in that direction.

Instead, led by business-minded engineers—including the influential secretary of commerce, Herbert Hoover—industry experts encouraged Congress to take an easier path. This path was shaped by several assumptions: first, that pollution-causing wastes were primarily a symptom of inefficiency; second, that firms had economic incentives to increase the efficiency of their facilities; and, third, that technological improvements would result in fewer pollution-causing wastes being discharged, eventually eliminating pollution as a concern. Industry experts asserted that regulations were unnecessary because companies were already making their operations more efficient. In the short term, they saw nuisance and damage torts as discouraging the most careless practices. In the long term, they argued, improvements in technology would reduce the flow of waste material entering waterways to acceptable levels. Hence, they called for education and continued technological innovation, not regulations.[30]

In the end, after initiating a major investigation by the U.S. Bureau of Mines and holding another wave of hearings, Congress passed a relatively weak piece of legislation, the Oil Pollution Act of 1924. This act prohibited ships from flushing their oily tanks within three miles of shore. Land-based facilities were not required to measure, monitor, or manage their effluents in any way. In essence, this piece of legislation, which brought closure to the first national debate over water pollution, simply shifted a major component of the oily waste entering the nation's harbors to deeper waters. After all, tanker captains and captains of oil-burning ships could still flush their oily ballast tanks in open water.[31]

Indirectly, though, the Oil Pollution Act of 1924 did have an effect on refinery operations, including those along the ship channel. In 1926, the Army Corps of Engineers issued a report required by the act and recommended that the prohibition of oily discharges from ships be extended to industrial facilities such as refineries. When a bill to amend the Oil Pollution Act was introduced in 1930, the directors of the American Petroleum Institute (API), the main lobbying group for the oil industry, acted quickly to convince legislators that the industry could regulate itself. Soon after the hearings, the newly formed API Committee on the Disposal of Refinery Wastes released the first volume, *Waste Water Containing Oil,* of a multivolume manual on the disposal of refinery wastes. The committee issued two more volumes, *Waste Gases or Vapors* and *Chemical Wastes,* in 1931 and 1935, respectively.[32]

In general, the authors of the API waste disposal manual subscribed to the notion that efforts to increase the efficiency of an industrial operation overlapped with efforts to eliminate pollution-causing wastes. For instance, given the importance of keeping acid sludge and spent chemicals out of the main waste stream, engineers on the API waste disposal committee emphasized the recovery of spent material as the obvious first step a refinery should take. Most refiners were receptive to this type of change. Although they did not always pay much attention to the details of what entered and exited their sewers, they were very aware of expenses associated with carting spent material away and replacing it with fresh stock.

For example, at the time of the Corps' report, refiners removed sulfur impurities from their products by washing those products with sulfuric acid. This treatment process generated large quantities of a waste product known as acid sludge, which consisted of the acid, the impurities it captured, and significant amounts of oil. The largest refineries were generating about 150 tons of acid sludge each day, and many refiners still disposed of this sludge without any effort to reclaim the material.[33] The API committee emphasized that extracting acid from this sludge and recycling it, instead of disposing of it as waste, represented an opportunity both to reduce the discharge of pollution-causing wastes and save money by not having to purchase as much fresh acid. They also pointed out that acid caused oil to emulsify, which made it difficult to separate oil from wastewater. Therefore, by keeping acid out of the sewer, oil and wastewater would not mix as easily and allow more oil to be captured before discharge.

The manual also pointed to a number of other such changes, including the recycling of spent caustic and filtering material, more extensive use of oil-water separators, and the separation of storm sewers from process sewers. Representatives from oil companies throughout the United States, including those that operated refineries along the ship channel, contributed to the committee's work and carried on waste-related research of their own. For instance, much of the work on the regeneration of spent filtering material was performed at the Humble Oil Company's Baytown facility.[34]

Although the API's efforts to create a waste disposal manual were motivated by a desire to forestall government-enforced regulations, the end result was more than window dressing. Indeed, in the 1920s and 1930s engineers could point to a number of cases in which efforts to make industrial operations more efficient overlapped with efforts to control pollution. For example, a typical refinery in the years just after World War I lost about 5 to 10 percent of the oil it processed. By the end of the 1930s, leading facilities had reduced their losses to about 1 percent. In addition, other advances, such as improved corrosion prevention and more effectively designed storage tanks, also resulted in significantly less pollution-causing material being lost to leaks, spills, and what today would be called fugitive emissions.[35] Similar advances were being made in other industries.

In the oil industry, faith in this efficiency ethic was also reinforced by developments in the management of oil fields, something with which many Houstonians were familiar. In the 1920s, as oil producers raced to extract as much oil as they could from newly discovered fields, they wasted tremendous amounts of oil and capital. First, in their effort to quickly extract as much oil as they could, they drilled far more wells than necessary. The resulting glut of oil not only reduced the price of oil but also overwhelmed the ability of producers to store their oil, forcing them to run it into open pits, where evaporation, fire, and seepage took a heavy toll. Second, in most fields, producers also treated natural gas as a waste product, flaring it as it flowed out of their wells. Underground, as the amount of gas dissolved in the petroleum dropped, the oil became more viscous and ceased to flow, resulting in most of the oil being left underground. Eventually, more water than oil flowed into wells, and disposing of this briny water presented producers with major challenges. In the end, only a small fraction of the available petroleum reached refineries, and oil fields were a thoroughly messy and dangerous place.[36]

By the end of the 1930s, however, much had changed. Producers were extracting more oil with fewer wells. Furthermore, the amount of salt water and natural gas being released to the environment had dropped dramatically. Instead, producers were injecting their water and gas back underground, which not only eliminated a pollution concern but also helped to maintain reservoir pressure. In turn, this allowed firms to extract more oil with less capital than ever before. Significantly, this effort to bring order to the nation's oil fields was performed under the banner of resource conservation and an agenda of increasing efficiency and eliminating waste. Many in the industry pointed to this success in the oil fields as validation of their faith in the gospel of resource efficiency, and nowhere would this ethic have been stronger than in oil centers such as Houston.[37] The fact that government regulations and strict enforcement, including martial law, were necessary to facilitate this change in oil field practices was conveniently ignored. In fact, refiners, who generally benefited from the inexpensive oil made possible by oil-field gluts, had long opposed efforts to regulate production.[38]

Hence, from the 1920s to the 1940s, a period of major growth for the industrial cluster serviced by the Houston Ship Channel, a pollution-control strategy was in place. Engineers and technical leaders generally believed that, in the long term, increasing the efficiency of production would result in less

Oil pollution in the Houston Ship Channel. Courtesy Houston Metropolitan Research Center, Houston Public Library.

pollution. And from one perspective, this strategy of controlling pollution by increasing efficiency and eliminating waste was working. Per unit of production, facilities were discharging fewer and fewer pollution-causing wastes. Furthermore, in the years after World War II, when firms constructed many new refineries and redesigned many older refineries, a de facto standard for pollution control, the API manual, existed. While lenient by today's standards, it represented a significant improvement from practices a generation earlier.

From another perspective, the strategy was an abysmal failure. The Houston Ship Channel, like many waterways throughout the United States, was still getting dirtier, not cleaner. Although companies had made improvements, facilities still discharged effluents with enough contaminants to overwhelm the waterway. The practices recommended by the API manual fell far short of eliminating all pollution concerns. In addition, the scale of production had risen dramatically, which overwhelmed whatever per-unit improvements in pollution control had been achieved. For example, in 1927 the refinery operated by Humble Oil processed 50,000 barrels per day, making it one of the largest in the world. By 1950, the facility was processing 260,000 barrels per day. In that same period, the Sinclair refinery grew from processing 16,000 barrels per day to 85,000 barrels. A refinery operated by Shell Oil, which did not even exist in 1927, was processing 100,000 barrels per day in 1950.[39] Facilities in other industries also grew in scale, resulting in an increase, instead of the hoped-for decrease, in the amount of wastes entering the ship channel. Furthermore, with the rise of petrochemicals, the toxicity of various wastes was also increasing.[40]

Another serious challenge was that by the mid-1950s, the most cost-effective improvements already had been identified and implemented. Some of these changes had even paid for themselves, if only by reducing expenses associated with nuisance and damage torts. Further reductions in the disposal of pollution-causing wastes meant investing significant amounts of capital for which there was little or no return. The notion that efforts to improve the efficiency of production overlapped with efforts to address pollution concerns was becoming less and less valid.

Shifting the Focus

By the 1950s, most people familiar with the problem of industrial wastes recognized that pollution-causing effluents and emissions would not simply disappear as firms improved their production processes. In most urban centers, sanitary engineers and health officials, who had significant experience managing domestic sewage, began looking more closely at industrial emissions and effluents. Their analyses were far more sophisticated than those associated with the debates leading up to the Oil Pollution Act of 1924. Among other things, they paid more attention to measuring and monitoring what was being emitted and more attention to the specifics of how to reduce those emissions.

In 1954, the Harris County Health Department, in hopes of better identifying and managing sources of pollution, hired its first director of Air and Water Pollution Control, Walter A. Quebedeaux. One of the concerns that triggered this search for a pollution control director involved problems with industrial effluents. According to a 1951 news article, chemicals being discharged were reacting with and consuming dissolved oxygen, crippling Buffalo Bayou's ability to break down bacteria-ridden sewage.[41] Although the focus was still on disease-causing bacteria, not the chemicals themselves, changes in perception were clearly taking place.

When Quebedeaux assumed his position as director, more than a hundred industrial plants, including three of the largest refineries in the world, lined the ship channel along its route to Galveston Bay.[42] The three main refineries—operated by Humble, Shell, and Sinclair—processed a total of 520,000 barrels of crude each day. Several other average-sized refineries contributed another 70,000 barrels.[43] Quebedeaux, though, was no stranger to industrial operations. Previously, he had been employed as the air and stream pollution director for the Champion Paper Company. Before that, he operated his own chemical analysis laboratory in St. Louis, where he provided what he described as "consultant and engineering services especially connected with litigation."[44] As the director of pollution control in Harris County, Quebedeaux immediately embarked on a campaign to quantify the concentration of various contaminants in the waterway and in the emissions and effluents of industrial facilities. One water quality survey performed in 1955 indicated that the channel from the "San Jacinto monument" to the "turning basin," a distance of about sixteen miles, was devoid of dissolved oxygen because sewage and industrial wastes consumed all the oxygen available.[45]

Quebedeaux was also active in efforts to control air pollution, both nationally and locally. For a time, he served as one of twelve directors for the Air Pollution Control Association, a national organization in which industry experts, health officials, and academic researchers met to discuss the latest developments on policy and technology relevant to monitoring and managing urban air quality.[46] At the time, the organization was focusing its attention on efforts to solve the photochemical smog problem in Los Angeles, and those efforts were revolutionizing the interaction between science, technology, and pollution-control policy.[47] Given that Los Angeles, like Houston, was a refining center and home to a rapidly growing automobile-dependent population, Quebedeaux paid close attention to developments in the field of air pollution control.

Before photochemical smog became a concern in Los Angeles, smoke-control experts had primarily been concerned with the sootiness of emissions from boilers and heaters. For example, major air quality improvements in St. Louis and Pittsburgh were achieved by encouraging residents to switch from burning sooty coal to burning clean natural gas in their home heating systems.[48] Such a strategy, however, would not work in Los Angeles, if for no other reason that few industries or homes in this oil-rich region burned coal. Other straightforward strategies, such as forcing refineries to reduce the amount of sulfur they emitted to the air, did not eliminate the smog either. In response, civic leaders in Los Angeles initiated a systematic program of research to understand and address their unusual smog problem. When funding from the federal government became available, scientists across the United States joined in the effort, studying the problem of air pollution far more systematically than scientists employed by industrial firms had. As a result, the pollution control research agenda shifted away from improvements in production efficiency and toward learning more about environmental interactions.

In Los Angeles, researchers discovered that the city's air pollution was the result of several contaminants, mainly nitrogen dioxide and unburned hydrocarbons, coming from a variety of emission sources and reacting in the sunlight to form ozone and other irritants. Automobiles and refineries were identified as key sources of the most problematic smog precursors.[49] In Houston, a similar mix of conditions—refineries, automobiles, and sunlight—existed, but concern about smog took several decades to become a significant public issue. In part, Houstonians articulated less concern because they placed less value on clear blue skies than paradise-minded Californians. In addition, the population of the city, meteorological conditions, topography, and the location of the ship channel cluster relative to the city core resulted in less of an air quality problem than that experienced in Los Angeles. Indeed, the 1950s measurements made by the U.S. Public Health Service indicated that Houston's air quality was better than that in Los Angeles. At the same time, Quebedeaux recognized that these measurements were slightly misleading, noting that "the samples were taken at points where the real pollution contamination could not possibly be sampled because of the prevailing wind conditions." Both of the sampling stations, he pointed out, "were in such positions that they did not receive a composite exposure to materials that are found along the ship channel."[50]

Although the level of concern in Houston paled in comparison to that articulated by civic leaders in Los Angeles, people in Harris County did express concern about the quality of the city's air. In the mid-1950s, for example, approximately 5,000 people from the community of Greens Bayou signed a petition asking that something be done to clean up the air in their area.[51] In the mid-1960s, the mayor of Houston admitted that he believed "Houston has a pollution problem . . . and a bad one . . . one that is growing worse." But he was also "extremely optimistic about the degree of control over air pollution that can be obtained through voluntary cooperation of the city and industry," and he did not believe aggressive action to be necessary.[52] In regard to water quality in Buffalo Bayou, most Houstonians accepted the stream's role as a transportation channel and as a sink for municipal and industrial wastes. Although people enjoyed industrial tours of the waterway, nobody looked to the channel as a source of drinking water or as a place to swim.[53]

Without widespread public support or sufficient resources, Quebedeaux, described by a local newspaper as a "prophetic voice crying in the wilderness," continued collecting and analyzing air and water samples.[54] He had initially hoped that nuisance law would provide him with the tools necessary to change industrial practices, going so far as to suggest that explicit pollution control regulations were neither necessary nor flexible enough in the face of changing technology.[55] However, by the mid-1960s he had met with numerous frustrations in getting unresponsive companies to change their practices. At the top of his list of things that frustrated him were firms that waited for lawsuits before responding to him, laws that prevented his staff from inspecting a plant at will, and cases lost in court by a prosecuting attorney "out of law school probably less than one year" going against an industry-hired "defense lawyer with twenty or thirty years' experience."[56]

At the same time, industrial practices throughout the country were changing, due in part to pressure exerted by health officials such as Quebedeaux. For example, in the late 1950s engineers at Humble Oil's Baytown refinery installed coalescing equipment and a new master oil-water separator to remove and recover more oil from their wastewater streams. They also added equipment to strip hydrogen sulfide from various waste streams, cooling towers to increase the reuse of cooling water, and a 20,000-gallon-per-minute pump at the point of discharge to increase the amount of dissolved oxygen in the water. In the early 1960s, the company also installed a three-pond lagooning system capable of holding water for forty-five days, with the extended retention time permitting enough bacteriological action to reduce pollutants in the effluent by an additional 70 percent. The engineers claimed that the quality of their effluent "showed more than a ninety percent improvement," making it "generally of better quality than that of its receiving body, the Houston Ship Channel."[57]

Discharging effluents "generally of better quality" than water in the Houston Ship Channel was neither a lofty goal nor a particularly useful standard, but determining what standards industrial effluents should meet was no simple matter. For one thing, industry officials resisted any effort to set enforceable standards and preferred to work with government officials in a spirit of "cooperative pragmatism," avoiding strict regulations that would require them to monitor and manage their emissions and effluents.[58] In addition, pollution-control officials and industry engineers alike recognized that the wide variety of wastes being emitted into the air and discharged into waterways prevented any easy analysis of the problem. Indeed, traditional measures of water quality were not sufficient to characterize water quality in a stream lined with petrochemical plants. Old measures of air quality were also inadequate in the face of pollution caused by emissions of sulfur dioxides and oxides of nitrogen.

Researchers were also putting new tools to use in monitoring pollution. For example, Texas A&M meteorologist Darryl Randerson began examining metropolitan areas from satellite photographs. Using photos from *Gemini-VII,* Randerson found that he could detect smoke plumes originating from facilities along the ship channel. He also noticed that some photographs of the area north of Houston appeared to be overexposed due to a diffuse, white haze that dispersed or reflected back much of the light.[59]

With industry stalling state-level efforts to produce air and water quality standards while public interest in air and water quality was growing, support for federal action increased. This support eventually led to two new pieces of federal legislation: the federal Water Quality Act of 1965 and the Clean Air Act Amendments of 1967. The first required states to formulate water quality standards for different classes of streams; the second required states to set ambient air quality standards for regional air sheds.[60] Although a new regulatory regime was clearly being constructed, these two pieces of legislation were relatively lenient. In many ways, they were the final test of an approach to pollution control that assumed some level of industry leadership. Neither act proved successful, and both would be replaced in the early 1970s with stronger pieces of legislation, significantly altering the landscape of industrial pollution.

Conclusion

In the three decades since the passage of the Clean Air Act of 1970 and the Clean Water Act of 1972, which together established the legislative foundation for the current system of pollution-control regulations in the United States, much has changed in regard to how firms manage their interactions with the physical environment. Point sources of emissions to the air and discharges to water bodies are now monitored more precisely than most people would have imagined possible in the first half of the twentieth century.

Although this regulatory system is neither perfect nor uniformly administered, governmental agencies are expected to take action when air and water quality fall below established standards. In heavily used streams such as Buffalo Bayou, agencies are now expected to manage the total amount of certain contaminants discharged, allocating specific amounts to individual firms. States must do the same for criteria pollutants emitted into regional air sheds that cannot meet federal standards. As a result, when locating facilities in heavily populated and industrialized areas, firms must treat their ability to discharge waste as a potentially scarce commodity that other entities might also wish to secure.

Land use and landscape also matter more than they once did. In many cities, significant lengths of waterfront have been redeveloped for public use. Even in Houston, people are looking at ways to better integrate portions of Buffalo Bayou into the city's landscape, with the focus primarily on a stretch of the waterway upstream from the navigable deepwater channel. Advocates who wish to see Houston become more bicycle friendly call that stretch of waterway a "squandered jewel" along which city planners have sited four large jail facilities. They also list ten reasons why Houston did not win its bid to host the 2012 Summer Olympics, with air pollution, traffic-causing sprawl, and the lack of pedestrian and cycling access figuring prominently in their reasons.[61]

By themselves, the added complexity associated with managing environmental quality in heavily industrial areas and increased public interest in landscape and access to waterways may not be enough to dramatically alter the siting of large industrial facilities. But other factors—the economics of operating in a global economy, access to raw material, and considerations of security—have also changed. Together, they are enough to suggest that industrial clusters such as the one served by the Houston Ship Channel are an artifact of the twentieth century, unlikely to be repeated in the twenty-first.

When the ship channel was constructed in the years before World War I, the United States had not yet answered the question of how to systematically manage the effect of industrial operations on air and water quality. Although some people in the United States were concerned about pollution, those concerns had relatively little effect on the development of the Houston Ship Channel. First, most Houstonians initially visualized the ship channel as a transportation channel only, not one that would eventually service a major industrial center. Second, nuisance and damage suits were seen as the main mechanism for addressing conflicts over industrial pollution. Therefore, even if anyone had anticipated the dramatic growth in industry along the channel, little could have been done to prevent potential concerns; the legal and institutional tools simply did not exist. Finally, even after the first few facilities were constructed and pollution concerns emerged, firms deflected challenges by emphasizing progress made in increasing efficiency and eliminating waste. Their response, and American society's general acceptance of this response, revealed tremendous faith in the ability of firms to eliminate pollution-causing wastes as they increased the efficiency of their operations. Only in the 1950s did this faith in the overlap between efficiency and pollution control break down, gradually leading to new legal expectations and industrial practices. By that time, though, the Houston Ship Channel was a mature industrial cluster.

In general, therefore, a guiding ethic for controlling industrial pollution did exist in the first half of the twentieth century, in the period when the industrial cluster along the Houston Ship Channel first emerged. However, it was an ethic based on increasing efficiency and eliminating waste, not managing environmental quality. Until that first ethic failed, as it did in the 1950s, any movement toward a regulatory system based on defining and managing environmental objectives was easily undermined by arguments based on the gospel of efficiency. In Houston, several other factors—the convenient clustering of industry along the ship channel, acceptance of the ship channel as part of the industrial landscape, meteorological conditions that generally moved air contaminants away from the urban core, and a general faith in self-regulation—continued to make efforts to identify and manage environmental objectives seem less important than elsewhere, delaying a significant public response until the 1990s.

CHAPTER 3

"Bad Science"

The Politics of Ozone Air Pollution in Houston

ROBERT FISHER

Houston has a serious air pollution problem that compromises its quality of life and promotes an ambiguous image of the city. On the one hand, Houston is a leading center in the world for both energy and health, a primary site for trade and transportation, and home to the most expansive petrochemical industry in the nation. On the other hand, both its image and quality of life are tarnished by the city having some of the worst air pollution in the United States.[1] Often the response to air pollution was to deny its existence and extent and to emphasize that it was a private, not public, matter. The history of air pollution in Houston is complicated by being politically charged, filled with scientific complexities and intricacies, and characterized by too little transparency. With that caveat in mind, this essay argues that part of Houston's historically weak record on ozone air pollution derives from a vehement and continuous opposition to both public environmental initiatives and chemical industry regulation.

What is not in question is the existence of severe air pollution. Industrial toxic air pollution rose from 108 million pounds in 1997 to 110 million pounds in 1998, the most in the nation.[2] A 1999 inventory conducted by the Texas Natural Resource and Conservation Commission (TNRCC), appointed by then-governor George W. Bush, underscored that the Houston area had 567 tons of volatile organic chemicals (VOCs) released daily into the air,[3] especially propylene, ethylene, butadiene, and benzene, ones most related to ozone formation.[4] This was dwarfed only by the daily release of 1,142 tons of nitrogen oxide (NOx).

Ground ozone, once broadly referred to as "smog," is not something emitted directly into the air. It is a complex atmospheric chemical reaction between VOCs and NOx that occurs in the presence of certain weather conditions—sun and heat above 70 degrees Fahrenheit. NOx is a product of high-temperature combustion, commonly from vehicles, incinerators, power generators, refineries, and other sources. VOCs include various organic chemicals that easily vaporize, such as solvents and gasoline, and are emitted commonly from oil refineries and petrochemical plants, as well as cars, trucks, and airplanes.[5] The combination of the high VOC and NOx releases with Houston's five to six months of hot, sunny, summerlike months results in the city having an "ozone season" longer by one or two months than even other Texas cities.[6]

Almost everything about Houston's air pollution problem—its causes, sources, extent, impact, and solution, and even whether the city and metropolitan area are making enough progress—is now a matter of significant debate. For example, on the question of the sources of ozone-related air pollution, while nearly 50 percent of the NOx and estimates from 30 to 75 percent of the VOC pollution come from the petrochemical and related industries, there is increasing acknowledgment that on-road vehicles (30 percent NOx and 26 percent VOC) as well as small area industries such as cleaning facilities and gas stations (2 percent NOx and 27 percent VOC) also contribute to the problem.[7] While all agree that high ozone rates in Houston are influenced by the city's unique weather patterns, Houston's air quality problem is heavily connected to petrochemical industry pollution. In most other cities, including Los Angeles, automobiles are the primary culprits.[8] In fairness to industry, throughout the United States it was long the sole target of environmental reform, to the exclusion of other sources of air pollution. Curiously, regarding ozone air pollution, automobiles rather than industry have been a primary target of reformers. Houston certainly has its share of air pollution from vehicles and nonindustry sources. But with more than a hundred petrochemical plants and refineries, industrial pollution is a major source.[9]

The comparison of Houston with Los Angeles has special significance. By the end of the twentieth century, Houston became known as the "smog capital" of the nation, surpassing Los Angeles. It did so in 2000, and this fact was quickly tied to the environmental record of George W. Bush, then the governor of Texas and Republican candidate for president. Houston's environmental record became a target for Democrat Al Gore's presidential campaign in 2000, and it brought much unwanted bad publicity to the city. During the prior fifteen years, Los Angeles dramatically reduced emissions that caused ozone pollution by 75 percent, whereas in the same period Houston achieved much more mixed results.[10] Since Houston surpassed Los Angeles, at least for a few years, as the most polluted ozone city, there seems to be broad public recognition, if not always progressive action, that Houston has a serious air pollution problem that requires public and private attention. There is still much debate about the scientific accuracy, extent, and solutions to address the metropolitan area's air pollution. A relatively new consensus on air pollution has developed, however, which recognizes that there is a problem and that it needs addressing.[11]

The Environmental Opposition

The historical opposition to pollution control in Houston is consistent with the broader environmental history of the United States. Samuel P. Hays emphasizes that a major contributor to the tensions and conflict that characterize environmental history is the role of "the environmental opposition."[12] For Hays, any analysis of the environmental opposition must include discussion of the roots of the opposition in economic, social, and political sources; the various forms the opposition takes, including the terms of argument and strategy; and the change that occurs over time in the opposition. Not surprisingly, what Hays and others assert for the environmental opposition since 1945 across the nation corresponds with Houston's air pollution experience. In a nutshell, he proposes that one central oppositional root is expanding manufacturing centers, especially chemical industries, which consider any form of regulation a threat. Land developers, opposed to zoning and regulation of land use, compose a secondary oppositional root. In terms of forms of environmental opposition, Hays asserts that on air and water pollution, the "direction and intensity" of the opposition have not changed very much since the 1940s. While the opposition will accept some policy innovations, in general they seem to be opportunistic and temporary responses, basically opposed to public regulation, which yields renewed opposition when openings develop to reverse policy.[13] Moreover, discourse from the environmental opposition remains tied to ideological critiques about "command and control" and "market strategies." It also continues to use a language of dispute that seeks to retard environmental advances by criticizing environmental research as "flawed science," "bad science," or "junk science."[14]

Such responses reflect another long-standing theme in the history of air pollution: the politicization of science. From doing their own studies to opposing those of others, industry sources and the environmental opposition, as well as environmentalists, have politicized the science of air pollution.[15] Lastly, in terms of change over time, Hays offers that the environmental opposition has evolved through stages to become a permanent part of public affairs. Jacqueline Switzer concurs, proposing that the biggest change in industrial environmental opposition has been the replacement of a "reactive stance" with an "active stance" of forming "coalitions to enhance their overall effectiveness."[16]

Houston's response to air pollution is also embedded in the city's political culture of "privatism," which reinforces and legitimizes the claims of the environmental opposition. Gregory Squires defines it thusly: "the central tenet of privatism is the belief in the supremacy of the private sector and market forces in nurturing development, with the public sector as a junior partner whose principal obligation is to facilitate private capital accumulation."[17] Some commentators characterize Houston as a "private city" in which an elite private sector rules and government historically serves as part of a "pro-growth" coalition led by the private sector.[18] For example, during much of the twentieth century, the political adage had it that if you wanted to know what city hall would be working on in any given year, you needed to check the priorities of the Chamber of Commerce the year earlier.[19] With the decline of the chamber in the late 1970s and the emergence of the Greater Houston Partnership (GHP) in 1989, the institutional leadership and players changed.[20] The chamber represented real estate developers and speculators, while the GHP leadership comes from multinational companies. To be sure, Houston is a much more complex, diverse, and pluralist place than it was in the 1980s, let alone the 1950s, and the GHP reflects the diversity of the new economy. Current corporate leaders could not and do not rule with the power of business elites in earlier eras. As this essay demonstrates, the public sector has developed into a more independent force in the city, not always in partnership with, let alone subservient to, business prerogatives. Nevertheless, despite this growing complexity of the city and the shift in power from the chamber to the GHP, the approach to addressing urban problems such as land-use planning and air pollution has remained rather consistent. It is best to let the private sector handle issues in private, without public interference or public scrutiny.[21]

This essay focuses on ozone air pollution, which is the primary form of air pollution that Houston has had to contend with publicly. Over the past fifty years, "probably more money and effort have been spent to control ozone than on all other air pollution problems combined."[22] It situates that history in a context of both powerful "environmental opposition" groups such as the Chamber of Commerce and the GHP as well as a long-standing political culture of privatism. To emphasize both factors, it concludes by examining a modest effort in the mid-1990s to develop an ozone forecasting system. The struggle over implementing this ozone alert system underscores tensions between environmentalists and their public allies, on the one hand, and the petrochemical industry and its big business allies, on the other.

Early History of the Politics of Air and Ozone Pollution in Houston

When as early as the 1940s Houston citizens complained about "smog," air pollution was still considered a local responsibility, and it stayed that way, more or less, through the end of the 1960s. In 1953, to address pollution concerns, Harris County created a Stream and Air Pollution Control Section and a nuisance law to bring polluters to terms.[23] To help the problem, in 1955 the U.S. Department of Health Education and Welfare authorized the Public Health Service to conduct air pollution research and to provide technical assistance to the city. The Houston Chamber of Commerce responded in 1956 with a two-year contract with the Southwest Research Institute to study air pollution in Houston. It was to serve as a benchmark for future studies and to determine what, if any, plans needed to be made. The study found air pollution concentrated in the Houston Ship Channel area, with sulfur dioxide and hydrogen sulfide most frequent, along with higher than average dust, chlorine, sulfate, and nitrogen dioxide counts.[24] But it concluded that "it seems safe to say that if any community-wide health hazard exists in Houston due to air pollution, then an equal or greater hazard exists in many other cities in the country."[25] Curiously, at the same time, commercial pilots were reporting a mile-wide plume of industrial haze streaking out from Houston.

From the outset, many in the region, which was a center of the oil-refining and petrochemical production, were concerned that bad publicity and public regulation might accompany closer scrutiny of the city's air pollution problem. It is important to keep in mind that in the 1950s and 1960s, industry still had a strong faith in being able to solve any problem with technology. Air pollution was certainly no exception. In fact, since then, advances in technology have often proved them right, but technological advances have continuously been trumped by economic and political priorities. To no small degree, the history of air pollution has been more about political will than technological adaptation.[26] Denial of the problem was not limited to Houston or the petrochemical industry; into the early 1960s, for example, automobile manufacturers denied that cars posed any threat to air quality outside of southern California. Before 1970, at the least, it was very common for industry to deny the problem and argue against interventions without more scientific knowledge.[27]

In 1963, Congress passed the pioneering Clean Air Act, partly in response to the appearance of photochemical smog in cities as disparate as Los Angeles, San Francisco, Philadelphia, New York, Chicago, and Miami.[28] This modest act continued to rely on the states to issue and enforce regulations on air pollution, but it also pioneered in terms of bringing the federal government into the debate on air pollution policy.[29] In response, Houston initiated a two-year local study on air quality. In 1966, the University of Houston hosted a Clean Air for Urban Environments Community Action Seminar focusing on the 1956–1958 study and one undertaken in 1964–1966. The seminar director concluded that Houston did not have a citywide pollution problem. He suggested that Houston air quality was better than the Northeast's, which used coal as its primary

Severe air pollution in the late 1960s. Courtesy Houston Metropolitan Research Center, Houston Public Library.

fuel. Houston used natural gas. He emphasized that Houston was better because of favorable weather patterns in which prevailing winds caused rapid dispersion and dilution of pollutants in the atmosphere.[30] The 1964–1966 study revealed that NOx was the worst contaminant, but concluded that there had been no real changes in air quality since the 1950s, only a little more dust and pockets of smog that "neared the level of eye irritation."[31] While such thinking was not uncommon throughout the United States, Walter A. Quebedeaux, the activist pollution-control officer for the county, claimed that the 1966 report was a "whitewash" for industry.[32]

In the late 1960s, due to the threat of increasing federal pressure and in an effort to stave off increased regulation of the petrochemical industry, the state responded with its own regulatory plan. In 1967, the state passed the Clean Air Act of Texas, establishing the Texas Air Control Board, whose membership was dominated by industry representatives. For example, one board member was John Files of Merichem Company, the same man whom Quebedeaux in 1966 had dubbed "polluter of the month" at the head of a "dishonor roll" based on the number of complaints received.[33] A pattern was developing: regulatory control of air pollution in Texas meant industry self-regulation. Addressing the problem of air pollution always had to be weighed against its short- and long-term costs to industry. As with the civil rights movement of the period, it would take federal initiative for Houston and Texas to take air pollution more seriously. By the end of the decade, not only in Houston but nationwide, industry was adopting a more combative stance against federal initiatives.[34]

Because members of Congress were concerned that states were doing too little regarding air pollution, because they concluded that air pollution was more than a local problem, and because of mounting pressure by environmental activists, Congress amended the Clean Air Act of 1963. The 1970 act created the Environmental Protection Agency (EPA), which would set and enforce federal air pollution standards.[35] It established national ambient air quality standards that were to be achieved by states in no more than seven years, at which point the EPA could step in and propose a federal plan.[36] The act included transportation control measures in several cities, including Houston, in order to fight "the urban pollution problem with every alternative possible to move people more efficiently in fewer cars."[37]

Pressure mounted in the early 1970s as Houston's ozone problem gained increasing visibility in federal EPA reports. In stark contrast to Houston's two prior air pollution studies in 1956–1958 and 1964–1966, in 1970 the EPA documented that Houston had one of the worst ozone problems in the nation.[38] City and industry leaders attributed this problem to the increase in automobile pollution. Quebedeaux argued that industry was the source because pollution did not decline on weekends when there were fewer cars on the roads. Whereas Texas began its first air study as early as 1952, "extensive air quality monitoring did not occur until 1971."[39] Simultaneously, in 1971 the Stream and Air Pollution Control Section became independent from the Harris County Health Department and was renamed the Pollution Control Department.[40] Reflecting this new concern, shortly after the passage of the 1970 Clean Air Act, the *Houston Post* started a daily feature indicating expected levels of particulate matter.

Measuring Ozone Air Pollution Over Time

Measuring the extent and impact of Houston's air quality has never been an easy matter or taken a singular form. The data, as with so much of the heated debate on this issue, is not always available, consistent, or easy to decipher. There have been numerous bodies (federal, state, county, city, and private) that measured the incidence of pollution, and they do so over time with varying technologies, forms of measurement, and standards. In a nutshell, due to this variety, there is no single source of reliable data on ozone air pollution over time. For example, data on the number of days per year that Houston's air was unhealthful, referred to as ozone exceedance days, is complicated by various factors. First, there's the number, quality, and placement of monitors. Where the monitors are placed has a huge impact on pollution reports, and the more monitors out there, the more exceedances recorded. One senior scientist at the TNRCC suggested that "if we added 20 more monitors we'd probably add 20 more exceedance days" (see table 3.1).[41]

Also complicating the data are the forms of measurement, that is, whether to use a one-hour or eight-hour standard and whether it is best to develop an average index for air pollution. Since 1997, the one-hour standard at 125 parts per billion (ppb) is to be used for nonattainment areas such as the Houston-Galveston-Brazoria area; the longer eight-hour standard at 85 ppb is applied to all other communities. Accordingly, this essay uses data from a one-hour-averaged design values index (see table 3.2) to illustrate the broader historical trend because this index best complements the qualitative analysis evident each year in the press, public reports, and public debates.[42] While the averaged design values index is not helpful as an indicator of the actual number of bad ozone days in any given year or years, it is valuable, though not perfect, as an indicator of the historical trend in ozone air pollution.[43] Because of an interest in change over time, because the exceedance index data is viewed as being

Table 3.1. Ozone Exceedence Days with Number of Monitors

Year	*No. 8-hr Exc. Days*	*No. 1-hr Exc. Days*	*No. Monitors*
1972	27	26	1
1973	32	23	2
1974	26	25	5
1975	49	41	5
1976	80	64	4
1977	43	43	4
1978	102	84	13
1979	112	89	11
1980	98	94	14
1981	109	92	15
1982	181	121	15

Source: Texas Commission on Environmental Quality, personal communication with author, 2004.

Table 3.2. One-Hour Design Values Index, 1982–2003

1982	260	*1988*	190	*1994*	202	*2000*	199
1983	290	*1989*	220`	*1995*	189	*2001*	185
1984	250	*1990*	220	*1996*	202	*2002*	175
1985	250	*1991*	220	*1997*	199	*2003*	175
1986	200	*1992*	210	*1998*	196		
1987	200	*1993*	200	*1999*	202		

Source: TNRCC, Personal communication with author, 2004.

a better predictor of trends, and because the one-hour data was more consistent with observed trends, this essay utilizes the one-hour-averaged index over other data available. In summary, Houston's air quality was severe in the 1970s; improved by the end of the 1980s; generally flattened, without much overall change, in the 1990s; and improved again in the opening years of the twenty-first century, though not without mixed results.

Another way to understand this historical data is to compare Houston's one-hour-design-trend values to those of other major cities. In this regard, the Houston metropolitan area could have made more progress over time, which is a central point of this chapter.

Ozone Air Pollution, 1970–1990

In the early 1970s, ozone air pollution became more of a public issue in Houston. In 1972, for example, 39 percent of sampling days between May and October had ozone levels greater than .08 ppm. In 1973, this figure fell to 24 percent, but in 1974 it was 32 percent.[44] As noted above, the debate about the primary source of Houston's pollution—industry or automobiles—was evident even in these early years. Industry, fearing regulatory costs, charged that automobiles and other sources were the culprits. If automobiles, rather than industry, were the primary cause, then the costs of the solutions would be more dispersed, more shared with Houston citizens. This concern about how much cleaner air would cost and whether it was worth the expense continually and consistently informed local and state policy analysis and planning, certainly more than questions about public health. All of the tensions and debates in the early years of the fights around air pollution were complicated by the "relatively meager" knowledge at this time of ozone photochemistry.[45]

At the federal level, the EPA under Republican president Richard Nixon weakened the 1970 initiative, postponing many compliance deadlines. It was facing much local resistance from business groups nationwide, feared prior goals were too difficult to achieve, and wondered if the regulations might cause economic disruption. In 1973, the EPA extended environmental transportation plans for Houston and other cities.[46] In 1977, the newly amended Clean Air Act under President Jimmy Carter was also a cautious piece of legislation. It extended compliance with EPA regulations by ten years, giving cities such as Houston a ten-year "breathing break" during which the federal government did not ask much from industry. In response to federal withdrawal, localities that had been more cautious in their opposition to prior regulations became strident. In 1979–1980, a study from the Houston Chamber of Commerce, a primary source of both the environmental opposition and the culture of privatism, concluded that Houston smog presented little health hazard. Shortly thereafter the director of the city's air quality program asserted that Houston's pollution did not harm people much because "I don't go downtown and step over dead bodies there."[47] In contrast, a study done in 1981 by researchers from the University of Texas School of Public Health, the EPA, and the Baylor College of Medicine underscored the link between asthma attacks and ozone exposure.[48]

In the 1980s, under modest and understated federal pressure and industry initiatives, Houston did improve its ozone air pollution. Measures such as the installation of seals on storage tanks and the beginnings of leak-detection programs helped decrease exposure patterns. From the late 1980s to the early 1990s, new emission controls reduced VOCs average ambient concentrations, though NOx releases did not decrease during the same period.[49] Industry and opposition groups still resisted public pressure. Air pollution was seen as a private, not a public, matter. As the pressure heated up again, with a mandatory EPA deadline approaching in 1987, Houston's leaders complained vociferously. This time, they argued, such public intervention was not fair given the dramatic economic downturn in Houston.[50]

For most of Houston's history before the mid-1980s, the city had been a veritable boomtown of rising profits and economic gain. In a context of such economic growth and high profits, business and city leaders argued that regulation was unnecessary, that bad air quality was overstated, and that industry could take care of it. These were classic positions of both the environmental opposition and the culture of privatism. Publicly, at least, they did not oppose regulation on the grounds that it was too expensive for struggling companies and the local economy to bear. They did threaten that Houston workers would have to choose between jobs in the petrochemical industry and clean air, arguing that the petrochemical industry could not afford both labor and environmental costs, which seems a remarkable argument to make given industry profits in the 1960s and 1970s. Now in the mid-1980s, with the first major economic downturn in recent times, industry leaders complained that regulation was an unfair economic burden. They could not afford the heavy costs of regulatory legislation at a time when the Houston and Texas economies were buckling under the decline in crude oil prices driven by the Organization of Petroleum Exporting Countries (OPEC) cartel.

As the 1990s opened, Houston was still averaging approximately sixty-two days per year that exceeded federal ozone standards. While, in private, corporations were making some initiatives to mitigate certain aspects of air pollution, publicly the city and its business leaders, according to the National Commission on Air Quality, continued their "combative" and "minimalist" strategy toward the EPA. The new Clean Air Act of 1990 went beyond the 1970 and 1977 acts—for example, classifying Houston as a "severe" noncompliance area for ground-level ozone and giving it fifteen years to meet federal standards.[51] The city was not alone. In 1990, approximately 100 urban areas continued to exceed the ozone standard, many, like Houston, by a significant margin.[52]

Ozone Struggles since 1990

At the opening of the 1990s, Mayor Kathy Whitmire's administration, which was liberal on social issues, echoed a two-decade old argument of groups like the Ozone Task Force and American Petroleum Institute that the EPA's "theory" of reducing hydrocarbons to reduce ozone smog "simply doesn't work."[53] However, dissent against keeping air quality a private matter was mounting, and in the late 1980s and early 1990s it was becoming more public. Before his death in August 1989, Congressman Mickey Leland distanced himself "from the cautious—critics says obstructionist—consensus on air quality issues prevailing among many of Houston's business and political leaders for decades."[54] Texas, Leland said, had suffered the most from the failure to control toxic air pollution from industries such as Houston's petrochemical complex. An Exxon company memo at that time expressed fear that "environmental activism might spread from working class neighborhoods already and blatantly polluted to more affluent ones that had more political clout."[55] In 1990, Gary Mauro, then Texas land commissioner, agreed. "Houston is the only city in the United States where the city leadership is still trying to say they don't have an air pollution problem. They're still trying to define away the problem instead of fight the problem."[56] In the same year, George Smith, a Houston resident and chair of the Air Quality Committee of the Texas Sierra Club, attacked the Texas Ozone Task Force. He accused this committee, which was led by representatives from the Greater Houston Chamber of Commerce, of playing down health risks and proceeding too slowly on smog control. In response, Larry Feldcamp, a lawyer who chaired the Ozone Task Force, said that "more study of smog is needed before 'disruptive' steps are taken." Feldcamp, who also represented the Texas Chemical Council, an industry advocate, said the city needed more scientific study "to avoid disruptive and draconian controls" in Houston.[57]

In response to pressure from clean air advocates from within and without his administration, Mayor Bob Lanier was forced to enter the fray. Something needed to be done, but what? A modest consensus developed at the time around a proposal for Houston to implement a relatively innocuous public-warning system. This forecast alert system, in place in other states, would give Houstonians information about expected ozone pollution levels.[58] Los Angeles, for example, had implemented a forecast alert system thirty years earlier. Even before that, in 1955, the Los Angeles Chamber of Commerce, committing something close to "civic heresy," resolved to stop any new industries that could not comply with smog-control devices from settling in Los Angeles.[59]

Houston's corporate community, reflecting the opinion of leaders from the petrochemical industry, opposed such developments. In March 1994, the GHP opposed a program that would ask Houstonians to ride buses and take other pollution-cutting actions on days when forecasters identified high smog levels would be likely.[60] They said state meteorologists' forecasting techniques were too inexact to target the days when violations of the federal smog limit were probable and that it could be "deceptive to the public, hurt efforts to attract new business and prove costly to industries that take voluntary measures on those days." The science of prediction "is not there yet."[61] Critically, this was a minority opinion. Certainly George Smith and others at the Sierra Club, Jane Elioseff and others at Galveston-Houston Association for Smog Prevention (GHASP), and Lanier's environmental aide Mary Ellen Whitworth did not agree. Any consensus on air pollution in Houston was breaking down. Tensions were erupting in public, and divisions were becoming evident.

Initially, the GHP consented to publicize smog forecasts on local television and radio stations but objected to their being sent electronically through National Weather Service transmissions because those would be sent to news organizations in other cities, which might hurt Houston's efforts to attract new businesses. The GHP reluctantly consented to a fax system sent to local media outlets. Five months later the smog alert program was stalled because of business fears the forecasts might be misunderstood. Curiously, other Texas cities with less severe air pollution, including Dallas, Austin, San Antonio, and Corpus Christi, already had such a forecast warning system. George Smith, angry about the continued undermining of this modest progress, declared, "I guess it's all right to expose people to air pollution but not to tell them what they can do about it. It's a crying shame."[62]

In November 1994, the TNRCC, no fan of government intervention, agreed it would start issuing advisories next year, broadcast the night before by the local media in Houston. But in May 1995, as the ozone season beckoned, the joint task force of government, industry, and environmentalists working to issue smog-alert forecasts in Houston the coming summer fell apart due to "the continuing hesitancy of business leaders to participate."[63] The Regional Air Quality Planning Committee for the Greater Houston area collapsed. Dewayne Huckabay, the city official who chaired the air quality committee, emphasized that local business leaders' reluctance to participate in the forecast program forced it to be disbanded. The TNRCC backed off quickly, saying it would not deliver forecasts to Houston without there being collaboration in the city. Marilyn Browning, the GHP's environmental affairs executive, said that the GHP "does not believe the state agency's smog-predicting capability is accurate enough, yet." "This is just a smoke screen," responded Jane Elioseff of GHASP. "They didn't want the negative publicity, because if we predict 40 days and then have an additional 20 days with high smog levels, it just looks bad for business."[64]

Environmentalists were infuriated by GHP stall tactics. Elioseff lobbied Mayor Lanier to have the city issue state prepared forecasts if GHASP organized a public education effort. Later, in June 1995, one of the largest environmental coalitions in recent Houston history called for local government officials to issue smog forecasts to help protect public health.[65] The newly formed Smog Action Task Force, consisting of forty-two environmental, health, legal, and medical groups, requested that the city release smog warnings based on state data. Lanier was not convinced yet, but he, too, seemed to be tiring of industry tactics. In August, under pressure from his own staff and environmental organizations, Lanier approved the City Health Department requesting information on predicted ozone levels from the TNRCC and then disseminating it to local news organizations. This seemed to end more than two years of debate on the issue.[66]

It was not as though air quality forecasting was new or untried. Early forecasting began with "intermittent control strategies" in the 1930s when Canadian smelters linked sulfur dioxide emissions to weather forecasts. Later, through the 1970s, "dispersion models were developed and coupled with meteorological forecasts to predict air quality concentrations."[67] In addition, forecasters tried deterministic photochemical models that produced estimates of air pollution concentrations. Currently, most of the models used to forecast regular ozone and particulate pollution are statistical. Sophisticated multivariate regression equations are used to forecast peak ozone and particulate concentrations. A combination of deterministic modeling and regression analysis seems to work best.[68] The point is that while air quality forecasting is less than an exact science, as is weather forecasting or any other prediction of future events, the methods are highly sophisticated and absolutely "indispensable for those concerned with public health and the management of air quality, and for the public."[69]

Back in Houston in the summer of 1995, the forecasting system in its first season proved even more accurate. Through July of that year, Houston was having an average air pollution year, with at least twenty-six days of unhealthful ozone levels during the first seven months.[70] The program of ozone alerts began the week of August 15, 1995. Most but not all the major local television and radio stations agreed to run the forecast spots. During the month following the commencement of the ozone-warning system, after having had twenty-six high ozone days for the first seven months of 1995, Houston had twenty-two unhealthful ozone days between August 15 and September 12. During that period, twenty-three ozone warnings were issued. The state's new smog forecasts were accurate approximately 81 percent of the time, issuing twenty-six forecasts in that month, twenty-one of which were correct. The system obviously was not perfect, but from the initial experience it certainly had a high degree of accuracy, and the alerts were certainly not unduly alarming people. The GHP said that "this is not a good test summer."[71] Nevertheless, it was on the losing side of the argument. The alerts were successful and modest enough to be implemented. So, in September 1997, after two years of a successful ozone-alert forecasting system, it was agreed that the Texas smog forecasts for Houston would be issued via the National Weather Service, "an arrangement that once worried some local business leaders."[72] From 1995 through 2002, TNRCC (now the Texas Commission on Environmental Quality) data reveals, the forecasts for one-hour exceedance levels of 125 ppb have an average accuracy of 77 percent.[73]

The implementation of the much-opposed forecasting system should not be misinterpreted as a change in GHP strategy. In late August 1997, one could hear GHP-sponsored ads on "news only" KTRH radio opposing revisions to the Clean Air Act in Congress, charging the act was based on "bad science."[74] In 1997, the EPA was trying to strengthen standards for ozone and fine particulates. The GHP, then-governor Bush, and the TNRCC (the group that was supposed to regulate pollution but in reality consisted of many political appointees from the petrochemical industry) all opposed as "bad science" an EPA initiative for more stringent ozone standards.[75] As the governor of Texas, George W. Bush was a staunch opponent of government regulation of industrial pollution and an avid proponent of industrial self-regulation. Barry McBee, chair of the TNRCC, reflecting the "bad science" position, said, "The EPA needs to do its homework before imposing this on the states, communities, and people of Texas and the U.S."[76] Environmental lawyer Jim Blackburn saw this as consistent with Houston's long history of denial and obfuscation. "What the hell is the Houston Partnership doing making comments on a Federal Standard to protect the health and welfare of the people? Their only interest was to make sure it didn't interfere with economic development in this town and they did it on the basis that they didn't think the documentation was very good."[77] After years of resistance, the GHP in 1997 was willing to go along with modest ozone alerts, especially since they were accurate and helpful. It was another thing, however, to actually want to do something about lowering ground-level ozone pollution, implement public policy to do something about it, and publicize its unhealthful consequences.

Other critics also continued to doubt whether there was really much change in the position of the GHP, now the leading voice of the environmental opposition and the culture of privatism. On the one hand, its position had always seemed counterintuitive. While Houston remained heavily dependent on the petrochemical industry, the GHP is an organization of some 130 business representatives, most of whom do not benefit from weak regulation of air pollution.[78] While the Houston economy is still very connected to the petrochemical industry, the economy of the city, especially since the economic downturn that resulted from the decline of oil prices in 1982, has become much more diversified. The city and the GHP take pride that Houston is no longer a single-industry economy such as Detroit, which fared poorly in the 1980s and 1990s. Still, the GHP's stonewalling related to air pollution and environmental regulation always seemed, as with the Chamber of Commerce, to have been unilaterally written by the petrochemical representatives, with other members who did not benefit from air pollution and were even being economically and physically hurt by it deferring to them. Informal conversations with midlevel managers and professionals in the industry underscore a diversity of views on air pollution and a more honest discussion on issues than the culture of privatism suggests. But that seems more true for the midlevel professionals and less so for industry leaders. From the latter, one hears an almost conditioned response reflective of the environmental opposition and the culture of privatism. It is therefore not surprising that even in 1997, critics such as Terry O'Rourke, a former assistant county attorney, doubted whether the acquiescence on air pollution reflected a real shift in GHP policy. For O'Rourke, drawing a comparison between business and political elites in Los Angeles, the overriding ethic in Houston's petrochemical industry remained "we keep the air as dirty as we can get away with, when it should be 'how do we make it as clean as we can afford.'"[79]

Postscript on Ozone Politics in Houston

Unlike many other cities, Houston failed to achieve significant sustained improvement in ozone levels in the 1990s. The early years of the decade showed improvement; the later years did not. The downward trend of the 1980s "flattened" in the 1990s.[80] The Texas Chemical Council concluded that improvement resulted from a 67 percent reduction in total air toxins released in Texas between 1987 and 2002.[81] Houston's experience was not consistent with most other metropolitan areas. During the 1990s, improved air quality characterized most of the more than 260 metropolitan areas in the United States. "But of the 34 areas with increasing pollution, most were failing in ozone. Ozone also was the culprit for most of the bad air days in the 94 largest metro areas in the '90s."[82] Houston, perhaps the poster city for this trend, was no different. "The American Lung Association's 'State of the Air' report covering 1997–99 calculated that more than 141 million Americans live in areas that received an 'F' for ozone pollution." That same report emphasized Houston's lack of progress on air quality. On three-year-interval averaged data covering the period 1996–2003, for example, the American Lung Association reported that the weighted average of the number of unhealthful ozone days in this period fluctuated from 51 to 61.5 and 56 to 60.[83] There were improvements since the late 1980s, but after the late 1990s data fluctuations were probably more affected by weather patterns than Houston initiatives.[84] According to GHASP, the lack of initiative on air pollution, not to mention the opposition to progressive measures, resulted in Houston recording the highest readings in the nation in 1997, 1998, and 1999: 234 ppb, 231 ppb, and 251 ppb, respectively. The next year Houston was the only city in the United States to have three days above 200 ppb.[85]

Houston's experienced only slight progress on ozone pollution in the 1990s. Even though the total number of days decreased in some years, by the end of the decade Houston would pass Los Angeles as the worst city in the United States in terms of both the number of ozone days and the number of days with the highest concentrations of unhealthful ground ozone. Houston would hold that dubious and much-publicized distinction in 2000 as well. As recently as 1992, Los Angeles exceeded the ozone standard 250 percent (two-and-a-half times) as often as Houston.[86] The dramatic decline in ozone pollution in Los Angeles was due to policy reforms, which made it increasingly expensive to pollute. As an executive director of a Los Angeles air pollution program put it, "There's lots of things that California has done—that proved very cost effective—that, just from an outsider's viewpoint, you wonder why a place like Houston hasn't done."[87] Between 1985 and 2000, in just fifteen years, Los Angeles had been able to cut pollution by 75 percent "through stricter regulation of cars, industry, consumer products and other pollution sources."[88] Houston's leadership, on the other hand, preferred to address the issue in private and stonewall in public; air pollution was an industry matter, a matter of self-regulation and control, not a public issue to be addressed through public regulation or civic engagement. There is no doubt that progress has been made on ozone air pollution in Houston since the early 1980s. Since the early 1990s, however, the progress has been inconsistent and at a much slower pace than in other cities.[89] "There is no indication that air pollution has been reduced in Houston since smog overcame young athletes at Deer Park High School on October 7, 1999. Over the past three years [1999–2002] Houston has had more days with ground-level ozone exposures above the federal health standard than any other city."[90]

Anthropologist Janice Harper argues that the future of air pollution in Houston is unclear, just like its skies.[91] Things have improved, partly as a result of NOx reasonably available control technology (RACT) rules, which substantially reduced industry emissions. In addition, in general, automobiles and trucks have cleaner, more efficient engines.[92] Air pollution in Houston was pulled from the private to the public realm with both the national embarrassment of becoming the worst city in the nation in terms of ozone pollution and with the 2000 presidential campaign in which Al Gore made Houston the prime example of Bush's privatistic environmental policy. In 2001, the GHP received an Environmental Excellence Award from the EPA for its efforts to improve air quality in the Houston region.[93] Design values for the past few years are down. Air pollution is improving, though 2004 was not a good year. Still, debates rage about the extent and significance of air pollution. The environmental opposition continues to argue about the extent of air pollution and the scientific accuracy of analysts and regulators.[94] At the same time, environmentalists and public health experts warn against continued delay-and-deny tactics.

Of course, this ongoing conflict is not limited to Houston. Houston is an excellent case study for what is transpiring at the national level, both in terms of the role of the environmental opposition and an increasing national political culture of privatism, on the one hand, and the conflicts that continue with environmental and public health proponents. For example, while environmental opposition groups, such as the Foundation for Clean Air Progress, continue to emphasize the extraordinary progress that has been achieved on this issue, national studies of the problem document increased public health risks and economic costs associated with ozone air pollution. A recent study by researchers at Yale and Johns Hopkins universities found that ground-level ozone hits cities hardest, causing approximately 3,800 premature deaths annually in the nation's urban areas.[95] A related debate continues in Houston as well over the primary cause of ozone air pollution, and this debate heavily shapes proposed policy and reforms. The Houston Area Research Center (HARC), which is responsible for providing leadership and management for the Texas Environmental Research Consortium, proposes that while in the past petrochemical industry sources were the primary cause of ground ozone pollution, especially on the east side of the city, now automobile emissions are the primary culprits, especially on the west side of Houston.[96]

What are we to make of the future impact on air pollution of the environmental opposition in Houston and the city's political culture of privatism? Bill Dawson, who for years covered environmental issues for the *Houston Chronicle*, thought that one reason for the GHP's stonewalling in the mid-1990s was that the partnership was led by oil executives, such as the CEO of Exxon. Once the chairmanship of the GHP passed to the likes of James Royer, president and CEO of the engineering firm Turner, Collie & Braden, and once the light of national visibility shone brightly on Houston's polluted skies, at least the denial stopped.[97] Based on this historical analysis and recent events, it is logical to remain more skeptical, more concerned that the stonewalling on air pollution runs deeper in the political economy and culture of the city. The private sector remains strongly suspicious of public-sector intervention, and those in the petrochemical industry are among the most hostile. This suspicion extends well beyond Houston to the corporate boardrooms in other cities, corporations that control the policies and practices of many Houston-based businesses and are even less sensitive to local concerns and problems.[98] The issue is more complex than a change in leadership at the GHP.

There is also the element of enlightened self-interest, a well-chronicled characteristic of the Houston business community.[99] Clearly, since the bad national press and public exposure the city received in the 2000 presidential elections, Houston's business community has modified its public position on air pollution. Says Anne Culver, speaking for the GHP, "We've got to get it cleaned up. Now. We agree with that. It's a black eye. It's going to keep people from wanting to move their businesses here, it's going to keep smart young people from wanting to move here for good jobs when they graduate from college. . . . We'd like to be Number 1 on lots of lists, but not the list of dirtiest cities."[100] The GHP concluded that this distinction of being the region with the worst ozone pollution "marred the community's image and threatens its attractiveness as a place to work and live."[101]

To help remedy the problem, a subset of the GHP spent over $1 million in 2002 on a "Clean Air: It's Everybody's Business" ad campaign. The GHP Web site lists a number of its initiatives on air pollution, which include a GHP-created Business Coalition for Clean Air, the Ozone Science and Modeling Research consortium funded by state and federal dollars, creation of a statewide coalition, and participation in a national policy conference. The GHP expresses its commitment to addressing this critical issue for Houston and the Texas Gulf Coast and points to progress already made, a 40 percent decrease in ozone exceedance days since 1980.[102] In many ways, in the past half decade the GHP has become the major leader on ozone air pollution initiatives, becoming both the environmental mainstream as well as an oppositional force. This reflects not only the partnership's understanding of the critical importance of the problem but also a change in environmental politics since the 1980s. In general, as Dewey notes, since 1980, "the idea steadily advanced that environmental goals must be balanced against the sorts of economic goals that traditionally had always weighed more heavily in policy decisions."[103] The environmental opposition has changed its stance, and the political center on these issues has moved closer to them.

On the other hand, in the fall of 2004 the Houston area was beset with "the most widespread smog event in recent years."[104] Brad Tyer suggests that the GHP continues to use its old tactics; it continues to "hang its hat on the paramount importance of smart science." Take the recent case of the GHP suing the TNRCC for its faulty science in developing the Texas Implementation Plan, its most recent initiative of clean air regulations.[105] The GHP argued that "the complex computer modeling employed by the agency is not accurate enough to justify the stringent reduction that the TNRCC—which is charged with bringing Houston into compliance with the Federal Clean Air Act by

2007—imposed on industrial polluters in 2000."[106] John Wilson of GHASP emphasized that industry strategy would criticize the TNRCC plan as being based on "junk science."[107] According to a GHP proposal, a recalibrated model preferred by the GHP would "reduce the economic burden [of cleaner air] to the region by $9.15 billion."[108] It would also, according to GHASP, reduce a NOx control requirement from 90 percent to 80 percent, increasing NOx pollution some sixty to eighty tons per day.[109] Environmentalists agree that the TNRCC plan approved by the EPA is faulty. Everyone involved agrees that the science is imperfect. The environmentalists argue, however, that the standards are too weak. For them, the real issue is short-term profits over public health rather than a question of "bad science."

The basic reasons for GHP opposition include not only doubts about the science but also concern for the costs of public regulation of the oil industry. The GHP argues that it would cost the Houston region $13 billion to implement the attainment plan, and would reduce fuel production capacity at a time when the nation is struggling to increase it.[110] The group is not alone. Kenneth Chilton and Christopher Boerner, writing for the antiregulation Cato Institute, propose that the whole issue of ozone's harmfulness is overblown, especially when the imposed economic costs for prescribed EPA plans are considered.[111] Environmentalists counter that toxic emissions in Houston are six times (600 percent) more than actually reported.[112]

The struggle continues, but the proponents of a slower, less expensive, and not as thorough approach to ozone air pollution have most of the leverage. Throughout the history of the air pollution struggle, environmentalist groups have been overwhelmed by the sheer power of the petrochemical industry and its allies in the city, region, state, and nation. Barring a major industry disaster, however, Houston should experience incremental improvement in ozone air pollution, because the problem is more transparent and more public.

The reality and image of a polluted oil town no longer fit with Houston's new downtown and aspirations as a world-class city. Improvements will also continue to occur in technologies and science to lessen air pollution and evaluate interventions. But the pace, direction, and policies governing the modest improvements will be heavily influenced by the nature and extent of change in both the environmental opposition and the dominant culture of privatism.

CHAPTER 4

"The Air-Conditioning Capital of the World"

Houston and Climate Control

ROBERT S. THOMPSON

Houston is typically a hot and humid place. With summer temperatures often remaining above 90 degrees Fahrenheit and afternoon humidity around 65 percent, the city often seems unbearably uncomfortable. To transmute their city's climate, as the *Houston Post*'s longtime editor George Fuermann reported, Houstonians in the early post–World War II years decided to build "the world's most air-conditioned city."[1] The ensuing dedication to climate control has convinced contemporary Houstonians and many others that the city could never have grown into a megalopolis without cooling and evaporative technology. Houstonians used climate control to transform the cityscape, as the technology required the construction of environments in which weather conditions exercised few constraints on human comfort or economic efficiency. Houstonians' access to climate control, however, was a process that hinged upon one's ability to afford the necessary machinery. Furthermore, efficient air-conditioning demanded that houses be built for the new technology, effectively accelerating growth at the city's perimeter. Under the mantra of constant "progress" and "growth," two Houstons developed in the decades after World War II—one outside of the city and cool, and the other urban and hot.[2]

Houston's boosters have called their city "the air-conditioning capital of the world," "the buckle of the Sun Belt," and "the land of the big rich," presenting Houston as a place of wealth, urban superiority, and technological prowess. Situated at a crossroads between the South and the West, Houston epitomized the post–World War II boom image of both regions.[3] Building on their cotton and oil roots, Houstonians created a city based on domination of the region's physical environment. Houston's residents not only strived for mastery over their urban space and the region's resources, they also struggled to attain control over the city's hot and humid climate. To compete economically with northern cities, businesses in the South could not afford to be deserted during the hottest summer months. Commerce had to be conducted year-round without regard for weather. Houston's climate—similar to that in Charleston, South Carolina; Jacksonville, Florida; and New Orleans—had to be controlled if business leaders were to remake the city's image.[4] All-glass skyscrapers, the Astrodome, and climate-controlled malls would dot the city's landscape by the 1970s, as Houstonians transformed their urban and suburban space. Air-conditioning, always framed vaguely as "progress," demonstrated Houstonians' mastery over a "natural" obstacle and a major step toward Houston's status as a "modern" city. The six months of each year marked by temperatures exceeding 90 degrees Fahrenheit and high humidity would no longer dictate the pace of Houston's growth.[5]

With the appearance of air-conditioned theaters in the 1920s, Houstonians got their first taste of cooled summer air, and within two decades

Early air-conditioned theater, 1937. Courtesy Houston Metropolitan Research Center, Houston Public Library.

the idea of progress through climate control took firm root in the city. Air-conditioning meant expansion, and, in the American business climate, growth was always perceived as good. As George Fuermann wrote in the 1950s, "Houstonians uncompromisingly share at least one characteristic of most Americans. They regard progress as a baby regards candy: it may not be healthful, but it is desirable."[6] Thus air-conditioning represented an important part of the city's struggle for economic prosperity. Growth, however, often has losers, and the "progress" ethic espoused by the city's business leaders either pushed or left many Houstonians, like other southerners, economically struggling.[7] Suburban development had typically been confined to only the wealthiest of Houstonians in the first half of the twentieth century. According to the 1980 Houston City Planning Commission, the majority of upper-middle- and upper-class residents left the central city for the expanding suburbs in the 1950s.[8] By the 1960s, Houston's reformulated landscape was developed in a way that left the city's poor in racially segregated pockets of poverty and placed more prosperous whites in a suburban ring around the city. As a city dedicated to southern racial traditions, segregated housing was, and would remain, a tradition in Houston regardless of a person's wealth.[9]

White Houstonians struggled to maintain segregation as Houston's African American and Latino populations grew in the 1950s and 1960s. Challenges to racial separation in Houston's public golf courses, libraries, and schools, as across much of the country, remained mostly unsuccessful in the 1950s.[10] While the youthful and optimistic "Leave it to Beaver" image of the decade defined the 1950s for many white people, the massive and deadly white resistance to desegregation in the decade struck fear in the hearts of black southerners. Although the city escaped the open hostility that plagued other cities, deeply entrenched segregation still necessitated that Houston's image had to be different than the urban reality. As in many cities, Houston's planning process stemmed from business elites, so creating a positive urban image became paramount—profits depended upon it.[11]

Developing a modern, marketable image of Houston in the 1950s meant erasing much of the city's past and its present. Before air-conditioning, Houston thrived as the agricultural and oil center of southeast Texas. Hot weather was necessary for cotton and rice agriculture and did little to impede the region's growing oil industry. The slow pace and traditional mores associated with southern climate and commodity production, however, presented a challenge to Houston's boosters. Building a city of over a million people by the end of World War II while maintaining traditional social boundaries required skill, but creating a megalopolis would demand real ingenuity. As such, Houston's politicians, over the shouts of prosegregationists, struggled to erase the most obvious form of political backwardness, namely de jure segregation, while maintaining de facto racial separation as a reality.[12] By the 1960s, access to air-conditioning became a leading indicator of which people were the winners and losers in Houston's cycle of growth, and most of those at the bottom remained in the city's most economically forgotten neighborhoods.[13]

Air-conditioning companies and salespeople encountered a number of naysayers to the new technology. Just as Henry Miller had ruminated over the changes wrought by the standardization of American culture in the 1940s, many Houstonians feared that air-conditioning, along with a host of other electric technologies, threatened to end their particular way of life.[14] Houston's leaders still lived among a population of southerners from Richmond to San Antonio who celebrated the climate's "peculiar" effect on people. The tradition of slowing down and moving outside during the summer was ending; year-round continuation of business was becoming the norm. After the Second World War, Houston was transformed from being mainly a commodity trading and oil-refining city into a financial and manufacturing center that prized the standardization of climate—a sensation that Houstonians also began to want for their homes. As reported in the *New York Times* in 1955, however, people across the country were concerned about the effects of climate control: "Air conditioning has already done such assorted things to our way of life that, frankly, I am worried about what it will do the future. Hell, it may make us a nation of air conditioning worshipers, with the box on the windowsill a new sort of god."[15] God had apparently arrived in the South in 1951 in the form of a small and relatively inexpensive window-unit air conditioner.[16] Designed to control the temperature, pollen, and humidity in a small, confined space, these mobile boxes cooled individual rooms while being fairly inexpensive and requiring no costly ductwork.[17] Sales of room air conditioners took off in the 1950s, with sales at under 230,000 in 1951 increasing to 1.3 million by 1955.[18] Climate control, or residence-wide temperature and humidity control, was soon realized in Houston through building new houses to fit central-air technology or by adding ventilation work to older homes. By the 1960s, climate control became immensely popular in southeast Texas.[19] Whether selling window boxes or climate control, air-conditioning salespeople proselytized Houston, telling citizens of the coming of a new age of ease and comfort.

The sales of residential air-conditioning units in Houston first took off, as one would expect, among the wealthy, but such limited sales also brought smaller profits.[20] Marketing cool air to less affluent Houstonians, furthermore, required a touch of boosterism. Thus the southern tradition of slowing down during the summers had to be replaced by a more profitable ambition. Boosters argued that air-conditioning simply made people happier. As emotions defied quantification, the promise of happiness constituted the perfect sales pitch. Salespeople bragged "that the cooler Houston citizen, big or little, smiles more:

that he is kinder to his family and neighbors."[21] With increasing competition in the market for air-conditioning units and climate-control systems in the 1950s, prices for units fell, and more Houstonians could afford to feel the effects of cool, calming air.[22] Air-conditioning, therefore, carried the promise of bridging class divisions and fostering the peaceful and harmonious society that growing Cold War and racial tensions in the 1950s threatened to shatter.

Air-conditioning, some proponents argued, would also improve Houston's struggle with immorality. The poverty and lawlessness stemming from years of segregation and racial prejudice stung Houston, but cooler temperatures were trumpeted as bringing an end to the problem. Houston's ministers readily acknowledged the city's history of vice, but they promised to end the decadence associated with Houston since its founding in the nineteenth century.[23] "With such a headlong start in everyday sin," wrote Fuermann in the 1950s, "it is a wonder that the Houston of more than a century later was not a real cutup, a Barbary Coast with air-conditioning."[24] Air-conditioning, according to its proselytizers, could not coexist with vice and assured that Houston would stray from immorality. To ensure this result, air-conditioning engineers notified church leaders that their buildings needed to be cooler.[25] Air-conditioning's proponents argued that the technology would "mean a great upturn in religious interest and a consequent improvement in general morality."[26] The city's chilled theaters, which had been around since the 1920s, apparently did not threaten to spread Hollywood's perceived immorality.

In the years following World War II, downtown Houston became home to a new complex of climate-controlled edifices. Advocates of air-conditioning touted the innovative design of a new Foley's Department Store in 1947: "The revolutionary new store is windowless, except for main floor display windows on four sides of the block it occupies. This feature, plus electronic air cleansing and air conditioning, will make the establishment perhaps the cleanest in the world."[27] A growing number of climate-controlled buildings emerged in the downtown business sector in the early 1950s, but window units in office buildings remained quite popular and less costly than retrofitting older buildings or constructing new edifices with integrated climate-control systems during the early years of the decade.[28] As cooler buildings became the norm, however, Houston's downtown seemed destined for commercial success.

The marketing of climate control reached epic success in the mid-1950s, with Houston's climate clearly becoming an enemy to middle- and upper-class folks and the prices of installation falling to where business owners could afford to install the technology. Thus Houston businesses struggled to defeat the weather through establishing air-conditioned working environments. While in the 1940s climate control was primarily a luxury, to many a decade later, life without it seemed impossible. So many Houston office buildings had been retrofitted for climate control by the mid-1950s that ventilation engineers found that the office-building market "was solely in new construction and there is, of course, a great deal of this."[29] With this overarching attraction to climate control, Houston business owners could even find themselves held hostage to their own comfort. "Air-conditioning operators struck today in about twenty big downtown buildings in Houston," reported a journalist on a sweltering day in late July 1954.[30] The operators bargained for more pay and better overtime compensation, knowing that Houston businessmen would not tolerate working in a hot, humid environment.[31] The strikers, not the businessmen, were left out in the heat, however, as owners of the buildings quickly found other engineers to cross the picket lines.[32] One lesson from the strike centered on the traditional maxim that demanding better wages or benefits in Houston often meant losing one's job. Another lasting legacy for the city was also formed in those few days; businessmen would not spend one moment in Houston's hot climate if they had anything to say about it.

The successes of a commercial downtown failed to accomplish the one thing that many middle- and upper-class white Houstonians wanted—to live far from poorer urban residents. Coupled with the racial strife of the 1950s, white flight from the city created conditions where the physical distances between different races and classes expanded. An all-white, wealthier city had to be re-created far from the reaches of the urban masses, far from the pockets of poorer urban residents. The 1940s had already witnessed population shifts to the urban periphery. The population in the central areas of Houston fell by 13,658 people between 1940 and 1950, while the total population of Houston rose from 384,449 to 596,163 during the same period, indicating that many new Houstonians moved straight to the suburbs rather than into the older parts of the city.[33] Houston's population, as in other cities, may have been growing, but not in the city core. Nonwhites felt most of the negative impact of urban neglect, but poorer whites also lived in the economically forgotten neighborhoods.[34] As Houston's Planning Commission noted of the city's core in 1958, "the predominant housing characteristics of these areas are undersized single-family dwellings crowded together either upon small lots or with several dwellings upon one lot. Where larger dwellings exist, they too are frequently crowded together and are often occupied by more than one family. The environment created by this type of residential use is usually substandard."[35] This collection of generally old and dilapidated housing stock within the neighborhoods of the antiquated ward system shared few similarities with the climate-controlled marvels of the newer Houston suburbs.[36]

With the region's availability of cheap and abundant electricity, climate control served many social and economic purposes for Houstonians.[37] It made the region's weather more palatable, fostered economic growth, and, most important, allowed for the development of a more prosperous suburbia untouched by both natural and social undesirables. Being white, however, was not enough to move to particular suburbs. Having the "correct" ancestry in the 1950s served to exclude certain folks of European descent, as the "right" whites were encouraged to move to new neighborhoods by developers' proclamations of "100 percent Anglo-American" suburban populations.[38] As a bonus, the "modern" buildings that allowed human control of outside elements could be more easily constructed in open-spaced suburbia, which formed a ring around the city's urban core. These climate-controlled houses appealed to those who fit a particular demographic and could afford suburban space, and ensured that only people traveling to work ventured into Houston's downtown.[39]

To make space for Houston's housing and population shift, the city council generously applied its powers of annexation. Granted extraterritorial jurisdiction (ETJ) by the state, Houston could expand freely. Two major annexations occurred in the years following World War II: in 1949, 83.74 square miles around the city were annexed, and in 1956 another 183.80 square miles around the perimeter of Houston came under city control.[40] Consequently, the city's population density dropped from a high of 5,264 persons per square mile in 1940 to 2,675 by 1960, with 938,219 people living on 351 square miles.[41] With this tremendous growth into outlying areas, white Houstonians stymied the political strength of the urban core's growing minority population.[42] In addition, suburban development allowed for de facto segregation to flourish in the city. Perhaps 1960s suburban Houstonians could have it all—control of weather, control of business, and distance from poor, mostly nonwhite populations.

The physical growth of Houston meant increased distance between diverse places in the city, and distances from residences to businesses grew. The image of success in Houston, in large part, already meant constant access to climate-controlled spaces, thus air-conditioning in automobiles played a significant part in Houstonians making the choice to live at the city's periphery. In the early 1950s, a cooling system was a luxury item available in only the most expensive models but by the 1960s the technology had spread to a variety of low-priced cars. Chevrolet, for example, projected a 35 percent increase in air-conditioning installations between the 1963 and 1964 model years.[43] Cool cars made the move to the suburbs possible by the 1960s, as they allowed progress-minded Houstonians to live nearly any distance from the city core as they went from home to work to retail in dry, cold air.

By the mid-1950s and early 1960s, an all-out rush to the suburbs had begun. Gulfgate Mall and Meyerland Plaza were erected in the mid-1950s at the perimeter of the city, and in 1961 Foley's opened a new department store in the suburbs at the new and huge Sharpstown Mall, as Houston's new urban reality began to take shape.[44] Where only one suburban mall existed in the very affluent neighborhood of River Oaks before World War II, car-centered and climate-controlled malls were commonplace by the early 1960s.[45] Affluent white Houstonians, through their building of suburban climate-controlled houses, businesses, and commercial centers, could avoid both the region's climate and the city core, and the lack of cars and adequate public transportation kept many residents of older neighborhoods from traveling to the outskirts of Houston.[46] Those outside Texas had already begun to take note of the new way of life in Houston:

> Now the Houston man of affairs, and often of distinction, awakes after an untroubled night in his air-conditioned mansion or apartment, rides to his office in his air-conditioned automobile, goes through powerful and intricate brain work in his air-conditioned office, and eats lunch in his air-conditioned club. He is wholly immune to Houston weather except for the mercifully brief moments when he is ducking from house to car, from car to office, and vice versa.[47]

In the circular exodus from downtown, business followed housing, and housing followed business; prosperous whites who moved to Houston learned immediately of the advantages of new, climate-controlled suburban housing over retrofitting older housing stock in the central city.[48] The consumptive and pleasure-seeking lifestyle of suburban Houstonians demonstrated all that was good about the country to many Americans.

Working-class Houstonians in the 1960s likely felt they had little in common with Houston's "man of affairs." Residential air-conditioning was far from ubiquitous in poorer neighborhoods, and work environments for many unskilled laborers were not artificially cooled. Even when poorer Houstonians worked in cooled environments, factory owners were quick to note that the purpose of climate-controlled "factories in this area is primarily for increased production and efficiency which results in profits to the owner. The fact that conditions for employee comfort are maintained is coincidental."[49] Poorer Houstonians of the 1960s were familiar with air-conditioning through work and in commercial buildings, but it would be years before a majority of working-class Houstonians would live in cool homes.

Aside from working conditions, Houston's residential development served as a prime example of uneven growth by the early 1960s. Houston's urban core, save downtown, was becoming "largely Negro and poor" and Mexican American, while the surrounding areas were "predominantly white and affluent."[50] Air-conditioning became the standard in the homes of the city's wealthier residents, but it remained elusive in Houston's old ward areas. Although some homes in these areas were fitted with artificial cooling, by 1970 poverty affected over a quarter of all the households, the average housing stock was more than forty years old, and little more than one out of three residences had air-conditioning. African Americans were more likely to live without artificial cooling, but many white residents near downtown also found themselves without access to residential cooling.[51] Developers made several attempts to attract young white professionals to the city core in the mid-1960s, but few marketing strategies succeeded.[52] Downtown Houston was a place for affluent whites to work, not to live.

By the late 1960s, many white Houstonians deemed the city core unlivable, and without climate control, being there became almost unimaginable. Years of economic neglect had left even the most palatial of edifices in disrepair, and the local media perpetuated an image of urban horror. One of the grandest of downtown pre–World War II buildings, Union Station, was described by a local reporter as "a civic eyesore . . . it is dirty, it smells, the walls are grimy, the paint is peeling off the ceiling, people lie stretched out on the benches Skid-Row fashion, and it is hotter than the hell Dante visualized."[53] Vice and crime were tied to the city core, and the national political discourse of "restoring law and order" to cities appealed to suburban Houstonians who were prepared to make a racial connection regarding lawlessness.[54] Inner-area Houston's heat, economic depression, and "dangerous" black communities were set in obvious juxtaposition to the imagery of Houston's newest suburban neighborhoods.[55] For any of your minor troubles in suburbia, at least you were not in the heart of the city, in that Hades, with those people—you had everything you wanted—as long as what you desired came in an air-conditioned suburban building.[56]

Houston's suburban developers made sure that everything people wanted came in suburban, climate-controlled packages. Air-conditioning sales in the early 1960s grew rapidly, even as actual thermometer readings throughout the nation generally stayed the same, and a move by the National Weather Bureau's official temperature station from the breezy Gulf-side of town to the hotter northern section of Houston only heightened the sense of oppressive heat in the city, meaning a perceived two-to-three-degree air-temperature difference.[57] The thought of rising temperatures meant even more demand for climate control. "The gain in dollar volume has been far more spectacular than the rise in temperatures and the air-conditioning industry appears certain to set new records by a wide margin," wrote journalist Richard Rutter in 1964.[58] Marketing of air-conditioning created a world in which climate control, regardless of outdoor temperatures, was deemed necessary.

The quest for full-time climate control skyrocketed in the early 1960s, and installations of central air-conditioning in single-family homes surpassed all other applications by mid-decade. Houstonians devoured the new technology, as those with means rushed to live in fully climate-controlled houses, even as the city's Chamber of Commerce heralded the high quality of Houston weather to attract business.[59] Houston by the mid-1960s grew at a rate that had only been outpaced by the city's population expansion during the oil boom of the 1920s. By 1970, in the west and southwest areas of the city, the population was 99 percent white, most homes had been built in the previous ten years, and 98 percent of the homes had air-conditioning—of which 80 percent of those residences had central air-conditioning. This area of Houston was now a modern, technological locale, expanding with new climate-controlled public spaces to complement the area's cooled homes.[60] Houston was no longer merely a southern or southwestern city, but a place of growth, of comfort, of convenience. With the weather subdued, the only climate that would matter was the city's business climate.

Houston's economy, although spurred by proclamations of its residents' unique business acumen, benefited greatly from federal largesse after the Second World War. In addition to a national commitment to automobiles and freeways, from which Houston's oil economy benefited, the city received a huge economic boost in 1961 when the federal government established NASA's Lyndon B. Johnson Spacecraft Center just southeast of the city.[61] This complex represented the ultimate in human control of the environment, where even gravity was put to the wayside. With the addition of Mission Control in 1965, Houston became the focus of more than earthly progress; it also represented universal progress.[62] A major component of the U.S. space team and their families lived in six newly developed, climate-controlled sections of Clear Lake built specifically for their arrival.[63] Control of all environments seemed mastered in southeast Texas, and public dollars flowed into developing modern marvels of architecture.

In 1964, Houston unveiled its ultimate showcase of climate control to the world. Officially named the Harris County Domed Stadium, but dubbed the "eighth wonder of the world," the Houston Astrodome represented the city's absolute mastery of the region's environment. Finished in time for the 1965 Major League Baseball season, the building awed the public.[64] After the inaugural game between the New York Yankees and the Houston Colt '45s, a New York reporter exclaimed: "That huge greenhouse in Texas is so monstrous in size that it doesn't give the immediate impression of what it is actually offering—baseball in a gigantic air-conditioned room. But after a while eyes wander and the observer is jarred back to reality. He sees the girder, the back wall and the dome, some of it much like an armory. That's when he reaches the one inescapable conclusion. He is indoors."[65]

Never before had a city so controlled the environment in which sports teams played and the fan base cheered. As the Carrier air conditioner hummed along, no matter the outdoor conditions, Houston's professional baseball team always played in 74-degree weather with 50 percent humidity, and spectators regularly found anything from a hot dog and a cold beer to lobster and hot

The Astrodome and its parking lot. Courtesy Houston Metropolitan Research Center, Houston Public Library.

toddies as they relaxed alongside the dome's $4.5 million cooling system.[66] The spectacle of the Astrodome marked a continuing affection for huge, air-conditioned spaces. Promoting the "air-conditioned, all-purpose structure," Mayor Louie Welch believed the Astrodome would "greatly enhance the prestige of our area and prove a welcome boon to our already healthy and vibrant economy."[67] As a type of blessing on the city, even the Reverend Billy Graham heralded the edifice: "This is in truth one of the great wonders of the world . . . a magnificent dream coming true before our eyes . . . the boundless imagination of man transformed to reality."[68] As Houston planners specifically positioned the Astrodome for people to avoid the traffic and congestion of the city, suburban Houston became synonymous with ease and comfort.[69]

As Houston expanded, downtown businesses and real estate developers appealed to citizens to remain committed to the city's traditional commercial core.[70] With declining sales, boosters appealed to the new sensitivities of Houston's middle and upper classes. In 1966, a "Main Street Mall" was proposed that would extend from Pierce Street to the Buffalo Bayou. Spanning three levels from subterranean tunnels to second-level skyways, pedestrians could shop in air-conditioned comfort.[71] The calls for downtown shopping revitalization, however, went largely unheeded, as controlled public environments became more popular outside the old ward neighborhoods. What did gain business support was expanding the downtown tunnel system that originated in 1947. These tunnels expanded to 2.8 miles and connected fifty buildings by 1977, while offering air-conditioned comfort and "safety" to people who traveled to and from offices, retail stores, and parking garages.[72] The growth in this area of Houston tended to be upward and office oriented as modern, climate-controlled buildings replaced "slums," especially in the traditionally African American Fourth Ward, known as "Freedmen's Town," on the west side of downtown.[73]

By the mid-1960s, on the eastern edge of downtown, the Texas Eastern Corporation looked to take advantage of Houston's fast-growing economy by developing a "modern" complex of buildings encompassing thirty-two city blocks. Running up to nine blocks east to west and up to four blocks north to south, the gas transmission company turned real estate developer looked to capitalize on Shell Oil's move of its corporate headquarters from New York City to Houston. To attract Shell and other businesses, Texas Eastern formed a development plan called the Houston Center, which included plentiful parking, modern "people movers," plazas and walkways, and multiple high-rises with residential, shopping, and entertainment space. By 1970, without Shell or any other marquee tenants committing to occupy the development and receiving public criticism of the ambitious plan, Texas Eastern faced a difficult decision regarding the plans to modernize eastern downtown. The corporation decided to go ahead and break ground for the forty-story Two Houston Center in January 1972 amid questions of its profitability. With several major clients committing to leases, Texas Eastern moved forward and finished One Houston Center in 1976. The major vision of creating a thirty-two-block air-conditioned complex of buildings was altered, however, as management turnover and meager profits slowed further development on the eastern side of downtown. Although several buildings and a new convention center were completed by the early 1980s, Houston Center looked little like the original plan. The developers of Houston Center found that competition from complexes built on the edges of town made development of a controlled environment in downtown a risky venture.[74]

One of the ultimate controlled environments was built at the west end of Houston. The Galleria Mall, with its multiple levels and scores of stores, represented the ultimate in luxurious shopping for Houston's middle and upper classes.[75] Built in the late 1960s, the complex defied the sensibilities of many of its visitors and was presented as the archetype of future malls. As the temperature sailed well past 100 degrees Fahrenheit, skaters laced up and headed to a full-size ice rink at the center of the mall. Journalist William K. Stevens noted that heat waves did little to change the lives of well-to-do Houstonians, "they simply stay inside where the air-conditioning banishes discomfort."[76] To many Houstonians, skating and whiling away the hours in air-conditioned comfort while outside temperatures hovered around 100 degrees seemed normal. Others obviously did not think so. As many people struggled to cope with the heat and disrepair of the inner city, Houston's leaders continuously marketed their city as the friendliest climate in town—for business at least.

Houston's leaders hailed nothing but success and progress, and any of the city's problems were "deemed insignificant" in the push for economic growth. With the climate made nearly prostrate, recruiting top-flight businesses and professionals came down to the city's economic potential. "Problems such as physical climate, a humid heat that drives citizens into air-conditioned cocoons, industrial pollution that taunts allergy sufferers with a literally stinking haze, a meager public transportation system and a leading homicide rate were deemed insignificant" upon Shell's 1970 move from New York city to Houston.[77] Houston did offer the corporation proximity to the emerging energy center of America, but Shell employees and their families, like many other immigrants to Houston, would never have to see much of the city anyway; they found themselves in suburban developments built specifically for the city's well-to-do newcomers.

As long as Houston grew, city leaders acknowledged few urban problems. Although city planners and architects warned of urban decay, boosters argued that any troubles that the city encountered resulted strictly from the city's growth, which could only be portrayed in a positive light.[78] After five years of tremendous population growth, in 1977 Mayor James J. McConn exclaimed that "our success has caused us to strangle somewhat, especially in traffic . . . but remember, we are talking about success. Our problems are related to growth, not to a mass exodus of people that troubles other cities. And amen, I'll take problems of success at any time."[79] Believing in the mantra of growth in the city's mostly suburban business centers necessitated ignoring the deterioration of the residential inner areas of the city.[80] This development was underscored by white Houstonians in the 1970s desperately trying to remove their children from Houston's urban public schools in order to evade desegregation decrees.[81] Houston's dedication to privatized economic expansion created wealth at the edges of the urban area while funneling money out of the inner areas. This practice, backed by public policy and planning decisions, led to expanded political and economic power within the elite classes and greater social distinctions between the haves and have-nots.[82]

Houston needed its boosters to belittle any of the problems that plagued the quick-growing metropolis. The city's leaders, for example, blamed mosquitoes for outbreaks of encephalitis rather than the human agency present in the city's sprawl through the countryside. As journalist James Sterba wrote: "Since air-conditioning, Houston's population has increased from less than 400,000 in 1945 to more than 1.4 million [in 1976]. That is part of the problem. They have spread themselves over 520 square miles. Real estate developers continue to implant tomorrow's subdivisions in yesterday's bogs to accommodate newcomers."[83]

Houston's growth had always defied reality, however, even before air-conditioning. Since the city's founding in 1837, Houston's residents had dealt with the city's swampy terrain, bringing heat, humidity, and illness. By the early 1900s, the oil boom sent Houston's population skyrocketing, but the difference with postwar growth was that the age of climate control allowed many Houstonians to avoid being outside. Thus residential tracts could even be developed in the old swampy lowland rice and sugar plantation areas to the southeast and southwest of Houston. To perpetuate an image of progress, however, nature, rather than human action, had to be blamed for any speed bumps on the road to growth.

As early as the mid-1960s, a number of Houstonians began to challenge the growth ethos of Houston's elite, and advocates of slowing down the growth of the city were not limited to those in the burgeoning movement of environmental activism.[84] Urban sprawl and the accompanying urban decay, for example, challenged some architects' long-held ideas about community, personal interaction, and responsibility to fellow citizens.[85] Jack McGinty, local architect and developer, charged Houston's private sector with negligence in letting parts of the city decay. "Everybody is so fat, dumb and happy doing their thing that they can't see what's happening to the place," McGinty ruminated in 1977. The money being made in the energy industry, according to the developer, made the Chamber of Commerce's proclamations of unique Houston business moxie ring something less than true. The city, McGinty believed, was "successful because we're in the nerve center of the energy situation, and because of that, people are flooding in here at such a rate that anybody can make money."[86] For Houston's business elite, however, as long as money flowed in, the city needed to maintain its path of unlimited economic and population growth.

No one needed to tell a vast number of Houstonians that the city's rising economic tide failed to lift all ships. People outside Houston's glitz and glamour areas reckoned their conditions with the city's boom image. Although some African Americans had made their way out of the city to the nearby suburbs just northeast and southeast of downtown and lived in climate-controlled houses, most blacks remained in neighborhoods that were generally lacking in air-conditioning. Spanish-speaking Houstonians fared little better, as in 1980 the majority of nonwhite Houstonians lived in noncooled residences in the city's core.[87] "Just as there are two Americas, there are two Sun Belts," wrote journalist William Stevens in 1980, "one white-collared and affluent, the other working-class and poor." Access to air-conditioning became one of the indicators of health and welfare. Houston's rows of shotgun houses and dilapidated houses came to fascinate the national media. "On Houston's East Side, far across town from the Galleria and the sleek suburban precincts it mostly serves, blocks here and there are lined by virtually identical one-story frame dwellings," wrote Stevens. "In these neighborhoods live the poor and struggling of Houston. Few can afford air-conditioning, and misery prevails."[88] Regardless of the positive rhetoric surrounding Houston's economic growth, the city's weather affected poorer, urban Houstonians more acutely than it did those living in suburban, climate-controlled environments.

As early as the 1950s, scientists detected that large cities created their own weather, and Houston was no exception.[89] The city's sprawl encouraged a distinct urban heat island that raised air temperatures in the whole region. The huge swaths of pavement and black roofs kept the city's sunlight trapped near ground level. Although increased rainfall remains the prime indicator of regional climate change, Houston's rising heat was remarkable; by the 1990s, the city was regularly five to nine degrees Fahrenheit hotter than the surrounding areas during the summers.[90] Without the means to cool one's home, Houston's weather became even more oppressive. With these rising temperatures, by the 1990s access to artificial cooling heightened the social chasm between those who had and did not have access to air-conditioning.

With climate control so prevalent in Houston, and increasing geographic separation between those with means and those without so great, two distinct societies developed in the city—one in which heat and humidity were mere inconveniences, and one in which people coped as best they could. Houstonians had for decades believed that heat-related deaths were "natural." This consensus, however, began to change by the early 2000s. Averaging sixteen deaths from heat per year, nearly one-quarter of all of Texas's heat-related deaths, the "world's most air-conditioned city" obviously did not ensure climate control for all Houstonians. As heat waves remain a threat to Houston every year, critics mounted a campaign that blamed heat deaths on "poverty, isolation and substandard housing in neighborhoods where crime and lack of businesses or social centers keep people shuttered in their homes."[91] With houses closed to outside air, power outages and heat-moving fans endangered lives when temperatures and humidity remained high for days.

Fixed-income senior citizens found themselves most vulnerable to heat-related deaths. Past the age of working in or easily traveling to air-conditioned environments, seniors' access to cool temperatures remained tenuous at best. With the price of residential climate control out of reach of many Houstonians, some elderly folks had to make a choice between food and medicine or air-conditioning. The costs associated with electric cooling, and the market model of cutting service to those who paid bills delinquently, left consistent climate control out of the reach of those with limited finances. In addition, community assistance programs in Houston found their budgets for utility-bill assistance inadequate at best.[92] The opening of neighborhood cooling centers, with free transportation to climate-controlled buildings, emerged as a quick fix during Houston's hottest days. This development, however, offered no long-term solution to seniors who consistently battled with a built environment in which they found themselves trapped in hot boxes in the middle of an urban heat island.[93]

With most buildings designed for artificial cooling, a great number of Houstonians developed a habit of merely avoiding the heat and humidity of the city. The threat of widespread loss of life in the case of a long heat wave, however, remained ever present. With this looming specter, along with growing concerns over pollution, energy conservation, and long-term economic development, some Houstonians began to look toward more efficient building practices.

The possibility of energy crises and the high cost of energy prompted some Houstonians to reconsider their building and development models. Architects learned that "something as simple as positioning the building so that east-west exposure is limited could reduce cooling costs."[94] By 2003, Mayor Bill White joined the crusade to encourage "green" development in Houston. Noting the urban heat island and the energy applied to artificial cooling, he argued that "Houstonians can and should lower outdoor air temperatures by increasing vegetation and using paving and roofing materials that absorb less heat."[95] By using language that appealed to many Houstonians, such as the idea of "progress" through "green" architecture, the mayor struggled to maintain the city's economic momentum.

The idea of rejecting Houston's environment in a wholesale fashion seemed to falter by the late 1990s. Even the once amazing Astrodome could not compete with the national trend toward open-air or retractable-roof stadiums.[96] Houstonians still concerned themselves with an image of progress, and convertible stadiums represented the most modern of ballparks. Not to be outdone, Houstonians built two such stadiums—one for the city's Major League Baseball team and the other for its National Football League team. Surprisingly enough, it was the Hades of Union Station that was remodeled to become the entrance to the city's hallmark of Major League Baseball stadiums. The city's well-to-do citizens, unlike in previous decades, actually wanted to be outdoors and in the middle of the urban core, on occasion. Houstonians still, however, wanted their connection to the city and its weather to be on their own terms, and as some baseball purists predicted, the roofs on the new stadiums tended to be closed more often than they were open.[97] When temperatures got too hot, or if it rained, Houstonians quickly entered back into the cocoon of climate control—and some fans could even watch their teams from the luxury of their own air-conditioned suites, separating themselves even from the heated mass of the crowd.

Houstonians' affection for controlled environments seems unlikely to falter in the coming years. As energy costs go up, however, the divide between those with and without air-conditioning will likely remain present. As Janee Breisemeister, of the Consumers Union in Austin opined: "In Houston in the summer, air-conditioning is a necessity, like food and shelter."[98] Access to climate control in the city, however, just like other needs, did not always cross socioeconomic boundaries. Where some people could ill afford to purchase or operate air conditioners, others cared little for price or efficiency; as long as the indoor climate was comfortable, any cost was worth it.[99]

It is easy to credit technology alone with changing people's ways of life. Houston, without a doubt, changed as a result of climate control, but the transformation was neither inevitable nor universally beneficial. The growth of Houston, furthermore, cannot be so easily explained as a conquest of the city's climate through the increased use of air-conditioning. The city's growth stemmed from the business community's commitment to economic expansion, of which climate control played only a part. The losers in Houston's growth, furthermore, can still be seen on the streets of the urban core. William Stevens's observations in 1980 remain relevant: "As the heat drives West Siders inside, to cool comfort, it drives East Siders outside. And as the sometimes ramshackle streets of the East Side are consequently jumping with life, those of the West Side are often deserted during the day, as if a sudden pestilence had wiped out the population."[100] The winners in Houston's cycle of economic growth are hidden away in their well-sealed coves, depending on blasts of cool air to reassure them that they are in control and successful.

PART 2 GROWTH OF THE METROPOLITAN REGION

Although founded in 1836, Houston—and this also is true for the surrounding towns and cities—is a product of circumstances much different from urbanization trends east of the Mississippi River. Houston, therefore, is by no means typical of all American cities. It is an archetypal twentieth-century city, which came into its own with the popularization of the automobile. In its modern form, the metropolis is multinodal, decentralized, and expansive, much like Dallas–Fort Worth, Oklahoma City, Phoenix, and Los Angeles. Lying on the southern fringe of the United States in the Southwest, it has been regarded as "the golden buckle on the Sun Belt." Its long-standing image as a boomtown was rightly derived from its flush years as a center of the American oil industry. It has been referred to as a "freeway city," "strip city," "space city," and in, a most unflattering way, the "soulless Los Angeles of the Gulf Coast." Houston is regarded as a consummate suburban society, especially given its low density and the plethora of homes fronted by lawns and equipped with attached garages.

Like all cities, Houston has changed over time. In recent years, the city has become less the boomtown driven by oil. It has begun to develop a legitimate downtown, and has seen middle-class citizens returning to the central city to inhabit the numberless new townhouses and loft apartments. It has retained, nonetheless, its vast suburban flavor, its many and varied racial and ethnic ghettoes, and its penchant for sprawl. On a day-to-day basis, and upon quick observation, the relationship between Houston's rampant growth and its role as an energy-intensive metropolis may not be so apparent. However, the Houston metropolitan region remains a product of its oil and petrochemical heritage, its serious dependence on gasoline-powered transit, its intricate and extensive network of roads and highways, its climate-controlled buildings, and its natural gas for heating. Such a footprint is large.

The essays in this section discuss two essential networks of survivability for the city and region: its sanitary system and its freeways. The section also focuses on the most characteristic physical feature of the city—sprawl, resulting from utter dependence on automobiles and trucks and a rather liberal policy of land use. It ends with a comparison of two dissimilar urban centers, Houston and Galveston, bound together by their proximity and location along the eastern Gulf Coast.

Martin Melosi's essay examines the development of sanitary services—water, wastewater, and solid waste—and their transition from local concerns to regional challenges. The origins of these services were typical for most cities in the United States, but that the city had to confront their expansion and impact on a regional level is not typical—at least on this scale. Rampant growth became the most serious issue that city officials faced in providing sanitary services for the city. Going beyond city limits to seek water and to dispose of waste created a series of encroachments that did not have to be considered in earlier times. Essentially, changes in the delivery of sanitary services were caught between a stated goal of relentless economic growth for the city and the questions of sprawl and regional conflicts.

Tom McKinney's essay treats the development and impact of the first interstate highway in Texas, which helped shape Houston and Galveston and became a model for future national highway development. The Gulf Freeway most obviously demonstrated the region's growing dependence on the automobile, and, as McKinney concludes, it "was among the first steps in the creation of Houston's modern built environment, which is often characterized by its highway system."

Diane Bates explores the issue of sprawl through a case study of the Piney Woods north of Houston. Rather than primarily trace the steps leading to sprawl and the varied paths that it took in the region, Bates dramatically demonstrates in words and pictures the destructive impact of sprawl on an important part of the natural environment originally beyond the city's northern border—particularly in "four counties that comprise the majority of the San Jacinto River watershed." She, too, casts her story in a regional context, bemoaning the fact that "thirty years of individuals escaping to forested suburbs has done little to protect the region's forests, much less improve the overall quality of the San Jacinto watershed."

Finally, William Barnett compares the economic and environmental relationship between Houston and its old rival, Galveston. As Barnett suggests, "Their histories have always been interconnected, and in the nineteenth century the two cities followed fairly similar paths, but in the twentieth century Houston and Galveston took divergent routes." Clearly, Houston's commitment to industrialization—of which oil and refining played a central role—led to a striking economic transformation with a whole variety of residual impacts. Galveston's revival as a "tourist getaway" took it in a different direction, one that has yet to show the dividends of the city's expectations.

These essays complement those in the first section by demonstrating some of the key impacts of Houston's emergence as a dynamic economic enterprise, fueled by oil, natural gas, and a vast refining capability. The causes and consequences of rampant growth are complex, but the style and substance of urbanization in southeast Texas speak to the interrelationship of economic choices, physical realities, and timing.

CHAPTER 5

Houston's Public Sinks

Sanitary Services from Local Concerns to Regional Challenges

MARTIN V. MELOSI

Sanitary services—water supply, wastewater, and solid-waste collection and disposal—form the circulatory system of a city. Not only do they transport vital resources in and carry unwanted materials out, but they also play a significant role in preserving health. The timing of and commitment to establishing and maintaining sanitary services in Houston were not particularly unique. In most respects, city authorities followed a pattern of development in line with many cities in nineteenth-century America. These patterns were shaped by population growth, the onset of epidemic diseases, inadequate maintenance of private wells and privy vaults, and newfound municipal authority and taxing power.[1]

The perspective of Houston as a "free enterprise city" eschewing public spending not directly benefiting profit-seeking business leaders does not seem to apply here.[2] For the most part, the city invested in a water-supply system, underground sewers, and solid-waste collection and disposal operations irrespective of the vagaries of the business climate. Houston, however, was not a leader in offering adequate sanitary services to all citizens in a timely manner, and city authorities certainly took into account the value of services vis-à-vis the economic vitality of the community.

Despite its apparent unremarkable history, the development of Houston's sanitary infrastructure provides insight into the nature of the city's growth from a small nineteenth-century town deep in unsettled southeastern Texas to the fourth-largest city in the United States. Houston's population in 1837 was a modest 1,500. Maturing as an important regional commercial and industrial center spurred on by World War I and especially World War II, Houston's population reached approximately 700,000 by 1945, and the city covered an area of about 17 square miles. By the late twentieth century, Houston clearly had arrived as a major American city, topping two million people within the city limits and extending over more than 600 square miles. Beyond these gross statistics, the Bayou City had long achieved world-class status as the "energy capital of America," a major international center of medicine and medical care, and an archetype of the modern Sun Belt city.

The development of the city's sanitary infrastructure is an important underpinning to the city's evolution. Prior to World War II, the expansion of sanitary services was guided by localized issues, namely Houston's modest population growth and the city's physical assets and liabilities. After the war, Houston's rampant physical growth, its population explosion, and its rising economic status produced a metropolis whose regional impact was immense and whose sanitary services were linked inextricably to southeastern Texas and the Gulf Coast. The city's regionalization was central in seeking new water supplies, in producing runoff that polluted Galveston Bay, and in disposing of countless tons of garbage and rubbish.

Especially with respect to sanitation systems, there have been two Houstons—a pre–World War II emerging city, where local concerns and interests dominated the delivery of services, and a post–World War II metropolis, where those services reached beyond the city limits to have broad-scale environmental and political impacts. This theme provides the backdrop for this chapter, and helps to explain the quality, extent, and importance of the city's sanitary infrastructure.

Origins and Implementation of Sanitary Systems, 1876–1945

Before the 1870s, sanitary services in Houston were primarily the responsibility of the individual. The city did not develop its first public water supply, sewerage system, and regularized solid-waste disposal program until it emerged as a commercial center in the 1870s and 1880s. The establishment of citywide systems was spurred by growth demands, the need for fire protection, and health questions—typical for most cities. However, the types of systems developed in this period—especially in the case of water supply and solid wastes—depended heavily on the geologic and climactic conditions of southeast Texas. Potable water came primarily from extensive aquifers beneath the city, and much of the refuse produced by Houstonians was burned to keep it from putrefying in the intense heat.

The new citywide sanitary systems obviously were meant to serve local, immediate needs and requirements. They did so unevenly, however, by giving preference to commercial districts and incorporated areas, with modest attention to the less affluent and to marginal land uses. The new services also proved inadequate in the wake of the aggressive physical expansion of the city and the great economic boom that followed World War II. As Houston became a regionally dominant metropolis, its sanitary services had significant influence on places beyond its borders. Local needs would clash with regional effects and impacts—political and environmental.

Water Supply

Water supply was the first important public utility in the United States, and the first municipal service that demonstrated a city's dedication to growth. In the early nineteenth century, most American cities relied on streams, ponds, and wells; by 1900, public systems were the major source for residential and commercial consumption and fire protection.[3] During its frontier days, Houston depended upon underground brick cisterns, overhead cypress tanks, and private wells for drinking water and for fire protection. Water from the bayous was considered good for drinking. When fires broke out in residential areas, bucket brigades tapped cisterns and shallow wells. Between 1838 and 1895, volunteer firemen provided the only fire protection.[4]

Houston developed its first public water-supply system in 1876. By national standards, the city's commitment to a centralized water source and distribution network was typical for a city of its age, size, and location. The system was unique insofar as the city relied exclusively on groundwater from countless wells from 1887 until the 1940s. The water was drawn from the Chicot and Evangeline aquifers running southeast to northwest from the Gulf Coast through Harris County's western half and into Montgomery and Grimes counties. Until it sought to develop surface water in the 1940s, Houston was the largest city in North America to rely exclusively on well water.[5]

The public uproar over a major fire that scorched the business district in February 1859 led to the installation of the first public cistern in downtown Houston. After a second major fire broke out in March 1860, the city constructed additional public cisterns.[6] For nearly two decades, little was done to upgrade the modest water-supply service. In January 1878, Mayor James T. D. Wilson noted the pressing need for waterworks and sewers. Dependence on cisterns for fire protection had become unworkable as demands for water by a growing population intensified.

Episodic health problems linked to the water supply also raised questions about the need to change the source. In the wake of a yellow fever epidemic in 1869, Mayor J. R. Morris called for a survey of Houston's water supply and concluded that the city should change sources. He investigated using the nearby

Brazos River west of the city, but engineers believed the water was impure. He also considered the San Jacinto River to the east for diversion into White Oak and Buffalo bayous, but the panic of 1873 derailed further action.[7]

Throughout the early 1870s, several northern companies approached the city council offering to install a water-supply system, but the city's shaky financial status as a result of the panic precluded issuing a franchise. When the financial picture improved, the council entered into a twenty-five-year contract with James M. Loweree of New York and his associates on November 30, 1878. Loweree's credentials as an experienced engineer and builder of several other waterworks won him the franchise.[8]

The Loweree group organized itself into the Houston Water Works Company on April 15, 1879, and began operations in August. The original waterworks consisted of a pumping plant, which drew water from Buffalo Bayou and pumped it directly into distributing mains. The water supplied was used primarily for fire-fighting purposes. Many people found the company's water unsuitable for drinking and continued to depend upon private cisterns. Some of the larger manufacturing companies dug their own wells. Shortly after the start of operations, a major fire broke out downtown, and the new system provided sufficient water at adequate pressure to fight the blaze. A grateful council immediately granted the company a permanent franchise.[9]

Despite advancements in fire fighting, Buffalo Bayou proved to be a poor source of supply, particularly after a fire or after the mains had been put under pressure. Sand and mud often clogged the mains, clouding the water and reducing the flow. Company officials vowed to enhance the system, but improvements came slowly.[10]

In 1884, local business interests, headed by former mayor and wealthy property holder Thomas H. Scanlan, purchased the Houston Water Works Company, and in 1888 it drilled its first artesian well. While the water company and others appreciated the relative ease in obtaining fresh water from a source other than the bayous, they did not know that Houston sat atop what was one of the largest artesian reservoirs in the United States.[11]

The new supply was immediately heralded as the solution to Houston's long-term water needs. Between 1888 and 1891, the company operated fourteen wells, which supplied an area of seven square miles.[12] The 1890–1891 *City Directory* noted: "Houston is the only city but one supplied through her system of water works with artesian water, the purest, clearest and best water to be found. . . . Houston's supply of artesian water is considered inexhaustible, as she has now nearly one hundred wells, spouting up cool, pure water."[13]

The water company soon discovered that a supposedly bottomless pool of pure water did not solve all of the problems of operating a water system. Bayou water often was pumped into the city reservoir to meet increased demand when fires broke out, making it unfit to drink for several days. In 1904, the U.S. Supreme Court ordered the company to cease pumping bayou water into the mains, and artesian wells thus became the sole source for the Houston distribution system.[14]

As Houston's population grew, the company sank more wells. The legal constraints on the use of bayou water left the city with few alternatives, since no provisions had been made for tapping surface water. Some existing wells began to produce at very low pressure or failed to produce at all, resulting in short-term water famines. In some cases, sand clogged the strainers, obstructing the flow of water. As complaints multiplied about water quality, poor fire protection, or minimal efforts to extend distribution mains, officials tolerated few excuses from the water company.[15]

While confidence in the artesian water supply was wholehearted, faith in the company faded. In Houston, as elsewhere, municipal leaders increasingly argued that cities could run their services more efficiently and effectively than private firms driven simply by profits.[16] The water company, however, was unwilling to extend service without a new franchise. Mayor Hugh Baldwin Rice concluded that while he had been doubtful of public ownership of utilities, "when it comes to the question of water, the very life and essence of a community, it would be far better for the City of Houston to own and operate its own water system."[17]

On October 6, 1906, the city purchased the waterworks, reflecting the editorial judgment of the *Houston Post* that "so vital an element of life as the water supply of a city should be in possession of the whole people. It should not be a source of revenue to any person or corporation."[18] City officials were quick to congratulate themselves for rescuing the waterworks from inefficiency and inadequate service. An August 1909 issue of *Progressive Houston* proclaimed, "Some are expending millions of dollars and pushing pipe lines scores upon scores of miles out into the country, perhaps into the mountains, in efforts to procure such water as Houston is daily turning into every yard of the city."[19]

The debate over municipal ownership ignored the viability of a water system based solely on artesian wells, but the optimism about improved quality of service had some merit. The city soon addressed problems of water source and pumping capacity, and then turned to extension of the distribution system. Although the placement of new mains favored downtown consumers and the affluent, aggregate mileage of the distribution lines increased significantly, from 69 to 105 miles between 1907 and 1912. In the years between 1910 and 1920, the miles of water main more than doubled, from approximately 98 to 200 miles—slightly more rapidly than population growth for the period.[20] The extension of service also meant a greater draw on supply, and questions persisted about available water pressure and rates of withdrawal—problems that played a role in the company's loss of its franchise. In addition, Houston's system was a single network of mains, while some cities had two sets—one with high pressure exclusively for fire fighting. This meant that in order to maintain balance between both services, less pressure for fire fighting was available in Houston than in other cities.[21]

Dependence on well water went unabated, despite the city takeover of the system. Until 1915, the largest percentage of the wells in Houston were "free flowing," meaning that pressure decreased when withdrawal was heavy. Adding more wells did not address the issue of reduced pressure. Pumps had to be installed at the wells to discharge water into the pipelines leading to the reservoir.[22] Theoretically, the underground supply was abundant for many future generations; realistically, demand on the system increased so rapidly that productive wells were depleted or ceased to be free flowing. Population growth accounted for much of the escalating demand, but agricultural and industrial uses also were important. After the opening of the Houston Ship Channel (1914), pressure on the aquifer from industrial use rose dramatically. By 1940, groundwater sources supplied more than 140 million gallons per day (mgd) distributed as follows: rice fields, 45 mgd; metropolitan district and fringe areas, 81 mgd; and the ship channel, 35 mgd.[23]

The immediate response to growing demand was to sink new wells, add pumps to existing wells, build new pumping plants, and extend distribution lines—all of which increased supply but did nothing to discourage demand.[24] Metering water use at the point of consumption was an early way to control demand, prevent waste, and limit development of new supplies. Meters were unpopular with citizens, but nonetheless became important tools nationwide in administering the water supply in publicly managed systems by the turn of the century.[25] In Houston, these devices were first introduced in 1909, with the municipal water system completely metered by about 1914. Between 1906 (with flat rates) and 1914 (with metered rates), the average daily pumpage decreased from 11 million gallons to 5.5 million gallons—despite a 65 percent increase in the population served. To complement the meter system, the city had to implement better methods of detecting leaks.[26]

Metering did not create a conservation-minded public, although satisfaction with the artesian supply continued into the late 1930s. Periodically, a cautionary note was voiced about the city's growth outstripping its well capacity, but many held out hope that Houston could continue to develop additional artesian wells rather than utilize surface water.[27] The satisfaction with the water supply, however, tended to mask structural deficiencies in the system. By the mid-1930s, it was essentially a collection of water plants and distribution mains pieced together by expansion into subdivisions along the fringes of the city or through modest annexation.[28]

Poor fiscal policy only aggravated the problems caused by expansion. Improvements were made through bond issues, since net revenues from water charges went directly into the general fund. The result was a serious lag in improvements. In the mid-1930s, two important organizational changes occurred: in 1935 the Water Department separated from the Department of Public Works, and through a 1934 Texas Supreme Court decision, revenues of the Water Department could only be used to pay for improvements to the water system, especially for the costs of its operation, maintenance, and fixed obligations.[29]

Despite the corrective actions, neglect of public health responsibilities was marked. Between 1915 and 1929, the Texas Department of Health and the U.S. Public Health Service conducted surveys of the Houston water system. Purity of the artesian water at the wellhead was not questioned, but there were several potential threats before water reached the consumer. In 1928, the city had been given only "Provisional Certification," contingent on making recommended sanitary improvements. Cross-connections had to be eliminated or at least regulated because they allowed water from any source to enter the city's system. Other precautions included providing laboratory control over water quality, replacing or eliminating tile sewers and privies close to wells or on water plant property, and chlorinating the water.

Little action was taken until after a major flood struck in May 1929. Buffalo Bayou left its banks due to heavy rains, and twenty feet of water inundated the Central Water Plant. The loss of the plant caused a severe water shortage, requiring pumpage from private wells. State and federal authorities slapped Houston's supply with "Prohibitive Status," and the city had to initiate emergency chlorination. In June, city officials signed a "sanitary improvement agreement" declaring that Houston would carry out recommendations made the previous year. In July, Houston received "Provisional Certification," and in 1933 "Full Certification."[30]

The increasing challenges to the water system began to erode the city's heretofore unshakable confidence in the groundwater supply. Artesian well pressure had begun to decline as early as 1910, and for several years the water level dropped by an average of five feet annually. During the early years of the Great Depression, the quantity of water withdrawn declined by 10 percent. With increased industrial pumping along the ship channel for cooling and other purposes, the static level (depth to retrieve water) of the wells worsened. The deterioration of the wells—plus the lack of a wholesale rate for water—caused many industries and owners of commercial buildings to drill more of their own wells. By 1941, the public supply furnished less than 40 percent of the total demand of the metropolitan area.[31]

The decline in well productivity, in addition to growing independent action of commercial and industrial enterprises, led several experts to view the water problem as nearing a critical stage. In the 1930s and early 1940s, as many as forty reports on the water supply's condition were issued, but they were often contradictory and inconclusive. In 1938, the Mayor's Advisory Board (in concurrence with the National Board of Fire Underwriters) recommended that a new well field be developed west of Houston and that implementation of a separate industrial supply be postponed.[32]

Alternatively, Alvord, Burdick & Howson—a Chicago engineering firm retained by the city—issued a report in February 1938 that favored the use of the San Jacinto River as a single, inexpensive, and reliable water source. The San Jacinto was the nearest surface supply, with a drainage area of 2,840 square miles above the ship channel. A reservoir site was available only fifteen miles from the city's industrial district.[33] The study painted a poor picture of the existing system, which was lacking in capacity in wells, storage facilities, reservoirs, mains, and fire hydrants. "Houston," it stated, "presents an unusually aggravated example of uncoordinated efforts to solve the rapidly expanding water requirements of a thriving community without any comprehensive plan or continuing policy." Comparing the Bayou City with other cities, it noted that "Houston, located in the greatest well water field of the world and within a maximum distance of fifty miles of the four rivers draining two-fifths of all Texas, has a water supply system everywhere deficient, from source to its ultimate distribution to the consumer." The report added that the Houston system "is virtually a group of small town supplies without the distribution facilities or interconnections essential to the delivery of water for either fire or domestic use." Financially, the Houston waterworks existed on a "hand to mouth" basis, and the city's investment in it was less than one-third that of the average city of its size.[34]

While the debate over Houston's water needs persisted, the Water Department decided on a middle course between those advocating drilling more wells and those supporting a shift to surface water. In May 1937—before the Alvord, Burdick & Howson report was made public—the engineering staff recommended filing an application with the State Board of Water Engineers to appropriate water from the San Jacinto River to complement the groundwater withdrawal. G. L. Fugate, chief engineer of the Water Department, pursued the combined groundwater/surface water program into the 1940s, viewing the damming of the San Jacinto as a source principally to supply industrial demand.[35]

The onset of World War II finally pushed Houston toward surface supply. On September 10, 1941, the city filed an application with the Federal Works Administration for financial support to improve the water-supply system and to obtain a supplemental supply from the San Jacinto River. Wartime exigencies directed the federal government's interest in the project to the eastern portion of the city around the ship channel, designated as a "defense area." The industries there would employ an estimated 90,000 workers during the war.[36]

Fugate conducted a study in July 1942 and concluded that a dual supply from groundwater and a single dam on the San Jacinto would be preferable to drawing on the Colorado River (although federal engineers disagreed about the potential of the Colorado). That same month the War Production Board authorized the San Jacinto River Conservation and Reclamation District to build a dam and other facilities on the San Jacinto to supply water for war industries along the ship channel and in the Baytown area east of Houston.[37]

Political wrangling ensued among federal, state, and local governments over the construction project. The city protested a grant of authority to an outside agency to construct water-supply facilities in its jurisdiction. Because of the urgent need for the water, the War Production Board suspended the district's decision-granting authority in August 1942, and declared that the agency best showing the ability to deliver water to industries in the Baytown area would be favored. Neither the city nor the district would permit a grant to the other, and the federal government announced its intention of constructing the facility itself.

Ironically, since the city's preliminary plans and surveys were well advanced, the Federal Works Administration adopted the city's program and in November 1942 employed the city as its architect-engineer, with Fugate as contract engineer. Actual construction of the dam began in December 1942 by Brown & Root, for delivery of water to industries at Baytown and in the Pasadena area (to the east of Houston) in 1943. Two open canals from the river were constructed to serve the ship channel: the West Canal leading to Pasadena and the East Canal terminating at the Humble Refinery in Baytown. Despite the increase in available supply from the San Jacinto, distribution facilities still did not reach remote sites along the ship channel, and as a result wells served the increasing demand from new industries.

In June 1944, Houstonians voted for a $14 million bond issue not only to increase the amount of groundwater supply and for more mains but also to buy the West Canal from the federal government, to build a dam across the San Jacinto River north of Sheldon, and to construct a filtration plant. Because of the need for additional funding, Lake Houston Dam on the San Jacinto was not placed into operation until 1954. The new public water supply provided water for the city of Houston and the industrial complex from Houston to Baytown, and also supported local irrigation for various products, including rice.[38]

The debate over water in the 1930s, coupled with the acquisition of wartime industries along the ship channel, began Houston's transition from a city totally dependent on groundwater to one eventually dependent on a dual supply. From the city's early years, residents and industries alike relied heavily on the underground aquifers located in or near the city. While this pure, inexpensive, and abundant source of water contributed to significant urban growth, the almost blind faith in its ability to sustain the city diverted attention from weak links in the supply system, especially the extent and reliability of the distribution network, and the need to anticipate future demand by exploring new sources, sources that flowed primarily beyond the city limits. New supplies meant new possibilities, but also new challenges that would force Houstonians to deal with jurisdictional and environmental issues not encountered before World War II.

Sewerage System

The development of an underground sewerage system in pre–World War II Houston took into account issues of public health, but gave only passing attention to chronic problems of heavy rainfall and flooding. Houston had developed a sewerage plan in the late 1860s, but construction of a citywide sewerage system did not begin until a quarter of a century after the public water supply. Such a delay was not unusual. Northeast and Midwest cities began constructing sewerage systems in the mid-nineteenth century, but not until after World War II did all large cities have complete networks. Prior to the installation of underground sewerage systems, urbanites disposed of their wastewater by throwing it on the ground or in a gutter (which doubled as a carrier of storm water). Human wastes were deposited in cesspools or privy vaults.

With rising population growth and the piping of water into homes and businesses in the mid to late nineteenth century, old methods of disposal proved impractical and unsanitary. The great volumes of water used in homes, businesses, and industrial plants had to flow out as well as in. Without ample sewers, wastewater flooded cesspools and privy vaults, inundated yards and lots, and generally posed a major health hazard. Few cities simultaneously installed sewerage systems with their new water-supply systems because of the cost involved and the inability—or unwillingness—to foresee the necessary interconnection of the two systems. Extensive construction of municipal sewers did not commence in the United States until the 1880s.[39]

Houston confronted many of the same problems that growing cities without sewers faced in their early years of development. Members of the Houston Board of Health voiced concern over good sanitation, and an 1866 ordinance prescribed fines for citizens who did not keep their privies clean or randomly emptied filth onto the sidewalks and streets.

Besides health concerns, southern cities like Houston faced severe drainage problems, exacerbated by frequent downpours. Soon after the Allen brothers founded Houston in 1836, every structure in the new town flooded. Runoff increased with more hard-surfaced streets, resulting in swelling watercourses and flooding. Harold Platt noted, "In May, 1868, . . . the newspaper had been quick to praise the horsecar venture for grading Main Street from curb to curb. But the first torrential storm replaced the street's 'metropolitan appearance' with sights of little boys floating down it on jerry-built rafts."[40] Between 1836 and 1936, Harris County experienced sixteen major floods, some cresting at more than forty feet, "turning downtown Houston streets into raging rivers."[41] Destructive floods in 1929 and 1935 raised serious public outcries. The 1929 flood caused staggering property damage of $1.4 million. The 1935 flood doubled that amount, resulted in the loss of seven people's lives, and inundated twenty-five blocks of the business district and a hundred residential blocks. In addition, the Port of Houston was shut down for months because of submerged docks, destroyed railroad tracks, and a channel blocked up with tons of mud and debris. Countywide flood-control action finally was taken in 1937 with the creation of the Harris County Flood Control District (HCFCD).[42]

Before and for a time after the water-supply system was in place, building drainage ditches and sewers in Houston was piecemeal and unsystematic. Because of the cost and old habits, city leaders rationalized that relying on cesspools, privy vaults, and open ditches made underground sewers less vital than fire protection, a good water supply, or paved streets. Funds were allocated only for specific sewerage projects, not for an entire system, and those who could afford them constructed private sewers.

In 1866, civil engineer Colonel William H. Griffin developed the city's first sewerage plan under the auspices of the Board of Health. Griffin, who became city engineer the following year, made a case for three main sewer lines that would spill into Buffalo Bayou and drain the southern portion of the city, focusing on the downtown area. While not an integrated system, the plan offered a practical starting point. The city council supported it, but directed Griffin to broaden the area of coverage. Two drainage ditches serving the downtown were completed immediately, and only the Caroline Street brick sewer—originally constructed for storm-water drainage and the first underground sewer to be constructed in the state—was completed in 1874. Attention turned to other city needs, and the call for a comprehensive system dropped down the list of priorities.[43]

By the mid-1880s, most Houston residents still relied on private cesspools. Tolerance for the lack of adequate sanitation was wearing thin, especially since Buffalo Bayou was being utilized simultaneously as a source for drinking water and as a sewage outlet. Citizens complained that "tar water" flowed out of the pipes, human waste floated in the bayou, and fish died because of the dumping of creosote. Diarrheic problems sent adults and children to the local doctors. The commitment to artesian wells for the city's water supply quelled immediate concern over the need for a source of pure water, but the question of increased incidence of wanton pollution of the bayou highlighted the need for adequate sewerage.[44]

In June 1887, city engineer W. M. Harkness drafted a proposal calling for the examination of sewerage plans in other cities. With the support of the Citizen's Committee—a businessmen's group interested in infrastructure and service improvements—Mayor Daniel C. Smith and the city council agreed in 1889 to develop a citywide system.[45]

Selecting the best approach was more than awarding a contract to the lowest bidder. The timing of Houston's decision came amid a national controversy over whether cities should adopt combined or separate systems. A combined system handled both household waste and storm water in a single large pipe; a separate system utilized a small pipe for household waste and a large pipe or a surface ditch for storm water. The debate over the systems actually began in Europe, where the first comprehensive sewerage systems were constructed. But in the United States, only combined systems were built in the 1860s and early 1870s, largely due to cost and the lack of a successfully operating separate system. In the wake of the devastating yellow fever epidemic in the Mississippi Valley in 1878 and 1879, hard-hit Memphis chose a separate system in the hopes of avoiding a similar future disaster.

Scientific agriculturalist and drainage specialist-turned-engineer Colonel George E. Waring Jr. was the major proponent of the notion that "sewer gas" caused disease, and he recommended a separate system to ensure the health of those in Memphis. Once implemented there, the "Waring system" provided only for house sewage, not storm water. Nevertheless, Memphis's residents' health improved markedly after its implementation, but the debate over the separate versus combined systems intensified rather than ended.[46]

The Citizen's Committee promoted the separate system concept in Houston, playing up Waring's work in Memphis and elsewhere. In November 1889, Wynkoop Kiersted, a hydraulic engineer from Kansas City, was hired as a consultant to help city engineer C. W. Jarvis devise a plan for the city. Among other reasons, Kiersted was selected because he had helped implement a Waring system in Kansas City in 1883. The Jarvis/Kiersted plan involved establishing sewerage districts and constructing sewer lines in several wards. The anticipated completion date was 1893, but the project took substantially longer.[47] What infrastructure existed in Houston by World War I was, with a few exceptions, a rudimentary separate system on the Waring model. Jarvis and Kiersted's plan shifted the focus away from Griffin's emphasis on drainage—and thus flood control—toward sanitary sewers. Piecemeal development left many Houstonians with makeshift or incomplete sewerage or no sewerage at all.[48]

While most districts were connected with an intercepting sewer, the system was a hybrid of separate and combined pipes with insufficient storm sewerage.[49] The Board of Public Works, responding to the city's chronic drainage problems, exhorted the city council in September 1903 that a system of sanitary and storm sewers should be regarded "as being of the greatest importance to the City of Houston." "In fact," the board added, "we regard a more perfect drainage system as being of paramount importance to paving as under present conditions many of our paved streets . . . become perfect canals after every heavy rain."[50]

Beyond the drainage problem, the pumping station and the disposal plant were inoperable, and other portions of the sewerage infrastructure were damaged. In 1904, the Sewer Department was made independent from the Engineering Department and the budget was increased, but additional defects were still being noted. In most respects, constructing an effective sewerage system proved much more difficult than developing a water supply. The department's 1904 annual report concluded, "There are some residences in the sewer district that are not connected with sewers, and in unsewered districts there are thousands of houses that cannot connect." Noting improvements in the following years, subsequent annual reports made clear that rapid growth placed great demands on the city for more extensions. The 1922 annual report noted that in developing areas, private parties constructed three times the mileage installed by the city.[51]

Sanitary sewers appeared to be the most pressing need because of public health concerns. Drainage ditches and the bayou system were regarded as sufficient for handling normal runoff. Because of its heavy rainfall and flooding problems, Houston faced more serious storm-water threats than cities of comparable size. Civic leaders were rightly proud of the Austin Street storm sewer in the eastern portion of the city (constructed in 1909), touted as the largest of its class in the nation, but the more typical practice in the city was "to carry off storm water, where it collects, by whatever sewer was near it," rather than to add more sewers specifically designed to handle storm water. The storm-sewer bond funds were only spent in parts of the city where major paving operations were located. Economics dictated construction practices, and left the city less well protected from the chronic scourge of flooding.[52]

Despite Houston's commitment to new sanitary sewers, city leaders failed to give immediate attention to upgrading sewage disposal or developing a plan for sewage treatment. Disposal of sewage was first addressed in the mid-1890s, although more for the sake of local businesses than for residents. Buffalo Bayou had become a convenient depository for street runoff, storm water, raw sewage, and dead animals. The battle over avoiding the bayou as a water supply went hand in hand with concern over using it as a sink for sewage.

A concerted effort to clean up the bayou did not occur until 1895, when Major A. M. Miller of the U.S. Army Corps of Engineers was sent to Houston to inspect the site for the proposed ship channel. He informed local leaders that the city had to rid the bayou of its sewage and other pollutants if it expected federal aid for a ship channel. With support from the Houston Business League, the city council accepted the recommendation of consulting engineer Alexander Potter to improve the sewerage system and construct a filtration facility. Under the plan, wastewater would be delivered by pipe to a central pumping station in the Fifth Ward, and then to filter beds four and a half miles away. Heavy matter would dry on the surface and be removed, and smaller material would be filtered through layers of stone, gravel, coke, and sand. The final effluent—which Potter declared as "fit to drink"—would then be dumped into the bayou.[53]

With prodding from the federal government, Houston was one of only a few cities by the early twentieth century with a filtration program of this type.[54] While filtration met immediate demands, pollution of the bayou persisted. The plan made no effective provisions for runoff or storm water, and surface polluters—such as grazing animals—were not deterred. In addition, the upkeep on the filtration facility was dismal, the filter beds were often clogged, and wastewater formed a standing lake.[55] Although the facility was repaired and new disposal plants added, an estimated 70 to 80 percent of the sewage went into the bayou from public and private sewers in 1916.[56]

According to a 1915 state law, no untreated sewage was to be dumped into watercourses beginning in January 1917. At that time, Houston became one of the pioneers in utilizing the activated sludge process, which permitted conversion of dry sewage into fertilizer, and built the first large-scale wastewater treatment plant to use the method. The technique gained international usage and became a standard practice for effluent treatment and water purification worldwide.

Growing out of the study of sewage aeration processes, a report by Gilbert J. Fowler of Manchester, England, stated that sewage inoculated with oxidizing bacteria could be clarified and free from potentially harmful bacteria within a few hours of aeration. The first use of the activated sludge process took place in Salford, England, in 1914. The first American installation was built in San Marcos, Texas, in 1916, and soon was followed by a facility in Milwaukee, Wisconsin. Houston's application of the process was on a much larger scale. While the San Marcos plant treated .12 mgd and the Milwaukee plant 2.0 mgd, the Northside Sewage Treatment Plant built in 1917 along Buffalo Bayou treated 5.5 mgd. The south facility in Houston, completed the following year, treated 5.0 mgd.[57] In 1932, Houston operated six small disposal plants and the Northside plant, and intended to replace the small plants with a large treatment facility south of the city.[58]

Despite better handling of dried sewage and the development of a relatively elaborate sanitary sewer system, wastewater treatment failed to alleviate chronic water pollution problems through the mid-twentieth century.[59] Houston's aggressive annexation policy in particular resulted in the connection of several poor-quality sewage disposal plants into the citywide system, which already was becoming outdated. Public health authorities feared that the polio epidemics striking the city in the 1930s were traceable, among other things, to the polluted condition of the drainage network. Surveys of bayou pollution revealed serious contamination, and on December 19, 1946, the Houston Engineers' Council adopted a resolution that read in part: "It is well known that the sanitary condition of the bayous and streams in and around the City of Houston are intolerable. They are, in fact, in some instances on a par with those found in China and India." The council recommended that the city chlorinate all sewage, that it construct adequate main-sewer trunk lines and treating plants, that surface drainage connections be disconnected from sanitary sewers, and that plans be devised to contend with industrial waste.[60]

The city's separate system also continued to place insufficient emphasis on storm sewers. As table 5.1 shows, construction of storm sewers lagged significantly behind sanitary sewers. While storm sewers have a much larger carrying capacity than sanitary sewers, the disparity in miles constructed is not explained by that fact alone. Surface drains were called upon to carry some of the load, as were older combined sewers and, of course, the bayous and the streets. Yet the lack of balance in developing a sewer system that met both the needs of sanitation and runoff left Houston with a serious unresolved problem.

Like the artesian well system, Houston's sewerage system was a one-dimensional approach to the city's more complex disposal, drainage, and sanitation needs. Carrying the promise of improved health, it seemed to offer protection against the ravages of many communicable diseases, but it gave only secondary consideration to other significant environmental threats, especially water pollution and flooding.

Table 5.1. Sanitary and Storm Sewers in Houston (in miles)

Year	*Sanitary Sewers*	*Storm Sewers*	*Other*	*Total*
1902	24.558	5.537	7.248	37.345
1914	94.03	21.721	—	115.751
1920	170.221	57.992	—	228.213
1925	282.868	88.013	—	370.881
1930	523.223	148.0	—	671.223
1937	616.353	175.498	—	791.851

Solid Waste

While Houston had modest citywide programs of street cleaning and refuse collection and disposal in the late nineteenth century, they did not receive the same attention as did water and sewerage, as was typical nationally. In the 1890s, Houston officials considered burning waste in a crematory; a decade or so later the city turned to incineration as its major form of disposal for several years to come.[61]

The 1880 U.S. Census reported that Houston had a city scavenger who collected garbage and cleaned privy vaults in the city center, while householders disposed of refuge in other areas. Ashes that were collected were used for fill or as a disinfectant. A city force under the supervision of the Board of Health conducted street cleaning as needed, sometimes once a day. Sweepings were used for filling low places in the suburbs.[62]

By 1902, the city was divided into twelve districts, with a disposal cart assigned to each. A scavenger disposal plant was constructed at the sewage pumping facility to accept night soil. In 1905, the new city administration determined that the existing carts and wagons were unfit for further use and decided to contract for carts with drivers, rather than to purchase new carts and new horses. Successful bidders purchased carts from the city, and a price was set for payment to the contractees for the ten-hour shifts required for daily collection.

City leaders ultimately decided to discontinue "the unsightly and insani-

Early garbage dump. Courtesy Houston Metropolitan Research Center, Houston Public Library.

tary dumping grounds" and to replace them with a furnace—or crematory—to be erected at the pumping station. The crematory had a sixty-five-cubic-yard capacity, considered large enough to service the business district. By the city leaders' own admission, the operation was "very satisfactory," but the crematory itself was "a crude and inexpensive structure."[63]

By the late 1920s, collection continued to be carried out by contractors working under city supervision. Incineration was the primary means of disposal, although the city maintained some landfill sites as secondary disposal facilities. There were six garbage incinerators constructed and one added by annexation (the total was later increased to eight). In 1930, the city was incinerating approximately 382 tons of waste per day.

As with water supply, environmental factors played a key role in refuse services. The need for quick disposal due to the intensity of the southern climate was the primary justification for reliance on incineration. Decaying matter left in the streets and alleys was not only unsightly but odiferous and unhealthy. In the 1904 *Annual Report,* the health officer bemoaned the breakdown of the first crematory because "the indiscriminate dumping of garbage in gullies and out on open lots, with its rapid accumulation of decomposing animal and vegetable matter, is anything but conducive to health."[64] Similar feelings were expressed in the years between 1910 and 1920, when garbage and other refuse were being regarded as major culprits in the battle to improve city sanitation. When health conditions in the city improved in the 1920s, the installation of new incinerators and the closing down of some open dumps were given partial credit (see table 5.2).[65]

The relatively low cost per ton of incinerating waste may have strongly impressed city officials as much as the environmental rationale. In 1930, incineration cost $.45 per ton in Houston, which compared favorably with disposal costs in other Texas cities. Cost was such a crucial factor that, as one city official stated, "We make no attempt at salvage, believing the increased cost would more than offset profit."[66]

The dependence on incineration came with an immediate price and further repercussions. Those living closest to the incinerators suffered the greatest discomfort and potential health risks from air pollution. If combustion was incomplete in the burners or if the units ran beyond capacity, great amounts of ash were belched into the air. But the concern of city leaders over neighborhood complaints was minimal in these years because incinerators were situated in poor, nonwhite neighborhoods—areas with little political leverage. Of the eight incinerators operated by the city between the 1920s and 1975, six were located in black neighborhoods, one in a Latino neighborhood, and only one in a predominantly white area. The five city-owned landfills, which ultimately complemented the incinerators, were all located in black neighborhoods.[67]

Table 5.2. Houston Incinerators

Location	*Number of Units*	*Capacity (tons)*	*Year Built*
5th Ward	2	25–200	1916
Howard Street	1	40	1923
5th Ward	1	40	1921
Studewood	1	40	1921
Howard Street	2	80	1925
Velasco Street	2	80–200	1925
Patterson Street	2	150	1929
Magnolia Park	1	15–75	—

Source: "Successful Refuse Service," *American City* 52 (May 1937): 56.

Neighborhood issues, therefore, did not deter the city from relying on incineration for many years. The practice was guided by perceived cost savings and the assumption that dumping was a greater health and nuisance risk than burning. From the earliest implementation of incinerators in the United States in the 1880s to the present, the debate over the most efficient and least environmentally threatening form of waste disposal pervaded many city halls. Faulty designs and improper operation of crematories—often increasing smoke problems—dissuaded some cities from using this disposal option. Others found the method too costly. In the late 1930s, the use of incinerators began to decline sharply nationwide as the technique competed with what became the disposal standard for more than four decades—the sanitary landfill.[68]

Climate alone was insufficient justification for Houston to employ a disposal method with liabilities such as air pollution. Possibly because incinerators were tucked away in black and Latino neighborhoods, they were shielded from sufficient attention. City officials cannot be held accountable for failing to foresee future demands for recycling or to anticipate changes in the waste stream that added many toxic materials, but Houston's waste-disposal program offered little provision for assessment of incineration as an effective disposal option in an expanding community and placed the heaviest pollution burden on those with the least political voice. Attempts to defend incineration as a cost-effective method relied too heavily on operations costs and not enough on capital investment and community impact. When city leaders were forced to seek alternatives, the incinerators simply were abandoned—artifacts of an earlier era of inadequate planning.

Sanitary Systems after World War II

Between 1876 and 1929, Houston leaders had settled upon clearly defined paths for the city's sanitary services: artesian wells, sanitary sewers, and incinerators.

By the 1930s, the resilience of those systems was sorely tested, and some of the inherent weaknesses recognized. Confidence in the water, wastewater, and solid-waste infrastructure was shaken by structural, economic, and environmental problems that plagued all of the sanitary systems. By the 1940s, the impact of war and pressures of growth—through annexation, in-migration, and natural increase in population—challenged the notion that the services were adequate for future needs.

Houston's history has always been defined by growth. In the post–World War II era, the city's emergence as a dynamic, regionally expansive metropolis not only transformed its politics, its economy, and its demographics but also placed its sanitary services in a new context. Aggressive annexations and unrelenting sprawl tested the integrity and functioning of the city's water supply and its wastewater and solid-waste systems. Like never before, they were connected to and influenced by a variety of watersheds and Galveston Bay. Solid-waste disposal sites claimed land beyond the city's earlier borders. Houston's sanitary services absorbed adjacent systems or at the very least influenced those systems beyond the city's grasp. In so doing, Houston's sanitary systems became regionalized, with noteworthy impact on the environment as well as on municipal responsibility for delivering adequate service.

Outward growth in Houston after World War II was made easier by an automobile culture stimulated by cheap gasoline and by a booming economy that attracted businesses and workers from around the country and the world. But Houston also benefited from negligible municipal competition along most of its periphery and a liberal annexation policy made possible by Texas law. The Municipal Annexation Act of 1963 was particularly significant, since it guaranteed space for future expansion for major cities statewide and encouraged orderly growth. The former succeeded, while the latter failed miserably.

Under the provisions of the law, Houston was granted extraterritorial jurisdiction (ETJ) over adjacent land, which it could reserve for future growth, allowing the city to place all real property within five miles of its limits under ETJ. Only Houston was allowed to annex in this area, and no settlements could incorporate themselves without permission from the city or without a change in state law. Houston could increase its city limits by 10 percent each calendar year. If less land was annexed, the difference could be added to a subsequent year's annexation as long as the city did not increase by more than 30 percent in a single year. By 1999, the policy allowed Houston to reserve 1,289 square miles for future annexation. Only in the southeastern portion of the city—the Pasadena area—was Houston blocked from expansion. With half the density of Los Angeles, Houston still held more than half of the population in its metropolitan statistical area (MSA) within the city limits.[69]

Water Supply

Aside from unprecedented expansion, the major change in Houston's water-supply service after World War II was the transition from a one-dimensional water system to a dual system increasingly dependent on surface water. Prior to the war, reliance on groundwater had already begun to expose the limitations of the once-abundant resource. The shift to surface water raised new issues, not the least of which was potential environmental problems affecting the whole region of southeastern Texas.

The major forces leading to the increasing growth and dominance of surface supplies were many. Most persistent was the unrelenting drive for economic development, which went hand in hand with rapid population growth. As a Chamber of Commerce report noted in 1954, "Obviously, the community cannot attract new industries which require substantial quantities of water until such new supplies are assured."[70] Also significant was the growing disincentive to rely on groundwater because of severe land subsidence.

The development of the San Jacinto River to supply water for World War II defense industries along the ship channel resulted in a dual water system for urban, industrial, and agricultural users in the Houston metropolitan area. That system consisted of potable water for residential and commercial use drawn primarily from underground aquifers with wells owned and operated by the city, municipal utility districts, and the private sector (including industry)[71]; and a separate industrial water supply of raw—or untreated—water from the surface source owned and operated by the city.[72]

This pattern of water-supply acquisition and delivery underwent substantial change over the years, with surface water becoming an increasingly important component of the city's water system. During the war, the industrial water supply from the San Jacinto River was taken from two reservoirs, and the water was sent to the ship channel through open canals (the East and West canals). In 1945, the city acquired the pumping station on the San Jacinto and the West Canal; in the same year, the San Jacinto River Authority acquired the East Canal. With the completion of the Lake Houston Dam and construction of the new East Water Purification Plant in 1954, a portion of the water from Lake Houston via the canal was diverted to the plant, and the treated surface water ultimately supplemented and was combined with ground water in the municipal district system.[73] By 1970, treated surface water met about 40 percent of the demand of the municipal system. The remainder of the water went through the plant's Industrial Water Pump Station for delivery to ship channel industries, although provisions were made for treated water to be diverted to the industrial area as needed.[74]

Despite satisfaction with the important role of the San Jacinto River to Houston's water needs, other sources—especially on the Trinity, Brazos, Colorado, and Neches rivers—continued to be explored. After the completion of Lake Houston Dam, the city of Houston began developing a reservoir site on the Trinity River, which was initially meant to serve a nine-county industrial area (the Houston-Trinity Industrial Complex). Lake Livingston Dam along the Trinity was completed in 1968, with water rights owned by the Trinity River Authority (30 percent) and the city of Houston (70 percent). By the early 1980s, it represented more than 80 percent of the Houston area's available surface water supply. Lake Conroe Dam on the San Jacinto was completed in 1973, with the San Jacinto River Authority owning one-third of the water rights and the city of Houston owning two-thirds. San Jacinto River water supplied municipal and some industrial demand, while water from the Trinity was initially used to supply industrial needs only. Tapping the new surface supplies required attention to water conveyance, pumping, and all of the issues that had faced the San Jacinto River project. Continuing industrial demand for water led the Coastal Industrial Water Authority (CIWA) (created in 1968) to develop the CIWA Conveyance System[75] to acquire water from Lake Livingston for industrial use and ultimately to replace water formerly delivered through West Canal for the industrial sector east of Houston.[76]

Within the city, a distribution grid delivers water to residential and commercial users. In the late 1940s, it consisted of a central plant, with other plants located around it in a wheel pattern. The eight existing plants were interconnected by large or arterial mains (800 miles of mains and 300 miles of service lines). As the city grew, the grid frequently had to be extended.[77]

Outlying areas not incorporated within Houston often continued to rely on wells and might only be connected to the city's grid after annexation. A report of the Harris County Home Rule Commission optimistically noted in 1957: "The presence of the hitherto abundant ground water helps to ease the water problem of the suburbs by making the technical problems of special water districts relatively simple; hence, the large number of districts in the Houston area."[78]

In reality, the ability to serve a city population that grew not only by in-migration but also by annexation was complex. Developers and subdividers, who were the essential players in Houston's growth politics, worked with city officials to acquire main extensions into their properties. While they would have liked the city to bear all the costs of furnishing and installing mains and hydrants, they usually had to share them. In the late 1960s, for example, the city was responsible for the design of the extension and contracts for installation on a fifty-fifty basis with developers within their property lines, and the city might also provide the pipe and hydrants. For short distances, the city usually furnished and installed the mains; for longer distances, the city could supply the hydrants, but the developer would be responsible for the remaining costs.[79]

The primary mechanism for establishing water service in suburban areas was the water district. After the disastrous Galveston hurricane of 1900—which inspired rethinking about ways to finance public building and rebuilding projects—an amendment to the Texas Constitution provided for taxing districts to be created for conservation and reclamation improvements, with provisions of the amendment liberalized in subsequent years to streamline the process. The special districts were granted unlimited bonded indebtedness and taxing power. Two types of districts were particularly important for suburban development: the Fresh Water Supply District (FWSD), created in 1919, and the Water Control and Improvement District (WCID), created in 1925.

The FWSDs were severely limited in issuing bonds for construction of drainage facilities, and thus were of little value in developing new communities. The WCIDs, designed for irrigation purposes, demonstrated sufficient flexibility for water, sewer, and drainage improvements and could be formed prior to the development of a new suburban community. By the mid-1980s, there were more than 800 WCIDs in Texas, with approximately half in the Houston metropolitan area alone.[80] In the process of annexing parts of its ETJ, Houston also absorbed many special districts, and in so doing reduced their economic authority. Between 1949 and 1979, the city annexed fifty-three water districts and assumed their bonded indebtedness.[81]

Planning Houston's water needs before World War II amounted to calculating how many new wells would have to be sunk and when and where to extend mains within a relatively small area. During World War II, it became apparent that there was a "precarious situation confronting the water system," which was in desperate need of new wells to relieve extensive use of the groundwater supply, a new purification plant, strengthening the existing grid, and contending with the problem of extending service to "fringe populations." Materials and monetary resources to accomplish these goals were constrained by the prosecution of the war, and yet the immediate postwar years did not produce radical change.[82] Outlying areas, especially in poorer neighborhoods, were not likely to get adequate service. The shift to surface water most immediately benefited industry, rather than residential users. And the slow weaning from groundwater was made more urgent by a growing concern over subsidence.

Subsidence occurred when too much groundwater was removed from a particular location, resulting in the surrounding clay collapsing and then compacting. Below the surface, fresh water cannot replenish the structure. In addition, subsidence also may break pipes and clog sewers, encourage saltwater encroachment into the aquifer, and increase the potential for flooding. Areas where well fields and production facilities were concentrated experienced the most population growth in these years, and also tended to incur the greatest subsidence.

There was little public notice of subsidence prior to mid-1953. But in the ten years between 1943 and 1953, there was already subsidence of approximately 0.5 foot in downtown Houston and as much as 3.25 feet in Pasadena. In the roughly same period, the problem of saltwater intrusion into the aquifers increased from a rate of one mile in seventy-five years in 1940 to one mile in sixteen years in 1952. In the late 1960s, it was estimated that subsidence lowered the ground level of Pasadena's north side and in Baytown by almost 7 feet; by 3 feet south of Houston at Ellington Field; and by 2 feet along the shore of West Galveston Bay. By 1978, subsidence of up to 10 feet was measured along the ship channel area. In the early 1980s, more than 20,000 acres in the Houston-Galveston area were below sea level, and even Lake Houston Dam had sunk 2 feet since its construction in 1954.[83]

In time, the issue of subsidence became a central concern when planning for future expansion of the water-supply system. The Harris-Galveston Coastal Subsidence District was created in 1975 as a way to control regional subsidence through regulation of groundwater use and through formulation of plans to convert area water supplies to rely more heavily on surface water. But unease among local leaders and bureaucrats did not necessarily lead to effective action.

The view that there was sufficient surface water to exploit was seen as a valuable countermeasure to past practices. However, continual dependence on groundwater in developing areas north and west of downtown—where access to surface water was limited—meant that city leaders ignored a problem that had plagued the area east of the city for years. Between 1878 and 1987, the area west of downtown experienced subsidence of one to three feet, due largely to aggressive land development. In the early 1990s, regulations meant to arrest subsidence overall required Houston to reduce dependence on groundwater from 43 percent of its total consumption in 1992 to 20 percent by the year 2020, along with expansion of surface supplies. This process has moved slowly.[84]

Other environmental problems arose in the development of the city's water-supply system. The notion that dependence on groundwater protected the city from polluted sources of supply was unfounded and somewhat naive. Water quality and water pollution issues are a constant concern for any water system, anywhere. The confidence in the quality of Houston's groundwater supply had been tested in the past when well water was mixed with bayou water before distribution to customers. More recently, agricultural runoff, and especially runoff from developed land that is made impermeable, increased the flow and quantity of water draining off the land. Urban runoff, especially, carried large quantities of nonpoint pollutants—pesticides, oil and grease, salts, bacterial contaminants, hydrocarbons, and heavy metals—which find their way into various surface-water and groundwater sources, seriously increasing flooding.[85]

Water pollution regulation with respect to Houston's water supply has had a mixed record. While drinking water provided by the city regularly meets or exceeds the water quality standards of the Texas Commission on Environmental Quality and the Environmental Protection Agency (EPA), a National Resources Defense Council report in 2003 showed that Houston's drinking water contained significant levels of contaminants, such as haloacetic acids, arsenic, and coliform. Wells in and around Houston sometimes show high radon levels. These problems are attributed to an aging infrastructure and dated technology.[86]

The shift to surface water engaged a different set of environmental issues from groundwater use. The controversy over the proposed Wallisville project is a good example. Authorized by Congress in 1962, construction began in 1966 on Wallisville Lake (or Reservoir), located approximately 40 miles east of Houston and 3.9 miles above the mouth of the Trinity River. The reservoir, to be created by a 39,200-foot dam, was meant to store 58,000 acre-feet of water. Its construction had support from the Galveston District of the U.S. Army Corps of Engineers, the Trinity River Authority, the Chambers-Liberty Counties Navigation District, and the city of Houston. Visualized as a multipurpose project, proponents claimed that it would increase the industrial water supply for Houston, control saltwater intrusion into the river, improve navigation, and enhance the habitat for fish and wildlife.[87]

The Wallisville project, including the construction of the Livingston Reservoir, had been proposed in response to drought conditions in the 1950s, and was meant to maximize the ability to draw water from the Trinity River. As one study aptly noted, "Stripped of its bells and whistles . . . Wallisville is at its core a water supply project."[88] Almost 70 percent complete in 1973, opponents successfully stopped construction of Wallisville Lake. In close proximity to an environmentally sensitive coastline and fragile wetlands, the project made the Corp's Galveston District a clear target for protest and litigation in a period of rising environmental awareness. Opponents of Wallisville (conservationist and environmentalist groups and the Texas Parks and Wildlife Department) employed a new weapon—the environmental impact statement (EIS), authorized under the National Environmental Policy Act (1969)—to further their cause.[89]

Criticism of Wallisville focused on several potential risk factors, many of which related to the link between the Trinity River and Galveston Bay. Opponents argued that the dam would reduce freshwater, sediment, and nutrients flowing to the bay. This would increase salinity, parasites, and pathogens harming the already weakened oyster fishery and could reduce shrimp harvests by as much as 65 percent. Exclusion of marine organisms from wetlands upstream from the dam and loss of Trinity delta wetlands also were possible. In the vicinity of the dam itself, approximately 13,000 acres of marsh, cypress swamp, and marine nurseries would be flooded. Opponents added that other sources of water for Houston, such as the Toledo Bend Reservoir, had the capacity to meet long-term needs without constructing the Wallisville Reservoir.[90]

Wallisville proved to be a "legislative Lazarus" frequently resurrected by Congress and the courts.[91] In September 1971, environmental groups, including the Sierra Club and the Audubon Society, sued to halt the construction because the Corps had not filed a proper EIS. After the completion of the EIS, construction continued through 1972, but was stopped again in 1973 when the district court declared that an EIS for the entire Trinity River project had not been prepared. In 1974, a federal appeals court returned the case to the lower court, and additional review of the EIS created more gridlock. In 1977, the Galveston District announced alternative plans to the original Wallisville proposal, including a reduction in the size of the reservoir from 19,700 acres to 5,600 acres, and another plan that would install a temporary nonimpoundment barrier on the river. Not until 1981 was the EIS mandated by the court released, but with an accompanying report dropping navigation and fish and wildlife enhancements as benefits of the project. The new EIS spawned another round of heated debate, which intensified when the Corps added back the omitted benefits (navigation and fish and wildlife enhancements) in its 1982 Supplemental Information to the Post Authorization Change Report (SIPACR).[92]

One observer stated that the SIPACR became "perhaps the most controversial document ever produced by the District" by reinstating the benefits without explaining why and by not circulating the report for public review.[93] Despite the brouhaha and the continuation of the court injunction against the project, Congress appropriated funds for the scaled-down Wallisville plan in 1983. In turn, the district court continued the injunction in 1986. But in a stunning reversal, the Fifth Circuit Court of Appeals in New Orleans ruled that the project could proceed after the fifteen-year delay—now as a single-purpose water-supply project.[94]

In 1989, under the Houston Water Master Plan, Wallisville Reservoir was identified as the best of four alternatives for supplying water to Houston through the year 2030 (the other three were an inflatable saltwater barrier on the Trinity River, the Toledo Bend Reservoir, and Brazos River water transfer). While litigation and environmental concerns continued to delay the project until 1991, it was completed in 1999. The Wallisville Saltwater Barrier put in place mechanically blocks upstream movement of saltwater from Trinity Bay during conditions of low flow from the river.[95]

The Trinity River Authority more recently has taken the position that, from its inception, the Wallisville project was "constructed for the purpose of controlling the intrusion of saltwater from Trinity Bay during low river flow conditions," rather than the multipurpose aspirations of supporters evident in the documents going back to the 1960s. From that time forward, opponents consistently viewed the building of the reservoir as posing environmental risks. The most strident among them labeled the project "a fraud," "a monumental rip-off," "an economic blunder," and "a slowly growing cancer."[96]

Although regarded by proponents as a ready answer to the growing demand for water in the Houston area, and as a serious environmental misstep by opponents, the Wallisville controversy exposed the difficulty and sometimes the peril of sustaining urban development. A shift to surface water had been inevitable for Houston in the wake of its breathtaking growth. In the case of Wallisville, a public battle was waged that produced at best a temporary solution to the city's water needs, but at a price that some people were not willing to pay. Even in seemingly water-rich East Texas, water wars raged as if the engagement had taken place in water-poor West Texas. The question remains: how many more times would such controversies emerge as Houston continued to grow? The panacea of surface water did not magically solve the problems of subsidence and groundwater depletion, but shifted the challenge of meeting water demand from below the ground to above, and from a local concern to a regional issue.

Wastewater

Houston faced an additional challenge in its effort to upgrade its existing sewerage system to meet the demands of growth. As with water supply, wastewater issues were increasingly regionalized after World War II. The difficulty in providing adequate sanitary sewerage with effective delivery to treatment plants and maintaining greater control of drainage was constant in the central city. As sprawl increasingly typified the Houston metropolis, providing these services became more difficult. Not only were new customers to be served, but a relatively centralized urban sewerage system did not extend into the city's periphery. Unease over adequate treatment and problems of runoff increasingly had to take into account the flow of effluent into Galveston Bay. To a greater extent than before, storm water exacerbated flooding problems beyond the residential and commercial property of the urban core.

The storm-sewerage issue was emblematic of a serious problem the city faced by not constructing necessary infrastructure to keep up with rapid growth and by not heeding the vagaries of Houston's weather. The city's forty-eight inches or more of rain each year seemed to be reason enough to taking storm sewerage and flooding seriously. The expansiveness of the city's impervious surfaces due to development directly affected the rate of runoff even more than the volume of rain. In addition, the increased rate of runoff made the response to potential flooding much more difficult.[97]

The traditional approach to deal with runoff, storm water, and flooding in

Houston was to include natural features—drainage basins, floodplains, and the rivers and bayous—into the drainage system alongside human-made features—channels, streets, roadside ditches, storm sewers, and (eventually) detention facilities. Great faith was placed in the natural features like the bayous as well as in the streets to offset the shortfall in storm drains and detention facilities. Despite efforts at flood damage reduction by the HCFCD since its inception in the late 1930s, almost thirty serious floods occurred in the area, resulting in millions of dollars in costs. However, after the 1940s Harris County did not suffer massive regional flooding again until June 2001. In 1939, the HCFCD developed a flood-control master plan and cooperated with the U.S. Army Corps of Engineers in a flood-control program, beginning with the building of the Barker and Addicks reservoirs in west Harris County and part of Fort Bend County in the 1940s. These were the first and largest flood-control detention facilities in Harris County. At this time, the eleven independent drainage districts in Harris County were merged under the HCFCD.

The primary focus of the HCFCD and the Corps's flood-control program was more than 2,500 miles of cement-lined channels built along the bayous between the 1940s and 1970s. Unstable Brays Bayou was canalized and lined with

Downtown Houston flooded in the 1920s. Courtesy Houston Metropolitan Research Center, Houston Public Library.

concrete between 1955 to 1960. White Oak Bayou became the last bayou in Harris County to be completely lined with concrete.[98]

Relentless development and an increasingly aging and overburdened sewerage system worked against effective storm-water control, as did periods of extensive flooding. In June 1976, almost ten inches of rain fell in six hours in the Brays and Sims Bayou watersheds. The Texas Medical Center was most seriously damaged (exceeding $20 million in losses), and there was significant damage to Rice University, the University of Houston, and the Houston Museum of Fine Arts. The cause of the flooding was not attributed to the overflow of Brays Bayou, but to lack of capacity in the storm-sewer system. Urbanization on the west side of Houston exceeded design estimates and was a central contributor to the ineffectual flood-control technologies in place.[99] In 1980, the HCFCD initiated a "no downstream impact" policy for new developments, resulting in greater use of on-site detention facilities. In 1985, consultants recommended construction of four regional detention facilities on Brays and Sims bayous and flow diversion channels for Keegar Bayou. Cost and politics scuttled the plan. Luckily, since 1983, Brays Bayou has not flooded, but it did experience "near misses" in 1992, 1994, 1997, and 1998. New studies and additional flood-control projects continued through the period.[100]

In June 2001, Tropical Storm Allison—which proved to be "one of the most devastating rain events in the history of the United States"—dumped more than thirty-eight inches of rain on the Greater Houston Area between June 5 and 9, causing twenty-two fatalities and more than $5 billion in property damage, including damage to 95,000 automobiles and trucks and 73,000 residences. Approximately 30,000 stranded people were housed in shelters. Allison went on to spread devastation in thirty-one counties in Texas as well as in Louisiana, Florida, Mississippi, and Pennsylvania. In Houston and Harris County, the rate and amount of runoff overwhelmed the drainage system. The storm sewers and roadside ditches in the area were designed to handle a rainfall rate of approximately one or two inches per hour (the streets are designed to carry at least four inches of water flowing slowly), far less than experienced during Allison but also less than occurred during many other major floods in Houston's history.[101] With resignation, one observer noted, "Harris County doesn't have earthquakes . . . doesn't have blizzards . . . doesn't have avalanches. We have flooding."[102]

While such an extraordinary event as Allison made an adequate response almost impossible, much lesser events demonstrated the vulnerability of the drainage system. Between 1980 and 1994, Houston suffered extensive flooding nine times, with many more localized events. The Comprehensive Storm Drainage Plan proposed in the 1990s—before Allison—was meant to evaluate the flooding patterns and storm-water generation.[103] Allison helped add addi-

Suburban Houston flooded in the 1970s. Courtesy Houston Metropolitan Research Center, Houston Public Library.

tional resolve to that evaluation, but solutions—or at least serious mitigation of the effects of flooding—remain far from being achieved.

The Wastewater Treatment System Task Group of the Houston Chamber of Commerce concluded in 1983: "The major deficiency in the City's wastewater system is the lack of capacity in both its collection network and its treatment facilities."[104] This observation applies to much of the city's history. After World War II, it was clear that the sewer system was not keeping up with Houston's aggressive building program. As *American City* noted, "In few cities is complete sewage treatment more seriously needed." By the late 1940s, a new sewage-treatment plant and new sewer lines were under construction. It was the goal to consolidate the six existing plants—only one of which was satisfactory—into two larger ones. The one adequate plant was to be retained and a new facility, the Sims Bayou Plant, completed in 1948.[105]

In the 1950s, there was no centralized sewerage system in Harris County. Outside of the center of Houston, sanitary sewerage came mostly from privies, septic tanks, and a variety of other collection systems that were public and private. Septic tanks were prevalent in parts of Houston, in nearby villages, and particularly in unincorporated areas. Some engineers noted that the privies and septic tanks with drain fields did not provide adequate sewage disposal, "and with this increase in suburban building tremendous amounts of disease-laden sewage in our roadside ditches, on the top of the ground, and in the yards of homes could easily be visualized if these methods of sewage disposal continued to be used."[106] The problem with septic tanks in use was that they failed to function well due to the heavy rainfall, the low absorption rate of the soil, the high water table, and poor drainage. The same engineers who criticized privies and septic tanks were encouraged that subdividers and home builders in several larger subdivisions were installing new sewerage systems and favored increasing the number of treatment plants by the late 1950s.[107] This was not a strictly voluntary response. The Harris County Health Unit would not approve an individual sewage disposal method, and hence subdividers who expected to acquire Federal Housing Administration—and later Veterans Administration—approval for building homes had to connect those homes to an existing system or construct their own. Some developers soon realized that they could profit from operating the utilities themselves.[108]

Coverage varied greatly in the county, however. Bellaire, an incorporated enclave within the Houston city limits, was completely sewered, as was Jacinto City. South Houston was over 80 percent sewered, and Baytown and La Porte were 95 percent sewered. Storm sewers, as in the central city itself, were less well developed along the periphery. In fact, areas annexed by Houston in 1949 and 1956 were without storm drainage systems. Disposal of sewage in Harris County was uncoordinated, often using the bayous as receptors for effluent. Because the county did not provide sewerage service, developers and individual citizens utilized their own devices or sometimes relied upon the formation of FWSDs or WCIDs, which had the power to finance, construct, and operate sewerage facilities. Homeowners who simply bought land from developers, but were not part of a large subdivision, could be left without sewers, as could people who did not own their homes. In the metropolitan area, there were sixteen sewage disposal plants in 1945, with thirteen operated by cities, one by a water district, and two by private individuals. Estimates suggest that each plant served 39,400 people in 1945, improving to 10,000 people per plant in 1956—in large measure due to the existence of ninety-seven sewage-treatment plants then in operation in the county. In 1945, the population of the cities in Harris County with sewers was 60,000, increasing to 176,000 by 1956.[109]

The decentralized sewage-treatment system met with serious criticism. Breakdowns in the plants could result in no sewage treatment for homes and businesses. Little help was available because of the lack of any backup system or larger facility with a wider coverage area.[110] By the mid-1970s, criticism of the city and county's decentralized sewer system was unabated. In 1974, Houston had forty-two wastewater treatment plants, two major sludge disposal plants, 179 pump stations, and approximately 3,600 miles of collection and conveyance lines. Much of the system was built by the city, with the remainder acquired through annexation or purchase.

While the statistics give a sense of wide-scale city service, a closer look reveals shortcomings. The average treatment plant in the 1970s served approximately 30,000 people (several plants had very low capacity) and were unable to support a growing population in other parts of the city. In addition, many of the plants were antiquated, could not meet modern water quality standards, and exceeded permissible odor levels. By comparison, a single plant served the entire population of Fort Worth and its suburbs. Dallas had three treatment plants for the entire city. A 1974 EPA EIS suggested that a way to offset the problems of Houston's decentralized wastewater system was to regionalize wastewater facilities.[111]

The inadequacy of Houston's sewerage system and wastewater treatment in the early 1970s is best expressed in the sewer moratorium. Because of insufficient sewage-treatment capacity, Houston was forced to develop a plan to avoid an EPA citywide moratorium on new sewer connections. The city chose a compromise in 1974 that protected downtown development, while curtailing new building throughout approximately 70 percent of the city. Connections in the ETJ were not covered by the plan. With the completion of Plan 208 in 1977—which outlined a long-term strategy for the watersheds of the Houston area—the moratorium was lifted.[112]

What regionalization actually meant in practice was fairly muddled. For a city strongly committed to growth, some leaders took particular notice of assessments like a 1975 Houston-Galveston Area Council report that argued, "With the completion of major interstate highway segments . . . sewerage facilities are becoming the prime determinant of the location of new development." There was, however, a cautionary observation in the report about the impact of wastewater system development as a growth tool:

> The capacity of the interceptor line (the major line that connects collector systems to treatment plants) has a major impact upon land use. There is a tendency to overdesign these systems in order to accommodate future development, and the oversizing contributes directly to the amount of development that can and probably will occur. When these large interceptor sewers are designed to run for long distances between the existing service area and the treatment plant, vast amounts of land are then opened up to sprawl between towns. The low-cost of land on the urban fringe encourages developers to move to the end of the new line, which results in leapfrog fill-in development patterns. Haphazard construction of sewerage facilities, therefore, increases the difficulty of properly planning the timing and size of other public facilities and spreads the urban area out in a pattern that is wasteful of land and resources.[113]

Regionalization of the wastewater system, therefore, could be mistaken for future growth, and not understood as a tool to more effectively deliver service to the citizenry of the metropolitan area. Confusing the two was common.

By the end of the 1980s, there was unease about the number of small wastewater treatment plants in the metropolitan area and the ETJ. These plants produced a poor quality of discharge, which aggravated water pollution problems in the region, and were less cost effective than large regional facilities. Remedial action led to the construction of new plants or enlarging older ones. Between 1964 and 1989, seventy-one small sewage-treatment plants were abandoned. In 1986, the Greater Houston Chamber of Commerce—a powerful force in the city's growth machine—prepared a report entitled "Wastewater Regionalization Plan for Houston" calling for the development of larger treatment plants in the ETJ prior to annexation. It was endorsed by the Developers Council, the Association of Consulting Municipal Engineers, members of the legal community, some local interest groups, and the city's Departments of Public Works and Health and Human Services.

In 1987, Mayor Kathy Whitmire formed the City Wastewater Regionalization Task Force with the mission of developing a wastewater regionalization program. A year later, the city council approved a wastewater regionalization plan for the ETJ. The step away from a historic focus on decentralization occurred without seriously questioning Houston's plans for future growth, using sewers as a kind of stalking horse.[114] Since state law required Houston to provide water and sewer service to at least 75 percent of the residents in previously annexed areas and in all new areas within four-and-a-half years of annexation, growth had a price.[115]

Despite the flurry of interest in sewage treatment in the ETJ, capital planning and system maintenance in the developed portions of the metropolitan area came under question as well. The city's growth in the 1970s and early 1980s led to substantial expansion of the wastewater system, but with significant diversion of funds from system maintenance. Overflows became more common when storm water began seeping into leaky pipes in the older parts of the system. In 1987, the EPA and the Texas Water Commission required the city to address the "infiltration and inflow" problems—a major contributor to water pollution in the city—resulting in Houston having to spend more than $1 billion over the next five years for system improvements.[116] In a 1991 Public Works Department report, department management was criticized for insufficiently focusing on long-range issues and simply delegating responsibility to "consulting engineers." The result was inadequate treatment plant facilities, a poor collection system, and limited investment in preventive maintenance.[117]

Pollution from effluent was a chronic problem in Houston only partially improved by greater attention to the water-treatment system. Sludge—an unwieldy semisolid precipitated by sewer treatment—proved to be the most difficult constituent. Despite the successful prewar operation of a sludge-drying plant and lagoons for the disposal of raw sludge, the problem remained after World War II. The lagooning method had begun in 1916, but by 1945 there was little available space for expansion. The existing open-air system also proved increasingly odiferous for residents and dissuaded commercial development.[118] Finally, in 1950 a sludge disposal plant—designed to serve more than 776,000 people—was completed utilizing the heat-drying method to produce dried sewage sludge for fertilizers. An enlargement program was completed in 1960, with further expansion in sludge disposal in subsequent years.[119]

While the city confronted its sludge problem forthrightly, it faced more serious difficulties with bayou and stream pollution. The Engineers' Council of Houston on December 19, 1946, adopted a resolution that read in part:

> It is well known that the sanitary conditions of the bayous and streams in and around the City of Houston are intolerable. They are, in fact, in some instances on a par with those found in China and India. Their bacterial index count generally exceeds several million per 100 cc. This condition is due essentially to the lack of modern treating plants. . . . The main sanitary sewer trunk lines and gathering systems are so inadequate in capacity throughout the city that during and after almost every heavy rain raw human sewage flows freely through the manholes onto our streets.

The Engineers' Council submitted its resolution to member societies for their approval, with the intention of petitioning the mayor, city manager, city council, and city engineers to remedy the situation through new construction, chlorination, and better handling of residential sewage and industrial wastes.[120]

The resolution outlined an aggressive and comprehensive plan to significantly upgrade the city's sewerage system. Meeting the challenge was hard fought and only slowly addressed. Reference continued to be made to the chronic problems of sewage and chemical pollution derived from Houston, Baytown, and other nearby communities as well as the ship channel industries. On June 28, 1948, on the heels of a polio outbreak that raised concerns about water pollution as a contributing factor, the *Houston Chronicle* noted that "pollution of both types—sewage and chemical—in the bayous, the Ship Channel and the bays has been serious for a long time." A headline in the newspaper on February 1, 1952, seemed to sum up the pace of change: "City Attacks Sewer Stink, but Slowly."[121]

A Houston Chamber of Commerce report in 1972 reflecting back on the wartime and postwar history of water pollution in Houston noted that the city in the 1970s "found itself unable to keep pace with requirements for sewage collection and disposal." The war effort several decades earlier had complicated the city's ability to keep up "with the requirements imposed by the population growth," and set before the city "the Herculean task of cleaning up a massive backlog of urgently needed municipal improvements, not the least of which was that of sewage collection and disposal."[122] Public health authorities had suspected polio epidemics of the 1930s and 1940s as related in some way to the pollution in the bayous during those years, and continued concern about the correlation between polio and bayou pollution led directly to surveys of the bayou conducted by consulting engineer Frank Metyko and delivered to the Harris County Commissioners Court in 1947.

Early postwar efforts at pollution control had been stalled, some believed, because of a cut in the city tax rate in 1944, at a time when population was on the rise. The municipal government, the 1972 Chamber of Commerce report noted, made progress in correcting the sewage disposal problem with the passage of a bond program in 1946—most likely induced by the polio scare, increased population pressure on the existing system, and efforts at federal water pollution legislation—and especially through a "massive program" in the 1960s "to eliminate the city as a principal polluter of the bayous and, thus, the Ship Channel."[123]

Debate and controversy over the contribution of ship channel industries to the water pollution problem—as well as the city's role—persisted for several years. In 1974, the Texas Water Quality Board called for further cleanup of the Houston Ship Channel, but the city of Houston and several industries requested greater latitude in the amount of pollutants they could discharge into the channel. About the same time, the state district court ordered Houston "without delay" to solve a sewage problem in several East End subdivisions that had not been addressed for forty-seven years.[124]

The inability of the city to upgrade and expand its sewerage system during a period of rapid growth in the 1970s and 1980s gave rise to a host of problems in the 1990s. A 1988 report concluded that sanitary sewer overflow caused by infiltration and inflow—that is, intrusion of storm water and groundwater into the sewer pipes—was a major cause of water pollution in the Houston area. Other studies came to conflicting conclusions and placed the onus on improperly operated wastewater facilities. Nevertheless, many agreed that the sewerage system had badly deteriorated and that, among other things, increased water quality standards would make improvements more expensive. The Lakewood Heights subdivision on Lake Houston, for example, saw regular sewage discharges into the lake after a heavy rain.[125] Although there were plans to spend $7 million in the early 1990s to improve sewage collection in the subdivision, this was only a small fraction of what some believed to be the $2 billion needed to improve the entire system. Through an agreement with the Texas Water Commission and the EPA, Houston had until 1997 to complete the improvements that were necessary because of what one EPA official called "policies of benign neglect."[126]

The 1994–1998 City's Capital Improvement Plan included projects to provide service to areas without sewers and to reduce infiltration and inflow through extensive repairs. Under Mayor Bob Lanier, the massive $1 billion Greater Houston Wastewater Program was under way, but not without concerns that there were flaws in testing and inspection of the system.[127] The battle over providing an adequate sewerage system that operated as designed and did not pollute adjacent watercourses continued.

Solid Waste

Houston's solid-waste collection and disposal system faced the same pressures of growth as the water supply and wastewater systems after World War II. A 1983 report by the Houston Chamber of Commerce's Solid Waste System Task Group stated that "Houston's current solid-waste collection and disposal system had managed simply to 'keep up' with the problem of waste. This system is no longer satisfactory."[128] This was not a remarkable statement to make for Houston or any large city for that matter. Nevertheless, the postwar years in Houston saw many trying days. Municipal collection was increasingly expensive; incinerators had continually malfunctioned, producing serious emissions problems; and landfill capacity was getting scarcer.

Among the most striking changes in the 1970s was a major shift in disposal method from heavy dependence on incineration to almost exclusive use of landfilling. The advent of recycling, however, modified the collection system in the 1990s and provided some relief to rising disposal costs. The transition from incineration to landfilling was not immediate, and not without problems. The first high-temperature incinerator for the city was completed in 1949.[129] In the 1950s, four new incineration plants were built. The city added two more incinerators in the late 1960s—one close to Interstate 45 and another near the Holmes Road landfill. The Holmes Road plant, opened in 1967, was widely criticized for its odors and emissions, and was closed within a year for major renovations. It reopened in 1971, but never operated at capacity and had chronic air-emissions problems. In 1973, it was closed permanently. Turning to Consumat mini-incinerators in that year, the city hoped that the natural gas–run plants would avoid the problems of their older incinerators. The energy crisis of the 1970s led to soaring natural gas prices, and many of the mini-incinerators never opened. Ultimately, all were dismantled.[130]

In the area of solid-waste disposal, Houston had bucked the national trend in the postwar years that had led to the broad acceptance of sanitary landfilling as the most widely accepted form of disposal in the United States. The sanitary landfill was a breakthrough that became the primary disposal option in the United States between the end of World War II and the 1980s. While there was no mass scramble to build sanitary landfills in the 1930s and early 1940s, momentum slowly shifted in that direction. From the 1950s through the 1970s, the prevailing wisdom among solid-waste managers was that sanitary landfilling was the most economical form of disposal and, at the same time, offered a method that produced reclaimed land.[131]

By the 1970s, sanitary landfills became the primary means of disposal in Houston, with sites within or adjacent to the city limits.[132] In some cases, the waste was transported to public and private transfer stations before being hauled to the landfills. In 1982, the Solid Waste Department collected 810,000 tons of solid waste, and private contractors collected another 1.2 million tons; by 2003 the Houston-Galveston region was generating 4.5 million tons of solid waste each year (58 percent from residential collection).[133] In the early 1980s, the greater Houston area had seventeen operating landfills (all private) to accept municipal waste, with seven reaching capacity by 1989. Some experts believed that the system of landfills in Houston would face a capacity crisis by the end of the decade because of the city's rapid growth.[134] The best known of the landfills, the McCarty Road landfill, had reached two-thirds of its capacity between 1971 and 1981.[135] In 1992, Ulysses Ford, who had just stepped down as the city's public works director, stated, "We don't have a landfill crisis like they do in the Northeast, but we need to be very concerned." New projections placed landfill capacity at the year 2000, but since it took many years to choose and secure new sites, unease continued to mount.[136] Projections for 2004 suggested that the region's twenty-one landfills would reach full capacity by 2013.[137]

Disposal was a very public issue for the solid-waste department, but the department faced several other predictable problems as well.[138] The coverage area of collection services always seemed to be expanding.[139] In such a labor-intensive business, problems between workers and management were sometimes intense, especially when much of the workforce was made up of minorities and much of the management was white.[140] For example, collection crews in the summer of 1980 staged a wildcat strike to protest what they believed was poor administration and defective equipment. In 1986, 84 "disgruntled garbage collectors" staged a three-day wildcat strike because of a new route system that added about 20 percent to the collectors' workload. Initially, 161 collectors were laid off (29 were recalled) because of the new system. Mayor Whitmire at first fired the strikers, but ultimately suspended them for thirty days without pay.[141] Automation of collection and privatization of service have changed the work environment in recent years.

Probably the most serious problem that the solid-waste program faced was the chronic practice of siting disposal facilities in minority, poor, and working-class neighborhoods. As noted previously, this practice began as early as the

1920s, if not earlier. Ultimately, it gave ammunition to the emerging environmental justice movement in the 1980s.[142]

Abandoned landfills posed another problem, especially in minority neighborhoods. Of the 223 (or 236, depending who is counting) abandoned landfills and dumps in Harris County in 2003, 36 of them lie within Acres Homes, an African American neighborhood in northwest Houston. Collectively, the landfills and dumps in Acres Homes—an area of just nine square miles—make up of the largest clusters of abandoned solid-waste facilities in Texas. Many of the dump sites that operated in the neighborhood from as early as 1932 until the early 1990s never received permits from the state.[143]

While the placement of waste facilities represented the most egregious element of the solid-waste collection and disposal program in Houston, efforts at curbside recycling in the 1990s likely represented the most forward-looking feature. In 1987, the city began an earnest effort to develop an integrated solid-waste management system—under the 20-Year Plan—that took a holistic approach to dealing with collection and disposal. An Urban Consortium Energy Program Grant aided in the development of a cost-analysis model, which grew out of the examination of a variety of management scenarios. The emphasis of the Houston program would be a public/private partnership that included some sort of voluntary source-separation recycling and efforts at waste reduction.

The first recycling project (1988) focused on scrap metal recycling using three neighborhood depository sites. In 1989, a second project was initiated—the Westpark Consumer Recycling Center. Located at the site of an abandoned incinerator, the center accepted aluminum, glass, newsprint, and some tin and plastics. Citizens received cash from the privately run center, and the city received a royalty on the recyclables. Also in 1989, a pilot office paper recycling project commenced, with a local recycling company providing bins for paper. Several large companies followed the city's lead and implemented their own office paper recycling programs. Another component in the plan was the "Don't Bag It Program" meant to encourage limiting disposal of yard waste—a major issue in the South—and encouragement to compost leaves and grass clippings.[144]

A curbside recycling pilot program—utilizing plastic bins for recyclables—was implemented in three phases beginning in August 1990, with 7,000 homes in two subdivisions participating. In September, 7,000 more homes were added, and in December 13,000 homes in fourteen subdivisions were added. Browning-Ferris Industries concurrently developed a plastic bag (Blue Bag) co-collection system for 19,000 homes, but by 1991 the city took over the program, converting many of the homes to the city's bin collection system. The recyclable program was not self-supporting and drew revenue from the city's general fund. Champion International Paper Company built a de-inking plant in 1993, relying on the city's recycled newspapers and magazines as a source for making new paper. All of the recycling efforts around the city were being promoted and monitored by the nonprofit Clean Houston, operated by Houston Clean City Commission, Inc.[145]

Progress in recycling in the Houston area had been significant in the 1990s, with a 168 percent increase in the amounts collected from 1993 to 1999. However, between 2001 and 2005 recycling dropped more than 13 percent—from collection of 11,770 tons to 10,210 tons. And the city has yet to meet the ambitious goal of 35 percent recycling of solid waste set by the EPA in the late 1990s.[146] Also, a 2003 *Waste News* report of thirty cities indicated that Houston's municipal recycling program ranked near the bottom nationally for the amount of material it recycled.

There are many reasons for the more recent tepid response to recycling in Houston: recycling rates across the country were down in the first decade of the twenty-first century; Texas does not set mandatory goals for the amount of recyclables, so local participation in programs like Houston are voluntary; the recycling program in Houston, which peaked in 1998, has undergone some changes, including the decision in 2000 not to accept glass; the city does not charge a fee based on trash volume, and thus there is little reason to recycle based on economic incentives; and the high cost of recycling inevitably makes the program compete with other city services for funding.[147] While the success of recycling has been modest in Houston, the city's effort to develop an integrated solid-waste system represents a planning process that had been sorely missing from its history of developing sanitary services.

Conclusion

A July 8, 2001, story in the *Houston Chronicle* bemoaning the state of the city's infrastructure began, "Every time Houstonians turn on a faucet, flush a toilet, walk under a streetlight or drive down the city's mangled streets, they are at the mercy of the Department of Public Works and Engineering." The article went on to question how the city's aging systems would fare in the future, and if public works officials and city leaders were up to the task to provide necessary leadership.[148] Such concerns and criticisms plague almost every large city, but Houstonians are nevertheless justified in expecting their municipal services to deliver the promised benefits. Growth has been a relentless adversary in providing adequate sanitary services, and long-range planning for growth often has been missing in action. Whether the shortcomings of service delivery were problems of foresight, problems of scale, or the product of fiscal conservatism is not absolutely clear.

The initial development in the late nineteenth and early twentieth centuries of a citywide water-supply system, underground sanitary sewers, and incineration of wastes were positive steps in meeting service needs in Houston. Banking on groundwater to provide the sole source of water, building sanitary sewers without similar attention to storm water and runoff, and seeking solid waste disposal at the expense of one sector of the Houston community ultimately were exposed as inadequate once the city grew into a metropolis.

The regionalization of urban expansion also regionalized the operation of Houston's sanitary services. Rapid urban growth, as several municipal leaders argued, not only constrained the ability of the city to "keep up" in maintaining its service infrastructure but also changed the rules of the game, and the impacts of those services went well beyond the city limits. Developing an adequate water supply for Houston now meant seeking sources of water many miles from the urban core and creating delivery systems that impinged on non-Houston areas. Constructing sewer lines and sewer treatment facilities to handle voluminous effluent was not sufficient to protect areas of outfall all the way to Galveston Bay. The bay also became the depository for staggering amounts of runoff from Houston's ever-increasing property development. Mounds of solid waste could not so easily be burned or buried within the city limits. Aside from the sanitary systems themselves, finding solutions to service needs also challenged traditional political institutions, demanding more attention to multiple jurisdictions and multiple interests.

In essence, the transformation of Houston's sanitary services from local concerns to regional issues was a dramatic reflection of how the city was changing since its modest beginnings in the nineteenth century. The environmental implications of Houston's transformation were particularly striking, suggesting an even greater need for clear planning goals in the future. The mental leap is as significant as the evolving services themselves.

CHAPTER 6

Superhighway Deluxe

Houston's Gulf Freeway

TOM WATSON MCKINNEY

The construction of the Gulf Freeway represents the culmination of pre–World War II planning, federal funding, civic pride, modern engineering, and a deeply held belief that all levels of government could cooperate in order to solve emerging urban problems in postwar America. For the federal government, it marked the start of a golden age of infrastructure building, as the Gulf Freeway was the first freeway constructed in the United States after World War II. For the Texas Highway Department, the Gulf Freeway was an affirmation that its wartime policy of suspending construction in order to save money for postwar projects was the proper path to take. For Houston, the roadway symbolizes the endorsement of the automobile as the primary mode of travel, an embrace of modernity, and a corridor for rapid and sprawling urban growth. The Gulf Freeway also represents a clear focus on urban problems during the postwar era, as it serves an urban population.

The passage of the Federal-Aid Highway Act of 1944 forever changed the way that roads are planned and constructed in the United States. The act, which amended the Federal-Aid Road Act of 1916, dramatically increased the amount of federal money budgeted for highway construction, an amount that continually increased during the postwar era. More important, it marked the first time that the federal government provided funds for the construction of urban roads, as previous highway appropriations were focused on rural roads.[1]

The new law simply defined an urban area as any community having a population of 5,000 or more, and this stipulation was specific to urban projects, as the funding formulas for primary and secondary highways took into account state land area, farm-to-market road mileage, and other variables.[2] The government only examined population in determining an urban center's eligibility for federal highway dollars.

Previously, the federal government considered urban roads to be the exclusive domain of the municipalities they served, and earlier highway aid cuts focused on the construction and maintenance of rural roads. The new law extended the jurisdiction of state highway departments into urban areas.[3] By earmarking funds for the construction of urban highways, the federal government sought to eliminate congestion in urban centers by ensuring that financing for such projects was available to all.[4] This also served an economic function, as the American trucking industry experienced a boom during the postwar era.

While the Federal-Aid Highway Act of 1944 provided funds for urban projects, it also specifically called for the states to designate a coherent system of highways. This was a revolution in American road planning, as work was done previously on a project-specific basis. Rather than designate separate systems for state highways, rural roads, and city streets, the 1944 act forced states to examine their transportation needs as broadly as possible and create a balanced system of roads. Tying long-range planning to road development, the act forced state highway departments to seriously assess their state's transportation needs. This stipulation is implicit in federal legislation since World War II, and it is the most important idea in modern highway planning.[5]

The Federal-Aid Highway Act of 1944 provided for a 40,000-mile "National System of Interstate Highways" to connect "principal metropolitan areas, cities, and industrial centers to serve the national defense, and to connect suitable border points with routes of continental importance in the Dominion of Canada and the Republic of Mexico," but it did not provide the necessary funds for construction.[6] The Federal-Aid Highway Act of 1956 amended the 1944 act to rectify this problem, thus launching the construction of the U.S. system of Interstate and Defense Highways throughout the nation, the largest public works program undertaken in the United States.[7]

The Texas road system was a casualty of World War II. The Texas Highway Department, headed by Dewitt C. Greer, "virtually suspended" highway development in Texas during the war. What was constructed was severely limited because of the need to conserve war materials and was generally focused on defense.[8] In addition to total neglect by the Highway Department, Texas highways also suffered because of heavy use by civilian traffic and by increased military traffic. Beginning in 1940, the U.S. Army conducted a series of war games across the Texas-Louisiana border. The army discovered that "only one-fourth of the East Texas roads used met federal military standards, and hundreds of miles of roads and many bridges suffered damage from the heavy traffic."[9] The substandard road system of rural eastern Texas, as well as many other highways throughout the state, fell into disrepair because of neglect and military use.

Division engineer J. A. Elliot of the Bureau of Public Roads found that this was common in the United States. In 1949, Elliot commented:

> prior to World War II, the increase in the number of motor vehicles in service coupled with the increased use of the motor vehicle caused a public demand for a large mileage of highways usable immediately. The only way this mileage could be obtained was by sacrificing desirable standards. During the war highway construction aside from that considered essential to the war effort was practically nil. At the end of the war the States found themselves with an increased mileage requiring reconstruction and higher maintenance on all mileage.[10]

Despite the fact that the Texas Highway Department did not initiate many projects during the war, it did not sit by idly. Suffering from a severe lack of manpower and a shortage of materials, the department used the break in construction to focus on planning initiatives in hopes of a postwar renaissance. Governor Ross S. Sterling is credited with the original idea for the Gulf Freeway during the Depression when he was head of the Texas Highway Commission, but the capital for such an infrastructure project was unavailable at that time. Beginning in 1944, the department studied transportation problems in Houston, Dallas, Fort Worth, and San Antonio, and enlisted the aid of "nationally-known highway planners."[11] These planners concluded that the urban environment of these cities would not allow for the sufficient widening of streets to relieve the mounting traffic congestion caused by increased suburban development and automobile ownership. As one engineer pointed out, "a 200-foot street, as required for an expressway, ruins practically the whole block."[12] The department's study discovered that "nearly one-half the highways in the state were ten or more years old, while in 1940 only one-third had been. Aging and deterioration of structures and surfaces were increasing rapidly, since little had been done to renovate the system during the war years."[13] The federal government openly endorsed this policy, as it wanted nothing to interfere with the war effort. In order to ensure the compliance of all state highway departments, it drastically cut aid for nonmilitary highway projects. This virtually halted highway construction in the United States during the war.[14]

While overseeing the drafting of postwar highway plans, Greer also did all he could to ensure that funds were available for construction once the war ended. He invested much of the Highway Department's budget in treasury bonds, "for postwar reconstruction of the state's highways."[15] By the end of the war,

Greer's investment gave the department a $30 million nest egg that was drawing 7 to 8 percent interest and was available immediately for highway repair and construction.[16] This fact, coupled with a coherent statewide plan and renewed federal funding, allowed the state to proceed aggressively with its postwar highway program.

Greer created "engineer-managers" to oversee the construction of urban highways in Texas. These special posts existed in four major cities in the state—Houston, Dallas, Fort Worth, and San Antonio—and engineer-managers were given the sole responsibility for planning and overseeing the construction of urban expressways. The Highway Department empowered this group to make rapid decisions concerning urban expressways, which demonstrated that the Texas Highway Department realized the importance of these routes and the need to complete them rapidly.[17] J. A. Elliot believed that Texas's postwar highway program was highly successful, stating in 1949 that "progress is being made in other States but not on such a large scale as in Texas."[18]

Houston, like many Sun Belt cities, experienced explosive growth during and after World War II. The influence of the automobile made a definite impact, as 90 percent of the city's growth occurred during the automobile era.[19] The construction of the Gulf Freeway, which began in 1945 and was completed in 1952, not only had spatial ramifications for the development of the city but also had a tremendous impact on the way Houstonians and modern Americans live their lives. The freeway started a revolution for Houstonians, as many traded the congestion of neighborhood streets for the smooth Portland cement of the Gulf Freeway and a shorter commute to and from downtown. The road's 50.75 miles to the Gulf Coast secured Houston's position as a regional city by providing an automobile-based alternative to the Houston Ship Channel, a path of expansion through annexation, and a closer relationship with Galveston.

The Gulf Freeway was the first limited access highway constructed after the war in the United States, and as such it served as a prototype for the construction of future highways. The freeway also fostered booming suburban growth as well, leading to the sprawling urban form that many equate with the Bayou City. In essence, the freeway was so successful that it was replicated repeatedly throughout the city, creating more sprawl with each mile and an increasing dependence on automotive transportation. The construction of the Gulf Freeway, therefore, illustrates many of the urban and environmental issues surrounding such roadways in Houston, and in the United States.

Houston emerged from World War II as a booming center of manufacturing and refining. The city's economic base benefited greatly from the war, as the city provided high-octane aviation fuel, petrochemicals, ships, ordnance, and other goods to fuel the war effort. The Port of Houston also underwent expansion, and new industries constructed plants along the ship channel. The attraction of jobs in these plants also contributed to the growth of the city, as rural residents left the countryside to seek their fortunes in the Bayou City. This trend mirrored the loss of rural population in the United States during World War II.[20]

Houston's growth continued to break previous records in the postwar era, as the promise of employment continued to attract new residents. The Houston Planning Commission remained busy from 1945 to 1954, as 910 miles of new streets were provided for new subdivisions composed of 124,000 lots. During those ten years, the Planning Commission platted fifty-five square miles, and during 1955 alone it recorded subdivision plats containing over 20,000 new lots.[21] In order to envision the amount of street congestion that must have occurred during this era, it is important to remember that the Gulf Freeway was not in use until 1948. Before that time, Houston drivers were confined to major thoroughfares and neighborhood streets.

In response to the growing congestion on the city's neighborhood streets, William James Van London, Houston's manager of urban expressways, announced plans for the Gulf Freeway's construction on November 2, 1945. Congressional approval of the 1944 Federal-Aid Highway Act, which offered up to $1,500,000 in matching funds to each state, allowed Texas to move forward with the freeway project. Van London estimated that the Gulf Freeway would cost close to $18,000,000, of which $3,000,000 would secure the right-of-way.[22] While it is unclear whether Van London was addressing the costs for building the freeway inside the city limits, the total cost of constructing the road was $28,643,521.[23] Van London explained that it was the city of Houston's job to secure the right-of-way, while the state would construct the highway using state and federal funds.[24]

William Van London was born in Canada in 1893 and moved to the United States at the age of two. He enlisted in the army in 1917 and served for two years. Educated at the University of Utah, Van London joined the Texas Highway Department in 1922. He was described by one reporter as "red-faced, partly bald and he talks in a straightforward, yet cautious manner," and as being "completely absorbed in his work," evidenced by his inability to finish a cigar while working.[25] The Texas Highway Department gave him the task of designing the freeway inside the city limits, which was marked to the south at Sims Bayou. The total distance was fifteen miles, and nine of those miles were considered "urban." Van London began preliminary design work in 1943, although he later admitted that he had toyed with the idea for the freeway as far back as 1941. The Texas Highway Department assigned the project outside the city limits to State Highway Engineer Jim Douglas. Van London and Douglas formed a good working relationship, characterized as "a perfect synchronized team."[26] Both men "love to build, and in the Gulf Freeway each saw a monument which was a challenge to any construction man."[27]

Van London's design for the freeway was innovative (see figure 6.1). Beginning in the heart of downtown Houston at Louisiana Street, he created a system of one-way feeder streets. Pease and Jefferson streets were designated to carry westbound traffic, and Calhoun and Pierce streets were designated to carry eastbound traffic. The four streets would merge at Dowling Street into the freeway, with three lanes on each side of a four-foot curb and an outside emergency lane. The freeway would narrow to four lanes at Telephone Road and continue to its link with the Galveston Highway at Sims Bayou.[28]

The design of the freeway also eliminated grade crossings of railroad tracks or with any other street. Van London avoided this problem by creating a series of overpasses, one-way frontage roads, timed traffic lights, and ten entrance ramps where automobiles could enter or exit the freeway. These measures not only served as safety features but also as regulators of automobile speed. While this seems less than revolutionary today, the Gulf Freeway was the first road in Texas to incorporate such measures. Erik Slotboom's study of Houston's freeway system finds the inclusion of continuous frontage roads to be a characteristic "that would set the Gulf Freeway apart from most emerging freeways in the nation and would also set the standard for future Houston Freeways."[29] Van London's design also incorporated drilled shafts with spread footings rather than the more traditional driven piles.[30] This technique helped to maintain low costs.[31]

His reliance on overpasses to avoid grade crossings resulted in the Gulf Freeway's nickname, "Van London's Roller Coaster." In a 1950 interview, one reporter commented on the reception the design received when it was first presented:

> A few years ago, when he was planning the Gulf Freeway, some city official engineers and others were skeptical.
>
> "Too elaborate—not needed." "Looks like something out of Buck Rogers." "What are you going to do with that roller-coaster, Van—compete with Playland Park?"[32]

Despite the initial criticism of his design, Van London's vision was adopted by the Texas Highway Department (see figure 6.2).

While the idea for the highway had been around since the 1930s, the actual securing of the right-of-way was left up to the city of Houston, Harris County, Galveston County, and the city of Galveston. The legislation did not give this authority to the state because of the fear that it might force a highway through a city. In order to ensure the population was willing to support the construction

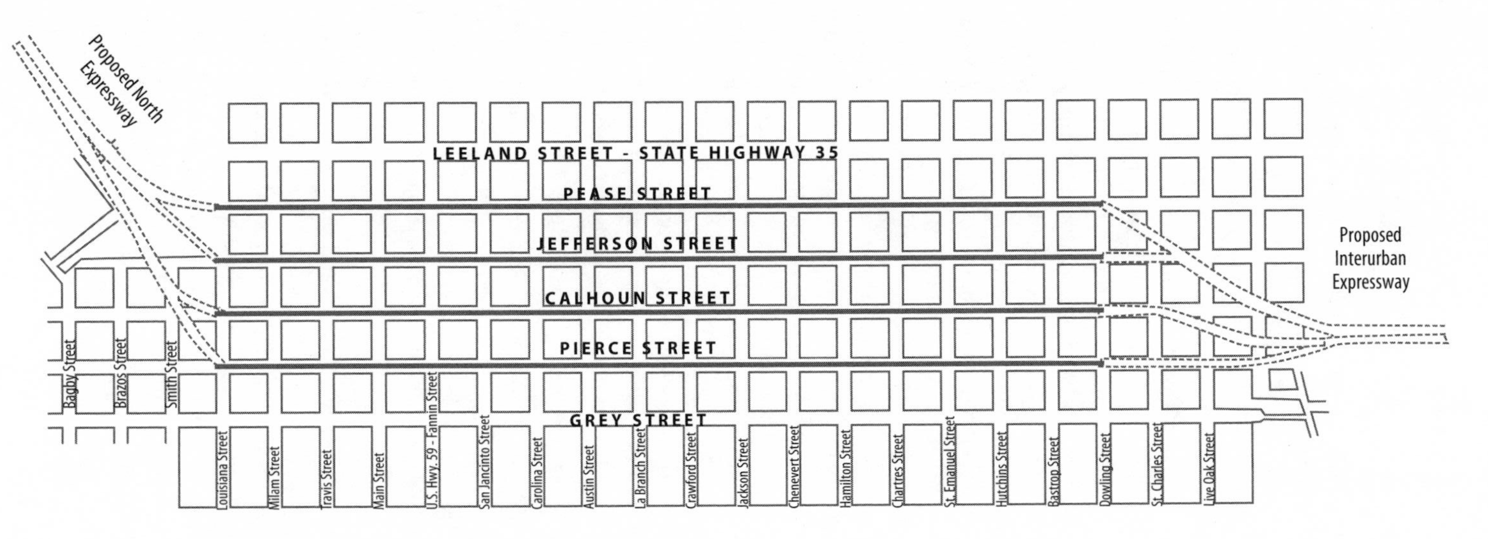

Figure 6.1. Van London's feeder street system for the Gulf Freeway.

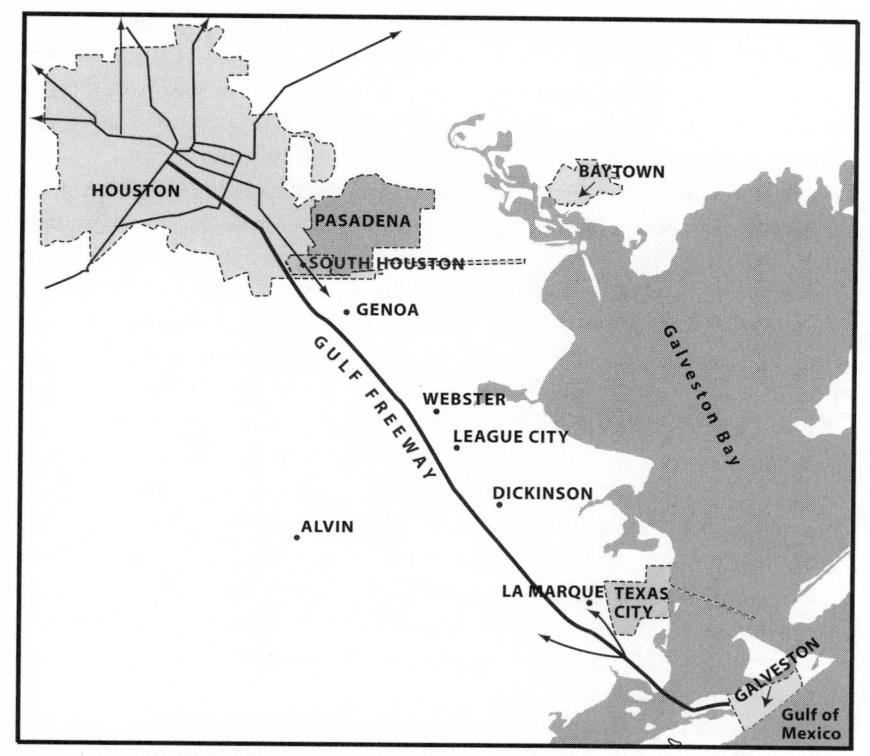

Figure 6.2. The Gulf Freeway.

of the highway, the law was structured to give the responsibility for the right-of-way to city and county entities. This law was changed in 1951 to give the highway department condemnation rights.

In the city of Houston, the task of securing the right-of-way fell to Mayor Oscar Holcombe and City Treasurer W. T. Collier. Holcombe's role in securing the needed land actually began in 1939, many years before the freeway route was designated. The Galveston-Houston Interurban applied to the city to abandon its tracks within the city during that year, and Holcombe purchased its right-of-way for the city, which would later provide part of the Gulf Freeway's route and enhance the reputation and popularity of the mayor.[33]

When Van London announced plans for the construction of the freeway, W. T. Collier was present, and he was questioned on the city's ability to secure the needed properties. His answer highlights one of the many difficulties in planning the first freeways in a metropolitan area: "I am not so concerned about buying the land as I am about getting 209 houses on it moved."[34] As previously mentioned, up to this point Houston had grown in a grid pattern without freeways. In order to accommodate the freeway, the form of the city had to be altered. This process included the removal of existing structures along with the widening and repaving of streets, the addition of storm sewers, and the installation of traffic control devices.[35] In short, the freeway required more than just a right-of-way. It required a restructuring of infrastructure.

Senior Land Appraiser W. T. Shannon had contacted many of the property owners by the time plans for the freeways were announced. He commented that "four house-movers—all we know of—have agreed to get to work on the moving as soon as we close the deals, and it will take about six weeks to move the houses."[36] City Manager John N. Edy said that if the house movers were unable to clear the property in time, he "may organize a house-moving division of his organization and do the job himself."[37]

By February 1946, three months after the project was made public, the city of Houston began to realize the headache caused by its efforts to secure the right-of-way, as city officials found it difficult to move some 200 structures in the freeway's path. All of them had to be removed before actual construction could begin. W. H. Bobbit, head of the city's land acquisition division, commented that local house movers were swamped, and that finding new home sites with sewers already installed presented still another problem yet unsolved. He, unlike City Manager Edy, thought that "the city doesn't want to go into the house-moving business."[38]

A month later, the city of Houston had still not finished securing the right-of-way. The Texas Highway Department nevertheless let contracts for work to begin, including a nine-mile stretch in Galveston County, and several sections within the Houston city limits. Brown & Root, Inc. secured the biggest of these early contracts—$2,053,015 for the construction of 1.2 miles of the expressway from St. Bernard to Wilma Jean (including two overpasses crossing the Houston Belt and Terminal tracks), and $957,000 for the widening of Pease, Calhoun, Pierce, and Jefferson streets, between Louisiana and Dowling, to carry their share of the expressway traffic. The company was also to trim and move 200 trees.[39]

In the summer of 1947, the city of Houston began to auction off the houses that stood in the right-of-way. This novel approach not only helped to clear these structures from the road's path but also helped the city to recoup some of its right-of-way expense as the money for the structures went directly to the city. This method also helped to expedite the removal of the buildings, as that became the obligation of the new owner.[40] Furthermore, the additional costs of moving the structures out of the freeway's path, such as those incurred by "the costs of cutting telephone cables or light wires where the house is too high to go under the cables; costs of crossing railroads where a railroad crew has to flag trains, if need be; costs of having plumbing and wiring worked over after the house is moved; of re-finishing inside, such as repapering," were all passed on as well.[41] The most notorious case involving the cutting of utility lines involved "one two-story brick house moved 2 1/8 miles the cost of moving or cutting such lines, and of moving an ice house in the way, ran to $1700. Highest such cost reported was around $5000 in moving a negro [*sic*] church 11 blocks."[42] Another example was the Humble Oil Company's donation of land for the freeway in the vicinity of the Friendswood oil field. As a result of this civic-minded deed, the oil company moved pipelines and storage tanks out of the path of the roadway in the ceded lands, estimated to be between 90 and 100 acres. This action reaped a large reward for the oil company—the remaining property became known as the Clear Lake area, the future home of NASA's Lyndon B. Johnson Space Center.[43]

One of the more interesting events that occurred during the construction of the Gulf Freeway happened at the home of James and Cora Tantillo on March 28, 1949. A group of sheriff's deputies and roughly 150 spectators arrived at the Tantillo home with a writ of restitution signed by Judge Ben Wilson, calling for the Tantillos' eviction. The Tantillos, however, refused to leave their home. Deputy O. C. Dawson, who was physically assaulted by James Tantillo during the eviction, described the events that unfolded at the home to a reporter:

> "When I got out there, the place was locked up tighter than a drum," Dawson remarked, "but I knew somebody was in there. So I got started removing a screen to get in."
>
> About that time the storm broke. Tantillo, an unemployed mechanic and owner of the house, hit Dawson, according to Deputy Sheriff J. E. Harrell, one of the reinforcements, who had been rushed to the scene.
>
> "Lock the screen," Tantillo shouted to his wife as he routed Dawson.
>
> Superior numbers won, though, and Tantillo, handcuffed, was taken to the Criminal Courts Building, while remaining deputies and workmen proceeded with the orderly eviction by carrying the Tantillo furniture outdoors.[44]

Despite the melee and arrest of her husband, Cora Tantillo remained defiant. She jumped into her bed and refused to budge, thwarting the best efforts of the deputies to remove her. She later relented to the eventuality of her home's destruction as workmen entered the house to remove doors and prepare the house for demolition. She walked out onto her front porch in resignation and surveyed all of her possessions being placed neatly in her front yard.

James Tantillo was later released on $500 bond. He had explained to the sheriff that he had not been paid for his property, and had no idea that he was to move. City employees who were interviewed at the time tell a different tale. According to the right-of-way department, Tantillo had requested that the freeway be rerouted in order to spare his five-room frame house and property. The condemnation commission did not grant his request, valued his property at $9,350, and deposited the same amount with the county court-at-law for him to pick up. According to the city, James Tantillo was notified in January of the condemnation of his home, and of the fact that he had five days in which to find a new home once the city secured a writ of possession.[45]

Whether or not James and Cora Tantillo received notice of the condemnation of their home, their story illustrates the conflict that can occur when repurposing land. As seen in numerous studies, humans have an attachment to place that is not easily understood or easily translatable into a political system.[46] With any major infrastructure project, there always seem to be both winners and losers, as some will benefit from the improvement more than others. In this case, the Tantillos were the losers, as they lacked both employment and political power. Houstonians were the winners, as the construction of their highways continued unabated. As time marched on, there would be resistance to the expansion of the Gulf Freeway, but these efforts would all be unable to stop it.[47] The only thing that stopped the construction of the Gulf Freeway was a lack of money.

The process of condemnation for the Gulf Freeway did not just affect land owned by nonelites, as land owned by Mayor Oscar Holcombe and his brother-in-law was also seized for the freeway. In April 1949, Holcombe told the city council that the only way the city would get his land was through condemnation proceedings, as it was the only way he, as the head of the city government, would get a fair price for his two-and-a-half acres of land.[48] The city council obliged the mayor, and voted to condemn the land.

The construction of the Gulf Freeway not only altered the built environment by forcing the removal of houses and other structures to make room for the freeway, but it also created the closest thing Houston has to rolling hills—freeway overpasses. The original urban section of the Gulf Freeway consisted of seven overpasses, which allowed the freeway to bypass grade crossings with major cross-streets and railroad tracks. Van London's design provided for a segregation of transportation modes so that neither would hamper the mobility of the other. His design also created soaring concrete traffic interchanges that would become a hallmark of the Houston built landscape, and these would also exacerbate the city's drainage problems.

By late June 1948, roughly three years into the construction, it was possible to drive to Galveston and not leave the roadbed of the freeway, as all of the grading, drainage, and laying of the shell substructure had been completed.[49] David G. McComb described a trip down this temporary roadbed in *Galveston: A History:* "My early memories, fragmented and distorted, include a ride to the beach over the unfinished Gulf Freeway in a 1946 Ford with my brother at the wheel and the windows rolled all the way down to circulate the oven-heat of the Texas summer. The white dust from the oyster shell foundation rose in a huge cloud behind us and drifted into the low trees along the right-of-way. They remained white until the next rainfall."[50] While this might seem like a fairly primitive way to get to Galveston, the incomplete roadbed was very similar to the former highway at the turn of the century.[51]

Mayor Holcombe led Houston into the freeway age on the evening of September 30, 1948, as he opened the first section of the freeway to traffic on the Calhoun Road and the Houston Belt and Terminal railroad tracks overpass. The event was viewed by 500 spectators, as well as Van London, Chairman John S. Redditt of the State Highway Commission, and District Engineer J. M. Paige of the U.S. Public Roads Administration. The 4.1-mile of roadway stretched from downtown Houston to Telephone Road, and cost $12,200,000. It was officially opened when Holcombe turned on the freeway's mercury vapor lights. "Happy Days Are Here Again," Holcombe's campaign theme song, was played from a portable phonograph, reminding those watching the spectacle who was responsible for the new freeway.[52] Traffic, which had backed up more than half a mile, rolled across the first freeway constructed in the United States since the end of World War II.

Oddly enough, the freeway had no name when it first opened to traffic, and it would remain unnamed until December 1948. Mayor Holcombe announced that a contest would determine it, and the winner would receive $100. Sara Yancey, an employee of the City National Bank, won the contest. When interviewed by a local paper, she commented that her winning entry, the Gulf Freeway, seemed "pretty obvious," as the road went to the Gulf of Mexico and because it was a freeway. Ironically, Yancey did not own a car.[53]

The Gulf Freeway had a tremendous impact on the way Houstonians drove, as evidenced by the fact that 42,000 vehicles per day were using it just five months after it opened. Traffic Manager T. E. Willier used pneumatic counters to determine that the traffic pattern of the southeast part of Houston had changed. While Willier was unable to make an accurate determination of how it had changed, he did comment that "traffic on Leeland, at St. Bernard, dropped 10,000 cars a day after the freeway was opened. It was 21,400 cars daily, and went down to 11,400. This traffic went somewhere, and obviously some of it at least went to the superhighway, but the picture is not complete enough to change traffic control arrangements."[54] Willier presented his findings to Holcombe in July, which indicated that the freeway was paying for itself in terms of time saved. According to his report, the completed section of the freeway cost $11,000,000, and the value of the saved time was calculated to be $2,688,684.[55] Dewitt C. Greer, after reviewing the report in August of 1949, predicted that the freeway would carry 70,000 or more vehicles per day. When pressed for the top number of vehicles the road could carry, Greer commented, "There is ample room for growth—everyone will just have to write his own ticket on the top figure, I guess." "Anyway," he said smiling, "it's doing a mighty big business already."[56]

The city of Houston continued to secure the freeway right-of-way, even though part of the freeway had been completed and opened to the public. On February 18, 1949, City Treasurer W. T. Collier predicted that his three-and-a-half-year effort to secure the path of the freeway soon would be finished.[57] The right-of-way cost the city of Houston an estimated $7,500,000, which it expected to make back from increased property taxes along the route of the freeway.[58] This seemed plausible, as land values skyrocketed along the freeway. The first sign that the values were escalating came when a consortium of businessmen purchased a piece of land measuring 640 feet by 150 feet for an unprecedented $72.60 per square foot, a price that was previously reserved for property in the downtown area. By the time the entire length of the freeway was opened in 1952, some land values had soared to 1,000 percent of their pre-freeway values.[59] While this is common for land values along freeways, it seemed somewhat miraculous to Houstonians at the time.[60]

Despite the fact the entire length of the freeway was not opened to the public until 1952, Willier released his report on the economic impact of the freeway on July 22, 1949. After praising Mayor Holcombe for his "foresight," Willier concluded his introduction by stating that his report "should convince citizens and governments [*sic*] officials that the large expenditures of public funds for modern-type roadways, such as the Gulf Freeway, may be warranted by assigning monetary values to time savings which accrue to motorists and operators of commercial vehicles."[61] His report noted that the freeway saved money and time for both commercial and private vehicles, and he estimated that these savings translated to roughly $2,500,000. He also commented on land values, noting that abutting lands increased in value up to 300 percent, but he also found that once-thriving businesses on neighborhood streets "have noted some losses in business because of reductions in traffic [on their streets]."[62] While any environmental component is absent in Willier's report, he concluded that, "a modern freeway is a 'necessity' and not a 'luxury' and that the relatively higher construction costs may be justified as a sound investment in the provision of efficient motor vehicle transportation."[63]

Willier's report offers a unique perspective on the rationale of the highway planner in the late 1940s. As head of Houston's Department of Traffic and Transportation, he was in charge of keeping traffic moving within the city. His primary concern was relieving congestion on neighborhood streets in a safe and efficient manner, which is precisely what the Gulf Freeway allowed him to do. This is evident in his report's mild consideration of neighborhood business owners who lost business as a result of declining traffic on their streets, as well as his lack of commentary on the environmental effects of the freeway. These were secondary concerns, as those of the "beautiful" freeway came first.[64]

Despite his priorities, Willier made two predictions in 1949 that came true. He predicted that the freeway would "aggravate" parking problems in the downtown area, and this would require some kind of "parking terminal."[65] Parking in downtown Houston did become problematic in the 1970s and 1980s, and the Metropolitan Transit Authority of Harris County, Texas (METRO) implemented the park-and-ride system in an effort to alleviate this situation. The service rapidly became one of the most successful METRO programs.[66] In order to escape the congestion and lack of parking associated with downtown office space, many businesses relocated their operations out of the downtown area entirely. The draw of undeveloped land, lower property taxes, and other benefits pulled many businesses out of downtown.[67] This trend led to the suburbanization of office space in Houston and to the development of such areas as the Galleria, Greenway Plaza, and the Woodlands.[68]

Willier also predicted that the majority of Gulf Freeway users would be "vehicles on local trips, including passenger cars of people living near Houston and working in Houston areas."[69] In 1973, the U.S. Census Bureau reported that 87 percent of all people in the Houston area, which includes Houston, Harris County, and the surrounding four counties, commuted to work, and 67 percent of that total drove alone.[70] The Census Bureau's study of commuting in America examined 125 cities, and, of those, Houston ranked first in the number of commuters. The fact that so many Houston commuters flooded the roads during peak traffic hours made the Gulf Freeway the most congested roadway in Texas by 1978.[71]

By December 1949, just four years and two months into the project, the final three sections of the Gulf Freeway were under contract, and work was progressing on two of them. Despite the fact that Van London originally projected the completion date of the urban portion of the freeway to be in 1951, he later pushed the completion date to some time in 1952 because of his concerns that the Texas Highway Department would be unable to secure financing for the completion of the rural portion of the freeway.[72]

Despite the financial woes of the Texas Highway Department, Van London announced his plans for the construction of the urban portion of the freeway leading to Dallas in January 1950. In order to accommodate this freeway, the original feeder streets of the Gulf Freeway—Pease, Jefferson, Calhoun, and Pierce—would all be lengthened and widened to accommodate the increased traffic load. In a manner similar to the original feeder design, these streets would merge together near Buffalo Bayou and connect with U.S. 75, the Dallas Highway, outside the city limits. The design of the freeway is similar to that of the Gulf Freeway, including one-way service roads and the elimination of grade crossings where possible.[73] The city of Houston began work on this route before Van London made his announcement. The process of purchasing the right-of-way was almost finished, and the widening of Buffalo Drive westward from its intersection with Heiner was under way. Buffalo Drive had already been widened in front of San Felipe Courts and Jefferson Davis Hospital, and the city let contracts for work between Taft Drive and Waugh Drive. Mayor Holcombe estimated the work would be finished by December.[74]

The extension of the freeway across the city before the road was finished to Galveston created the idea of perpetual freeway construction in the minds of many Houstonians. The original concept of the Gulf Freeway called for a road that extended from the heart of downtown Houston to the city of Galveston. There was no mention of a northwestern route to Dallas, and by adding

The newly opened Gulf Freeway and the Gulfgate Mall. Courtesy Houston Metropolitan Research Center, Houston Public Library.

other connections and modifications to the original concept and design, such as the 1967 Pierce elevated section, the idea that the freeway would never be finished emerged in the minds of many Houstonians. *Houston Post* reporter Mark Morrison expressed this idea best in his 1973 statement that the freeway "had been under construction for 27 years but highway engineers say it will be 1980 before planned work is complete."[75] *Houston Post* reporter Rick Barrs commented in 1982 that "because they have been asked so often for nearly four decades, [highway] department officials get a little sore when asked when the Gulf Freeway will be completed."[76]

The idea of perpetual freeway construction is based on a lack of understanding on the part of the population, as well as aggravating urban factors outside the control of highway officials. The most glaring problem is the fact that the original purpose of the freeway has changed over time. Van London's chief problem was to relieve congestion on neighborhood streets and to connect Houston and Galveston. His design was based upon these ideas, and he allowed room for expansion of the roadway as the population grew at a steady pace. He could not have predicted the rate of Houston's expansion, especially when one factors in the growth rate of the late 1960s and 1970s Sun Belt population migration. The construction of NASA's Lyndon B. Johnson Space Center at Clear Lake, south of downtown Houston, was also something outside Van London's thinking. As one highway engineer said, "When Houston stops growing at such an agonizing speed and we stop needing to upgrade the road, then we'll be through with construction."[77]

The developing automobile dependency of Houstonians was something that Van London vastly underestimated, but again, the extent of the city's growth was something outside his thinking. The *Field of Dreams* catchphrase, "build it and they will come," definitely applied to the Gulf Freeway, as Houston commuters flocked to the roadway. In just two years after its opening, 100,000,000 vehicle miles were traveled on the roadway. The first traffic count for the road yielded 28,800 cars per day in 1948, a total that jumped to 62,500 by July 1950.[78] By 1982, the Gulf Freeway, which was formerly the most congested highway in Texas, carried 150,000 cars per day.[79] The Houston Planning Commission rated its practical capacity to be 100,000.[80] While highway department officials once viewed these statistics with glee, by the 1960s they viewed congestion with as much dread as did the commuters who sat in traffic during peak hours.

In 1963, the Houston Planning Commission launched a study of the Gulf Freeway. It found that the freeway was the only road that stretched from the downtown area to the southeast part of Houston's metropolitan area. The study also revealed that the freeway served not only downtown Houston but also the cities of Pasadena, Deer Park, Lomax, La Porte, South Houston, and the industries and their facilities along the south side of the Houston Ship Channel. The commission concluded that too much had been expected from the freeway, the only major traffic corridor in the area, resulting in deficient transportation facilities and exceedingly high congestion. The report concluded that "because of the substantial lead time required to plan, finance, acquire right-of-way, and construct the additional freeways required, it is evident that severe congestion will develop long before any adequate and permanent solution of the problem can be realized."[81] When Van London designed the Gulf Freeway in the early 1940s, the entire population of Harris County was roughly 600,000; by 1960 the population was 1,243,158.[82]

The freeway also served as a stimulus for urban sprawl in Houston and in the towns that it ran through. Houston suffered from a shortage of housing during World War II because of governmental controls on construction and materials. The city's housing shortage was exacerbated by the fact that many of the war industries were located along the channel, and this gave federal housing authorities the impression that Houston had no housing shortage, as housing could be found in the central city. Paul Levengood described the effect that this lack of understanding of local conditions had on the war worker:

> In the pre-freeway days of the 1940s, a fifteen-mile commute across the city's traffic-choked roads was a far from welcome prospect. When coupled with wartime rationing of gasoline and tires, and an inadequate public transportation system, the daily task of simply getting to and from work became an exhausting, time-consuming ordeal for many. In reports detailing the alarmingly high rates of turnover for ship channel area war plants, the number one factor given for voluntary separations was a lack of adequate housing nearby and employee frustrations with living an inconvenient distance from their place of work.[83]

Levengood further argues that working- and middle-class Houston families began planning the construction of their own homes once the war was over. This nationwide trend began in Houston as early as 1943.[84] The stage was set for a postwar housing boom across the Gulf Coast centered on the Houston region.

The Gulf Freeway cut through large portions of vacant land, making it easily accessible to developers and postwar homebuyers. Proof of such development is rampant in "A 15-Year Study of Land Values and Land Use Along the Gulf Freeway In the City of Houston, Texas," conducted by Norris & Elder, a firm of consulting engineers, for the Texas Highway Department and the Bureau of Public Roads in 1956. This report found that much of the vacant land that lay alongside the freeway had been developed into residential subdivisions and apartment complexes since 1945.[85] The rate of population growth in the

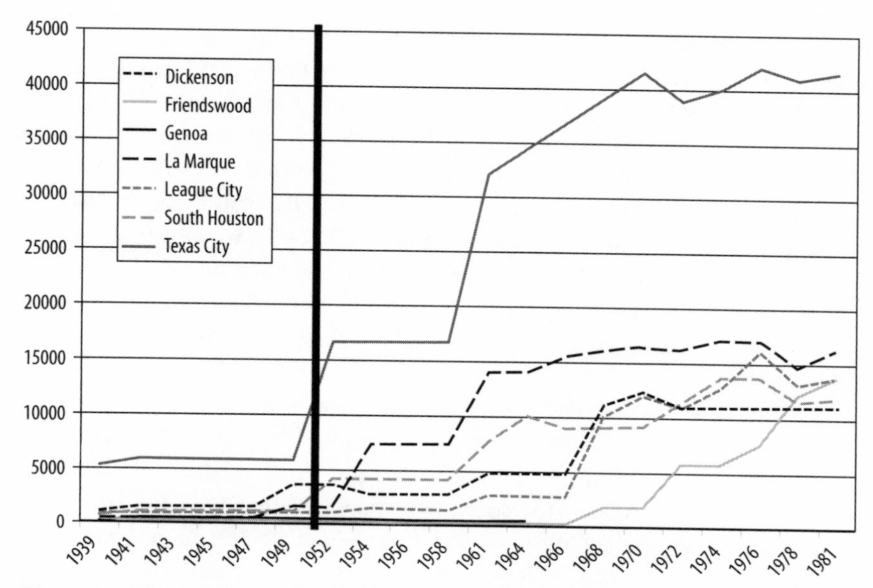

Figure 6.3. The population of Gulf Freeway cities, 1939–1980. The vertical line corresponds to the year 1952, the year the freeway was completed. Source: *The Texas Almanac and State Industrial Guide, 1939–1980* (Dallas: A. H. Belo, 1939–1980).

towns along the freeways also echoes the conclusions found in the Norris and Elder report (see figure 6.3).

Despite the future congestion that the Gulf Freeway would become noted for, sections of the road continued to be finished, and the entire road was opened to the public on August 2, 1952.[86] Dubbed a "50-mile Rhapsody in Concrete" by the Houston press, the opening of the entire length was marked by a lavish ceremony held on the Dickenson overpass. Driving in antique cars, officials from Galveston and Houston met on the overpass and shook hands in front of 5,000 spectators. Chairman E. H. Thornton Jr. of the Texas Highway Department officially dedicated the freeway "to the 7,500,000 people who call this great Texas empire home . . . and in tribute to their wisdom and foresight in planning for tomorrow."[87] At the end of the ceremony, the Galveston delegation played host to the other delegates by inviting the Houston and Highway Department officials to a lavish dinner, the Miss Texas Pageant, and a fireworks display over the Gulf of Mexico.[88]

Galvestonians generally welcomed the new freeway to Houston, as the congestion on U.S. 75, the old Houston road, grew steadily worse as the population in the small towns between the two cities and the number of vehicles increased.[89] Some thought that the Gulf Freeway would detract from Galveston's economy as the Houston Ship Channel had done previously. C. E. McClelland, editor of the *Galveston Daily News,* attempted to lessen these fears in an editorial published on August 2, 1952: "No, Galveston and Galveston County folks are not going to become the adopted orphans of Houston. Houston is merely going to be a closer neighbor, whereas Galveston and the county are relatives who have their own prosperity fish to catch and fry. The more relatives fish together the better the financial fish fry will be."[90] McClelland also commented that Galveston County towns such as La Marque and Texas City, which were rapidly becoming satellite communities of Houston, increased the county's property tax base and stimulated the local economy.[91] Other Galveston business leaders agreed with McClelland's commentary, calling the new road a "business stimulator."[92]

The impact of the new road was felt at once, as the volume of traffic between Houston and Galveston increased immediately and leveled off 52 percent above what had previously been normal.[93] Houstonians had always enjoyed forays to Galveston to escape Houston's summer heat and to relax in the warm water of the Gulf Coast. This fact is well documented in Steven M. Baron's *Houston Electric: The Street Railways of Houston, Texas.* The Galveston-Houston Interurban, an electric streetcar service that was completed in 1911, made Galveston a popular destination for Houstonians. Baron comments that visiting Galveston was so popular that sometimes the interurban's fleet of cars was overtaxed, and streetcars from Houston were used to handle the increased demand.[94] Despite a long-standing rivalry spurred on by the city's ports and transportation facilities, the Gulf Freeway helped to confirm Houston's and Galveston's roles as regional partners.

The Gulf Freeway offered tourists a quick and safe way to Galveston, as previous travelers had to endure the trip to the coast on U.S. 75, one of the deadliest roads in Texas. The freeway was touted as a new and modern way for Houstonians to reach Galveston. Mayor Herbert Y. Cartwright of Galveston saw the freeway as "not only a thickening of the golden stream of tourists and week enders [*sic*] from Houston and other inland points, but another step in the old queen's [Galveston's] struggle to overcome land transportation difficulties which have hampered her in commercial matters."[95] The road's impact spurred "the nearest thing to a boom it's had since shortly after the 1900 storm," by stimulating commercial development on Galveston Island, especially its undeveloped western side.[96]

Despite the benefits the freeway brought to Galveston, the city actually spent no money for its construction. All the city was required to do was to widen Broadway to allow for two extra lanes. Yet this action caused protest, since Galvestonians mistakenly believed that the palm trees that lined the route were to be removed. Once these misgivings were laid to rest, the citizens of Galveston enthusiastically supported the construction of the road, as it cost them nothing. Houston papers noted this with some irritation, as the city of Houston had spent millions of dollars on the construction of the highway. This irritation is apparent in a cartoon that was published in the *Houston Chronicle* on August 2, 1952. The cartoon shows the city of Houston, represented by a suit-wearing cowboy, laying down the freeway while the city of Galveston, represented by the Greek god Poseidon, looks on smiling and doing nothing. Entitled "Pulling the Two Cities Closer Together," the cartoon clearly shows Houston doing all the pulling.[97]

The freeway not only benefited Galveston, but the other towns lying in the freeway's wake also experienced growth. For example, South Houston, an incorporated bedroom community, grew tremendously. The city of Houston encircled South Houston by its 1949 annexation, which made Houston the largest city in the South.[98] Mayor George W. Christy thought that the addition of the Gulf Freeway would increase the safety in his city, as the old Galveston Road cut through the middle of town—a fact that attracted unwanted speeding traffic. It was the mayor's sincere hope that "the speeders . . . will take their races over on the freeway, if they must drive recklessly, and give our ears a rest here [from ambulance sirens]."[99] South Houston felt the effects of the freeway through physical growth rather than commercial sales as "the boys who whoosh through town don't spend a dime" on their way through.[100] Hundreds

"Pulling the Two Cities Closer Together." Published in the *Houston Chronicle* on August 2, 1952.

of new homes, including a new subdivision, were in the process of being built when the freeway was completed, forcing the city to expand its sewer and water lines to meet the increased demand for city services. The city also had thirteen-and-a-half miles of its twenty-two miles of roads freshly paved as a result of the growth. The continued suburbanization of Houston, as illustrated by South Houston, was an effect of the freeway's easy access to the area. This access continued to make South Houston, as well as many of the smaller communities along the freeway, a desirable residential area, much like Sam Bass Warner's streetcar suburbs—but with automobiles instead of streetcars.[101]

While the majority of the small communities along the freeway welcomed it as an economic and growth-producing boon, a few of them eyed it with caution. Friendswood, an old Quaker community on the border of Harris and Galveston counties, was one such place. While the *Houston Post* pondered the idea that Friendswood could become a suburb of Houston, the people of Friendswood desired to keep it "a quiet residential neighborhood."[102] Already a self-described city of commuters, the residents of Friendswood generally felt the same as the head of the school district when he said, "We don't have any honkytonks and we don't want any."[103] However, the community of 500 did indeed become a suburb of Houston over time.

While the Gulf Freeway offered a convenient means for moving about the region, it also contributed to Houston's air pollution. Houstonians, as well as many others from the surrounding towns, became increasingly automobile dependent in the post–World War II era. The increased amount of suburban housing development served to reinforce the necessity of a commuting lifestyle, as the distance between job and home continued to increase. The explosive suburban growth that accompanied the Gulf Freeway and other roads has given Houston the unwanted nickname of "the monster that ate East Texas," as wave after wave of construction washed over the Gulf Coast region in tsunami fashion, leaving behind a dizzying array of suburban houses, shopping malls, and commercial establishments. By fostering such growth to accommodate new residents, commute times to the city increased, as more and more cars were using the same roads from locations farther and farther out. This produced traffic congestion and, coupled with the industries on the Houston Ship Channel, created air so polluted that one could see it by the early 1960s.[104] As time marched on, Houston and Los Angeles would pass the title of America's smoggiest city back and forth. In the modern era, NASA scientists have found that Houston is "an ideal prototype [for studying pollution] in that it possesses a combination of the many potential sources that contribute to increased pollution: a growing population already in the millions, an enormous amount of automobile use and an abundance of chemical industry and power plants."[105]

Most of Houston's pollution-control efforts were focused on industry, as the refineries and petrochemical plants in the area contributed greatly to Houston's pollution problems. The role of automobile emissions cannot be overlooked in forming a complete picture of the city's environmental problems. While the state and federal governments implemented many programs to lessen the emissions produced by industry, as industrial pollution was viewed as the greater threat to the environment, automobile emissions were harder to curb. It is also important to note that these initiatives are modern in their origin, as Houston's pollution problem was not a public priority until the late 1950s. Until that time, Houstonians celebrated Highway Appreciation Week and each new highway opening with glee.[106]

In 1957, the Houston Chamber of Commerce sponsored an air pollution survey conducted by the Southwest Research Institute. This survey, for all intents and purposes, was the first of its kind in the Houston region. The report found that pollution was a localized event, as the sea breeze is favorable for the rapid dispersion of pollutants. It noted that the problem of heavy pollution in Houston generally occurred when the wind came out of the east or northeast, as this allowed the mixing of pollutants from several sources. On the whole, the report was seen as reassuring.[107] Upon reviewing the report, President Ben C. Belt of the Houston Chamber of Commerce said: "The only way to have no air pollution at all is to have no industry at all, or no city at all; so all our considerations are relative. We must compare the findings of the Houston survey with conditions in other industrial cities. From the standpoint of pollutants and dust in the atmosphere, we compare very favorably, or better, especially when it is considered that our problems are not community-wide."[108] Houston still remains an auto-oriented city, and while it has made great strides in reducing industrial pollution, auto emissions still remain a serious problem (see the essay by Robert Fisher in this volume).

A prime example of one of these mushroom communities is Clear Lake, which exploded as a direct result of its freeway location. The Clear Lake area, composed of several small towns, became home to NASA's Lyndon B. Johnson Space Center in 1961. Land values in the area skyrocketed up to ten times their previous values, as the Del E. Webb Company developed the area in a master-planned fashion.[109] Despite the fact the Clear Lake is twenty-two miles from Houston, Houston annexed the Clear Lake area by strip annexation, following the path of the Gulf Freeway in 1977. Clear Lake fought back by challenging Houston in court, effectively arresting the process until 1987, when it finally lost.[110] Houston became "Space City, USA." Perhaps this phenomenon was best expressed in 1962 by Houston city planner Ralph Ellifrit when he said, "The real axis of growth for the City of Houston is down the freeway."[111]

The Gulf Freeway, despite the politics that would turn it into a course for municipal expansion, was one of Texas's most touted achievements; engineers from around the country came to study it. Even before it was finished, the U.S. Bureau of Public Roads called it the most outstanding highway engineering development since World War II.[112] The bureau conducted a survey on the road, which it planned to use "to convince other cities of the nation of the value of expressways."[113] In November 1948, *American City* magazine examined the roadway and found it to embody several significant engineering trends.[114] On April 14, 1949, a group of Texas A&M University engineering students came to study the freeway and hear a lecture on it by Van London.[115] The American Association of Roadbuilders held their fiftieth anniversary celebration in Houston in January 1952, which included a viewing of the Gulf Freeway.[116] While it is unclear what effect these viewings and lectures had upon highway planners, the fact that they occurred is evidence enough that the freeway was seen as a significant piece of highway construction.

Regionally speaking, the Gulf Freeway was studied and replicated. Van London's influence can be clearly seen in the entire Houston freeway system, since he designed most of the first wave of major thoroughfares. Although he retired in March 1955, Van London's work included almost all of Houston's early freeway system, as he was the chief designer of the 1942 street plan. While it would be the job of A. C. Kyser, Van London's assistant and successor, and Houston city planner Ralph Ellifrit to construct the system, Van London left behind a tremendous legacy, which few know about.

It is also important to note that the Gulf Freeway is a part of the larger Houston highway system. While this essay has examined the road in a relatively isolated fashion, the Gulf Freeway was never conceived as a stand-alone project. The first wave of freeway construction, which lasted from 1942 to 1976, also included the North Freeway (I-45 North), Interstate 10, Loop 610, and U.S. 59, as well as a number of other state highways. State and local officials often touted the Gulf Freeway's success in their efforts to secure highway funds or convince voters to pass bond issues. It is not a coincidence that, in a plea for more highways in 1952, Mayor Holcombe referred to the freeway as "the beginning of a new era in modern highway construction and [it] brings into focus Houston's need in this direction which the city administration is moving to meet."[117]

It is also important that the population of Houston stood firmly behind Holcombe's pleas, as shown by their overwhelming acceptance of the Gulf Freeway and the perceived benefits it brought to the city, such as increasing suburbanization, additional industry, and commercial establishments. To Houstonians of this time, freeways were seen as a symbol of modern prosperity, as they increased land values and contributed to Houston's growth. While many of the features and benefits of the freeway are less than incredible today, one

Downtown Houston flanked by freeways, 1960s. Courtesy Houston Metropolitan Research Center, Houston Public Library.

must keep in mind that to the contemporary eye, the "golden lane" must have been seen as incredible.

The problem with any system is that it can be overwhelmed, and this is what later generations of Houstonians found, as the Gulf Freeway became the most congested highway in the state. Houstonians, like many of those in urban America, were frustrated with sprawl, pollution, hectic commutes, and traffic congestion by the late 1960s. Their perceptions of freeways were vastly different from preceding generations; the highway became the root cause of their problems instead of a blessing. This attitude expressed itself in the changing highway policy that gave citizens a larger voice in planning thoroughfares and made funding increasingly difficult to obtain.

The Gulf Freeway was among the first steps in the creation of Houston's modern built environment, which is often characterized by its highway system. With its inception in 1942, the Gulf Freeway served as a model for the city's entire freeway system. The built environment of Houston was remade to accommodate the roadway, as hundreds of structures were moved or demolished in the central business district alone. Future highway planners continued to replicate the Gulf Freeway throughout the region in hopes that the new freeways would have the same effects as the original. This action created a sprawling urban form and fostered automobile dependency as travel distances increased. The roadway provided the city with a pathway for the city's vigorous annexation program of the region's continuing suburban development. It also fostered automobile-related environmental problems and drainage issues, as well as Houston's rise to regional dominance. Houstonians once saw the Gulf Freeway as a panacea for all of the city's congestion problems on their neighborhood streets, but the fact remains that as the system aged and more commuters joined the ranks of the existing ones, the "superhighway deluxe" became just another roadway to wait on as traffic slowed down.

CHAPTER 7

Urban Sprawl and the Piney Woods

Deforestation in the San Jacinto Watershed

DIANE C. BATES

In form, Houston is an archetypal postmodern city. Houston's major expansions in size and population coincided with the redesign of urban environments that emphasize private automobiles and the decentralization of workplaces. Recent renewal inside Houston's "Inner Loop" cannot counter the reality that Houston can be best described as a sprawling, "edgeless" city with multiple hubs for employment, such as downtown, the Galleria, Greenspoint, NASA/Clear Lake, and the Woodlands.[1] The development in the city of Houston passes seamlessly into neighboring suburbs, whose borders with other suburbs are equally unremarkable. To drive from east to west or north to south through the Houston region takes longer than an hour on the interstate and much longer under normal traffic conditions. To study the environment of Houston, then, one must acknowledge that the city of Houston itself represents only a part of the story.

The Houston metropolitan region spans across three major natural ecosystems: the coastal plain, which continues from downtown south and southeast to the Gulf Coast; the prairie, which stretches from downtown west toward Katy; and the Piney Woods, a heavily forested region that extends north and east of downtown. Houston's urban expansion into these three ecosystems provides the wide variety of residential options and recreational opportunities found in the region, and also explains the diversity of environmental problems that the Houston area faces.

This chapter considers the environmental impact of urban expansion on the Piney Woods region found to the north and east of the Inner Loop (U.S. Interstate 610). In keeping with certain ideals within environmental research, it focuses on the most important watershed found in this quadrant, the San Jacinto River. By comparing satellite images from 1979 and 2000, the four-county area that includes the San Jacinto watershed is analyzed to reveal the replacement of forests with suburban development, especially upstream from Lake Houston. These satellite images also provide evidence that master-planned suburban developments that have maintained the aesthetic qualities of the Piney Woods nevertheless alter the characteristic properties of the land. As a consequence, forested suburbs are unlikely to provide many of the environmental benefits that an intact forest can provide, including flood control, erosion control, air and water filtration, and environmental cooling.

The San Jacinto Watershed

Rather than focusing on the entire Houston metropolitan area, which contains large areas of coastal plain and prairie, this study examines the four counties that comprise the majority of the San Jacinto River watershed: Harris, Montgomery, San Jacinto, and Walker counties. Although linked by the San Jacinto, these counties feature vastly different social systems. Comparing forest change across these counties leads away from the study of the city of Houston, but provides important points of comparison. Between 1979 and 2000, Walker and San Jacinto counties maintained much of their rural characteristics, while Harris County has consistently been the most urban. The county that has experienced the most social and environmental change during the study period is Montgomery, which transformed from a largely rural county in 1979 to a largely suburban county in 2000. A satellite image of the four-county region appears in figure 7.1.

The San Jacinto River itself is not a particularly large or important river on either a national or regional scale. However, its impounds at Lake Conroe (Walker and Montgomery counties) and Lake Houston (Harris County) provide both drinking water and major recreational facilities for area residents. After joining with Buffalo Bayou, which bisects the city of Houston, the San Jacinto has been artificially deepened to become the Houston Ship Channel, home to the nation's largest concentration of refineries and an important estuary for Galveston Bay, which in turn is vitally important to the Gulf Coast commercial fish and shrimp fleet. The San Jacinto watershed, however, provides an excellent case study to examine the impact of low-density urbanization (often called "sprawl") on forests in Houston's metropolitan region.

The rural counties in the study area provide a sort of environmental baseline for forest change during the study period. Statistically, San Jacinto and Walker counties recorded high population growth rates during the study period (95 and 48 percent, respectively; see table 7.1). However, because of the relatively small populations in these counties in 1980 (11,434 in San Jacinto and 41,789 in Walker), they have been able to preserve the rural character of land use.[2] San Jacinto County gained only 10,812 inhabitants, while Walker's population increased by 19,969 during the study period. Both the East and West forks of the San Jacinto rise in Walker County, which straddles the ecological boundary between the pine forests and the hardwood prairies more typical of central Texas. Walker County remains primarily rural, with a population of 61,758 in 2000 at a density of 78.4 people per square mile. Although low-density cattle ranching is common, 60.76 percent of Walker County was forested in 1992, including a large portion of the Sam Houston National Forest.[3] Also containing part of the national forest, San Jacinto County contains an even higher percentage of forested land: 79.44 percent of the county's land was forested in 1992. In 2000, San Jacinto County was home to 16,372 inhabitants at a density of 39.0 people per square mile. The human density in both Walker and San Jacinto counties was lower in 2000 than in Texas as a whole, which averaged 79.6 people per square mile.

In contrast to these more rural counties, the San Jacinto watershed also contains the highly urbanized Harris County and the increasingly suburbanized Montgomery County. Harris County surrounds and contains the city of Houston, the fourth-largest city in the United States. In 2000, 3.4 million people called Harris County home. Between 1980 and 2000, Harris County added nearly 1 million inhabitants, such that the county's population was 41 per-

Table 7.1. Demographic and Forest Indicators by County

	Harris	*Montgomery*	*San Jacinto*	*Walker*
Population 1980	2,409,547	127,222	11,434	41,789
Population 2000	3,400,578	293,768	22,246	61,758
Population Growth (percent change between 1980 and 2000)	41	131	95	48
Density 1980 (population per square mile)	1,393.6	121.9	20.0	53.1
Density 2000 (population per square mile)	1,967.0	281.4	39.0	78.4
Percent of Land in Forest, 1992	24.3	68.7	77.3	63.8
Acres of Land in Forest, 1992	267.0	460.5	282.9	320.9

Sources: United States Bureau of the Census 2004; Patrick E. Miller and Andrew J. Hartsell, *Forest Statistics for East Texas Counties—1992, Resource Bulletin SO-173* (December 1992), New Orleans, U.S. Department of Agriculture Forest Service, Southern Forest Experiment Station.

cent larger at the end of the study period than it was in 1980. Harris County's density of 1,729 people per square mile greatly exceeded the average for Texas and was second highest in the state (after Dallas County). Nonetheless, in 1992 nearly a quarter (24.3 percent) of Harris County contained forest, especially in the northeastern quadrant.

The county that has changed most dramatically during the study period is Montgomery County, which has taken on an urban (or at least suburban) appearance, more like that of Harris County. Between 1980 and 2000, Montgomery County's population more than doubled, from 127,222 to 293,768 inhabitants. Population density increased from 121 to 281 people per square mile. Notably, Montgomery County's population increased by a whopping 61 percent between 1990 and 2000, adding on average 106 people per square mile during the decade. More than 64 percent of the housing in Montgomery County was built after 1980. Although Montgomery County contains important employment and commercial centers in Conroe and the Woodlands, many of the county's residents commute to Houston. Eighty percent of workers over the age of sixteen in Montgomery County commuted alone in a private vehicle, averaging a 32.9-minute commute (another 13.2 percent carpooled in private vehicles). All of these indicators demonstrate the rapid expansion of development in Montgomery County—a trend easily confirmed by anyone who has lived in or driven through the county in the past two decades.

The San Jacinto River thus drains a highly diverse region, represented by the four counties that contain the greater part of its watershed. San Jacinto County is rural, forested, and sparsely settled. Walker County is rural with a combination of forest and ranching activities, but also sparsely settled. Montgomery County is rapidly changing from a rural to a sprawling suburban county. Finally, Harris County represents a highly urbanized county with significant residential, industrial, and commercial land use, although a substantial portion in the northeast quadrant of the county remains forested.

Deforestation and Suburban Sprawl

Since the last inventory of Texas forests was completed in 1992, there exists no systematic data to measure deforestation in the San Jacinto watershed.[4] Anyone driving along the major transportation corridors in the region (for example, Interstate 45 or U.S. Highway 59) cannot help but be struck by the large amount of frontage land that is being converted to nonforest use. This conversion is most apparent in Montgomery County and the forested northeastern quadrant of Harris County, both of which have seen dramatic increases in population since 1980. Residents and this researcher, a former resident, suggest that the primary reason for forest conversion is linked to the growth of suburban developments.

Conversion to suburban land use in Montgomery County and northern Harris County largely takes on the appearance of urban sprawl. "Sprawl" is a disparaging term used to denote extensive residential, commercial, and office development.[5] Typically, sprawl involves single-use areas, or "pods," that contain single-family homes with large lawns and garages, low-rise and low-density office parks, or retail big box and strip centers. Office and retail pods generally require large amounts of land for parking and typically provide limited access to housing, such that residents must leave the residential pod through a main exit to a major thoroughfare even to return to nearby offices and stores. Consequently, sprawling development is perhaps most characterized by dependence on the private automobile, both in terms of transportation and for the large proportion of land devoted to streets, highways, and parking lots. This sort of automobile-based urban organization tends to create major social and environmental problems, most obviously including traffic congestion and poor air quality.

Sprawling development has produced these effects in the Houston metropolitan region.[6] Sprawl "helped" Houston replace Los Angeles as the "Smog Capital" of the United States in 1999 and 2000 by registering the largest number of days in which at least one air quality indicator exceeded federal standards.[7] A report compiled by the Texas Transportation Institute found that 60 percent of freeway lane-miles and 50 percent of principal arterial road lane-miles were congested in the Houston region during peak times.[8] A Houston-based coalition for environmental organizations links sprawl with water quality degradation, reduction in freshwater inflows, erosion, wetland and other habitat destruction, and flooding,[9] concerns echoed in the Texas Center for Policy Studies' *Texas Environmental Almanac.*[10]

Residents' concerns about the ugly aspects of sprawl in Houston have encouraged the development of "master-planned communities" in which development is controlled and regulated by a private centralized authority and companion homeowners' associations. The two largest of these master-planned communities in the San Jacinto watershed are Kingwood (now a part of the city of Houston) and the Woodlands. In 1972, a former subsidiary of Exxon, the Friendswood Development Corporation, created Kingwood, "the Livable Forest," on property between the East and West forks of the San Jacinto River north of Lake Houston. Unlike other suburbs and zoning-phobic Houston, Kingwood attempted to maintain the aesthetic qualities of the pine forest while promoting suburban development within relatively strict zoning and beautification guidelines. Two years later, developer and oilman George Mitchell founded the Woodlands just north of Spring Creek, a boundary between Harris and Montgomery counties and a major tributary to the West Fork of the San Jacinto River. The Woodlands has tried to improve upon Kingwood's model with even stricter controls. After Kingwood was annexed involuntarily to the city of Houston in 1996, many residents perceive that environmental standards have changed for the worse.[11] Likewise, while the Woodlands has actively and successfully fought annexation by Houston, some residents suggest that environmental standards have worsened since Mitchell sold his interests to the Woodlands Operating Company, a partnership led by Morgan Stanley and Crescent Real Estate Equities. As of 2000, however, Kingwood and the Woodlands were frequently cited as models of "smart growth" in the region, and both communities continue to attract residents who prefer to live in (and can afford) prestigious wooded communities over other options in the Houston metropolitan area.

Despite the presence of these two well-known suburbs in Houston's periphery and the negative hullabaloo about sprawl throughout the region, few people have directly identified deforestation as an environmental problem linked to sprawl. Forests are natural filters for air and water pollution, provide habitat for wildlife, control floods, prevent erosion, and add other intangibles to the human environment, and the lack of connection between forest loss and sprawl represents an important oversight for people concerned about Houston's environment.

Background on the Pine Forests of East Texas

Although most people do not think of Texas as being a major forested region, in fact the eastern Piney Woods region contains vast acreage of pine and bottomlands forests. This same forest belt stretches north and east along the Appalachian Piedmont and ends in the southern Pinelands of New Jersey. In 1992, the U.S. Forest Service (USFS) conducted an inventory of the forty-three wooded counties of East Texas and found that 11.9 million acres were covered in forest, amounting to 55.09 percent of the total land area.[12] Forests were most concentrated in the USFS Southeast Texas region, between metropolitan Houston in the west and the Louisiana border in the east and from the Gulf Coast in the south to the northern borders of Sabine, St. Augustine, Angelina, Houston, and Leon counties. USFS surveys for southeast Texas reveal a slight decline in timberland from 1935 to 1992, which contrasts with both the northeast region and the eastern United States as a whole. Between the two most recent forest inventories, however, all but four counties in the forty-three-county region either had gained timberland or changed less than 20,000 acres. Most gains came from the reversion of agricultural land to forests, consistent with the gradual movement away from agricultural activities in the modern United States. Loss of forested land is generally attributable to local land-use decisions made by private landowners and has not been systematically analyzed.

The forests in East Texas consist of two primary types: loblolly-shortleaf pine and oak-pine. Loblolly pine (*Pinus taeda L.*) represents the most common pine species over this entire region, due to longleaf pine harvesting in the late nineteenth and early twentieth centuries and subsequent fire suppression. Loblolly pines are fast-growing, commercially valuable pines that tolerate the high levels of heat and rainfall common to the area, as well as thrive in the region's soils.[13] Loblolly pines also are a part of much of the coastal flatwood wetlands in East Texas, which are notable for ground saturation and slow drainage during the winter and early spring.[14] The many streams and bayous of this region are surrounded by bottomland hardwood forests, containing oak, cypress, and sweet gum, which are often flooded during heavy seasonal rains. Toward the western edge of the Piney Woods, oak and hickory species begin to replace loblolly pines as the dominant tree species because of the drier climate and changing soil type.

Commercial forestry in Texas is concentrated in the southeast region, where 44 percent of timberland is owned by forest industry firms, particularly in counties that contain the Neches River and Sabine River basins. Forest industry–owned timberland tapers off from east to west, where private ownership of forestland becomes more important in the San Jacinto River watershed. According to the 1992 USFS forest inventory, the forest industry owned only 10.38 percent of timberland in Harris, Montgomery, San Jacinto, and Walker counties. In contrast, 16.02 percent of timberland in these counties is owned by municipal, county, state, or federal governments, and over 70 percent of timberland is owned by private individuals and nonforestry corporations (39.49 and 31.47 percent, respectively). Forest conservation and deforestation in this region must thus be understood outside the realm of commercial forestry. Changes in forest cover reflect plans and decisions of private landowners and not of professional foresters.

Measuring Land-Use Change with Satellite Imagery

Different land uses, such as forests, fields, and urbanization, reflect light and heat differently. Visual light and infrared reflections are recorded by LANDSAT images. These images can be classified using "remote sensing" software designed to sort pixels based on similar reflection signatures. This study uses three images to distinguish land use in the four-county area: one from 1979 and two from 2000. Two images were required for 2000 because the four-county area was not available in a single satellite image; thus two images (one from July and one from September 2000) were "mosaicked" together to create a single image. Mosaicking matches pixels in the overlapping area of two images based on similar reflection values to produce a single image. The satellite image from 1979 and the mosaicked image from 2000 were then "clipped" to include only the four-county study area. Because the original satellite images excluded a very small area in northern Walker County and a very small area in far southern Harris County, these areas cannot be analyzed.

To evaluate land use, a computer program classifies pixels into twenty-four different categories based on similar reflection signatures. By using aerial photography, visual satellite images, and knowledge of the region, these categories were then manually reclassified into four broad land-use categories: water, forest, field, and highly reflective surface. Both the 1979 and 2000 images were reclassed so that pixel values reflected these four categories. While the first three categories are self-explanatory, the last requires some clarification. Urban areas are highly reflective because pavement, rooftops, and concrete tend to reflect back the light and heat that reach them, while vegetation tends to absorb light and heat. The 2000 image contained some cloud cover in Walker County, which produced a reflection signature essentially indistinguishable from that produced by urban areas in Houston. Exposed riverbed in the Trinity and San Jacinto rivers produced a similar signature. Thus "highly reflective surfaces," instead of "urban land use," is used in the following examination of the San Jacinto watershed. Nonetheless, most of the highly reflective surfaces indicated in the classified satellite images represent urban land uses.

In 1979, water made up 3 percent of the land area of these four counties, forest covered 45 percent, fields covered almost 40 percent, and highly reflected surfaces accounted for 12 percent of the land area (see table 7.2). In 2000, water still covered about 3 percent of the area, but forests had been reduced to 40 percent, and fields had been reduced to 37 percent. In contrast, highly reflective surfaces increased to more than 19 percent of the land area. While some of the increase in reflective surfaces can be explained by the presence of clouds in the 2000 image, careful inspection of the images confirms that this error does not substantially change the overall increase of highly reflective surfaces and a decline in forest and fields. Forest loss was concentrated in the central third of the image, containing northern Harris County and nearly all of Montgomery County, where there is no cloud cover in the 2000 image to distort the reflection signature (see figure 7.2). Likewise, pixels that were not classed as highly reflective surfaces in 1979 but were in 2000 are concentrated at the southern

Table 7.2. Percent of Land Area of Four-County Watershed in Specified Uses in 1979 and 2000

	1979	*2000*	*Change 1979–2000*
Water	3.00	3.14	+0.14
Forest	45.05	40.19	−4.86
Fields	39.70	37.22	−2.48
Highly Reflective Surfaces	12.26	19.45	+7.19
Total	100.01	100.00	

third of the image (central Houston), where there are no clouds distorting reflection signatures. These highly reflective surfaces clearly expand northward along the Interstate 45 and U.S. Highway 59 corridors, again into areas in the satellite images that are not distorted by cloud cover (see figure 7.3).

Comparisons of land use in 1979 and 2000 demonstrate that the proportion of land in both forests and fields has declined, so there is no evidence to suggest that forests are being converted for agricultural use. In contrast, this study found an increase from 12 to 19 percent in land that has been classified as highly reflective surfaces, mainly pavement, concrete, and rooftops. Although cloud cover in the 2000 image inflates the growth of highly reflective surfaces, there is enough evidence to conclude that a significant proportion of this watershed has been converted from forest to urban use during this time period. Overall forest loss has been moderated by reforestation in rural Walker and San Jacinto counties and the relatively stable forest in Sam Houston National Forest. In other words, rural San Jacinto and Walker counties' forests demonstrate the same pattern of recovery typical of the southern forest belt, but forests in Montgomery and Harris counties have experienced considerable removal over the study period.

This finding is consistent with another remote-sensing analysis conducted by American Forests in 2000, which found that heavily forested areas in southeast Texas had declined by 16 percent between 1972 and 1999. In the four-county region in the American Forests study, however, forest cover changed from 45 percent in 1979 to 40 percent in 2000.[15] The discrepancy between the American Forests study and the current one is most likely methodological: the American Forests study measured "heavily forested" as land that had more than 50 percent tree cover, while the current study used reflection signatures and ground-truthing to determine classification as forest or nonforest. Ground-truthing involves visiting sites throughout the study region, recording actual land use, and connecting that information to the satellite image using global positioning system (GPS) coordinates. Ground-truthing may be a more subjective measure of land-use classification, but it reflects the actual use of land rather than simply tree cover, which can misrepresent "forest" in wooded suburbs such as Kingwood and the Woodlands.

Indeed, the current study finds that even environmentally conscious master-planned communities such as Kingwood and the Woodlands have not retained the characteristics of the pine forest, at least in terms of reflective signatures. While distinguishable from more highly urbanized locations such as downtown Houston, the reflective signature of Kingwood and the Woodlands was much more similar to urban areas than to natural forests, such as those protected in the Sam Houston National Forest. Moreover, large patches of forest were removed between 1979 and 2000 in the areas within and adjacent to both of these communities (see figures 7.4 and 7.5). In effect, satellite data suggests that wooded suburbs like these may preserve some aesthetic qualities of the forest but fundamentally change the character of the land away from forests.

Why don't master-planned communities in Houston's periphery protect the forest environment? Part of the problem has been the emphasis on preserving an aesthetic forest instead of the forest ecosystem. In driving through the Woodlands or Kingwood, it may "feel" like traveling through a forest, but strips of forest buffers along roads do not offer the same habitat, hydrological control, or heat-absorbing properties as a forest that extends more than a few yards before meeting a lawn, parking lot, or structure. Degradation occurs when as little as 15 percent of a watershed is covered in impervious surfaces—a proportion much lower than the typical "footprint" of buildings and driveways in a residential lot, even without considering the substantial additional amounts of land dedicated to roads, parking lots, and office or retail centers.[16] While the Woodlands made commendable efforts to protect the Spring Creek watershed and other floodplains, the satellite image demonstrates that the overall effect of its growth has been forest loss and fragmentation. When considering the large scale of residential and commercial development in the Woodlands and Kingwood in the past few decades, the potential and actual deterioration of the forest habitat (including the San Jacinto watershed) can be nothing but a major concern for residents.

Another way that these forested, master-planned communities contribute to the deterioration of the forest system is through the replacement of native plant species. Both the Woodlands and Kingwood have promoted the conservation of native pines, but they have been uneven about the protection of underbrush in forested areas and tree borders. Aesthetic and fire-safety concerns have privileged nonnative species and large lawns. Lawn maintenance and trees that cannot withstand the extreme drought-flood weather cycles typical of the San Jacinto watershed may eventually contribute to water shortages. Chemical fertilizers and pesticides, necessary for grass and nonnative species, are a major nonpoint source of water pollution, especially nitrogen.

Moreover, as long as deed-restricted communities in the Houston area continue to cater to people in higher income brackets, the likelihood of reducing pressure on the region's forests is unlikely. The sanctity of the personal automobile among better-off Americans and, more recently, the "right" to drive whatever type of automobile one can afford are critical barriers to reducing the environmental impact of any new development in Houston. Even while the city of Houston unveiled its METRORail in 2004 and the city's voters have approved another 72 miles of light rail by 2008,[17] outlying suburbs remain dependent on and fiercely protective of automobile culture. And while residents of both Kingwood and the Woodlands point to the unsightly car-based strip-center sprawl along Farm to Market Road 1960, there is apparently little political will to move suburbanites to alternate forms of transportation. The walking paths in both communities are designed more for recreation than functionality—the space between the "town centers" in the Woodlands may be miles from the housing developments they are designed to service. Even the new Woodlands Waterway, modeled in theory after San Antonio's Riverwalk, is surrounded by parking lots and chain stores, with distances between freestanding restaurants, malls, and centers more on the scale of driving than walking. Without demonizing our dependence on automobiles too much, it seems clear that as long as we base development (housing, commercial, or offices) on automobiles, land-extensive development will dominate the Houston regional landscape.

At a regional level, master-planned communities allow a more privileged fraction of the region's residents to enjoy an aesthetically pleasing environment, while the region as a whole suffers from the systematic degradation of the area's forests. Aesthetic forests may comfort people who dislike concrete and skylines, but they offer few of the environmental services of a genuine forest, such as the filtering of airborne and waterborne pollutants, flood control, erosion control, or wildlife habitat. Air quality suffers from the increased commuting of people from more distant (even if wooded) suburbs. With forest conversion comes greater potential for runoff, and the San Jacinto watershed has experienced devastating flooding in 1994, 1998, and 2001, with most of the damage downriver of both the Woodlands and Kingwood. Deed restrictions also push undesirable land uses from master-planned communities, as noted by members of the unincorporated Atascocita neighborhood, south of Kingwood.[18]

Conclusion

The preservation of forests on the fringes of urban areas cannot be accomplished through wooded suburban development such as the Woodlands or Kingwood. While master-planned communities might represent a qualitative shift in the right direction in terms of aesthetics, they fundamentally alter the forest through fragmentation and conversion to nonforest uses. Even a relatively large number of native ornamental trees cannot replace the environmental services of an intact natural forest. This reality is indicated by the continued worsening of air and water quality measures and increased incidents and severity of flooding in the San Jacinto watershed. As the region's forests shrink and become more fragmented, the habitat for wildlife also changes. Opportunistic wildlife, such as raccoons, coyotes, and deer, proliferate and become a public nuisance, while more specialist species decline or are regionally eliminated, as in the Houston area with wild cats and wolves.

This study demonstrates that despite the overall trend in the southeastern United States toward reforestation, we cannot ignore how suburbanization affects the forests adjacent to cities such as Houston. This is especially the case given the land-extensive, sprawling nature of suburban development in the region. Because of the environmental services provided by forests on the urban fringe, loss of these forests can be expected to exacerbate many urban environmental problems, such as flooding, drought, air quality, and urban heat islands. Suburban sprawl has already been linked to a multitude of environmental ills, and this study provides preliminary data that should add deforestation to the list.

The forests of the Piney Woods service the environmental health of the entire Houston region. They represent a great natural resource, not in terms of the wood that could be sold or the land that could be converted to other uses, but as a major contributor to the regulation of water quantity and quality, air quality, wildlife management, and overall quality of life. Regional forests are protected by some public management, most significantly the National Forest Service in Sam Houston National Forest. However, few municipal or state parks exist to protect the forest itself. For example, Lake Houston State Park, one of the largest publicly owned forests in the San Jacinto watershed, has already been considered for reduction in size and bisection by the Grand Parkway. Given the potential environmental risks associated with deforestation, further removal of the pine forests in the San Jacinto watershed must be addressed within a regional environmental plan. Area residents must come to recognize forests as part of the regional environmental infrastructure and not merely as aesthetic amenities to suburban living.

Until master-planned communities, even those with environmental goals, coordinate with the city of Houston and other suburbs and unincorporated territories, they will at best remain islands of green in a sea of concrete. Houston's refusal to institute comprehensive zoning contributes to this process, as the region's more affluent residents seek out environmental and aesthetic amenities in the urban fringe. Recent activities in the region, including the METRORail, urban infill development in downtown and midtown, and the heated debate over the expansion of the Grand Parkway, suggest that Houstonians are not unaware of these problems, nor are they unwilling to change behaviors to promote "greener" objectives. This vigilance, however, must be regional and deliberate in orientation; thirty years of individuals escaping to forested suburbs has done little to protect the region's forests, much less improve the overall quality of the San Jacinto watershed.

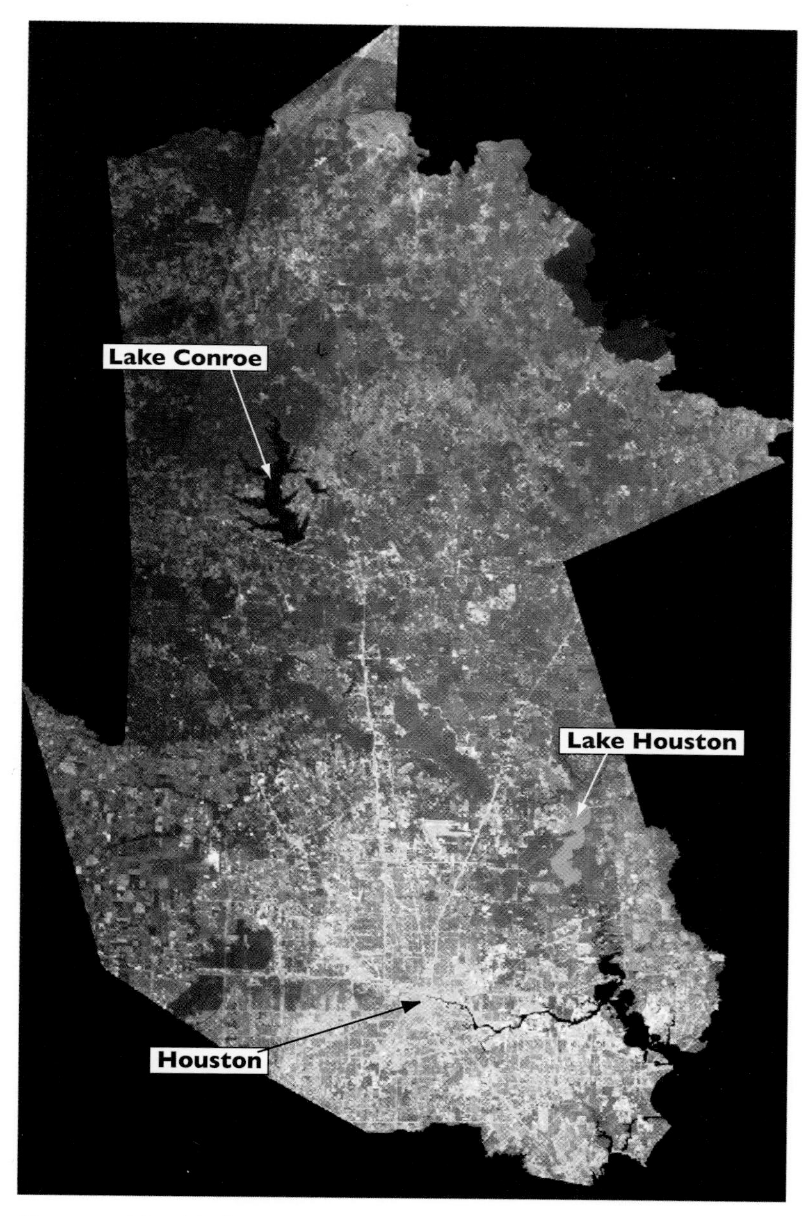

Figure 7.1. Mosaicked LANDSAT Image of Walker, San Jacinto, Montgomery, and Harris counties, 2000. Red indicates vegetation, and light blue indicates highly reflective surface.

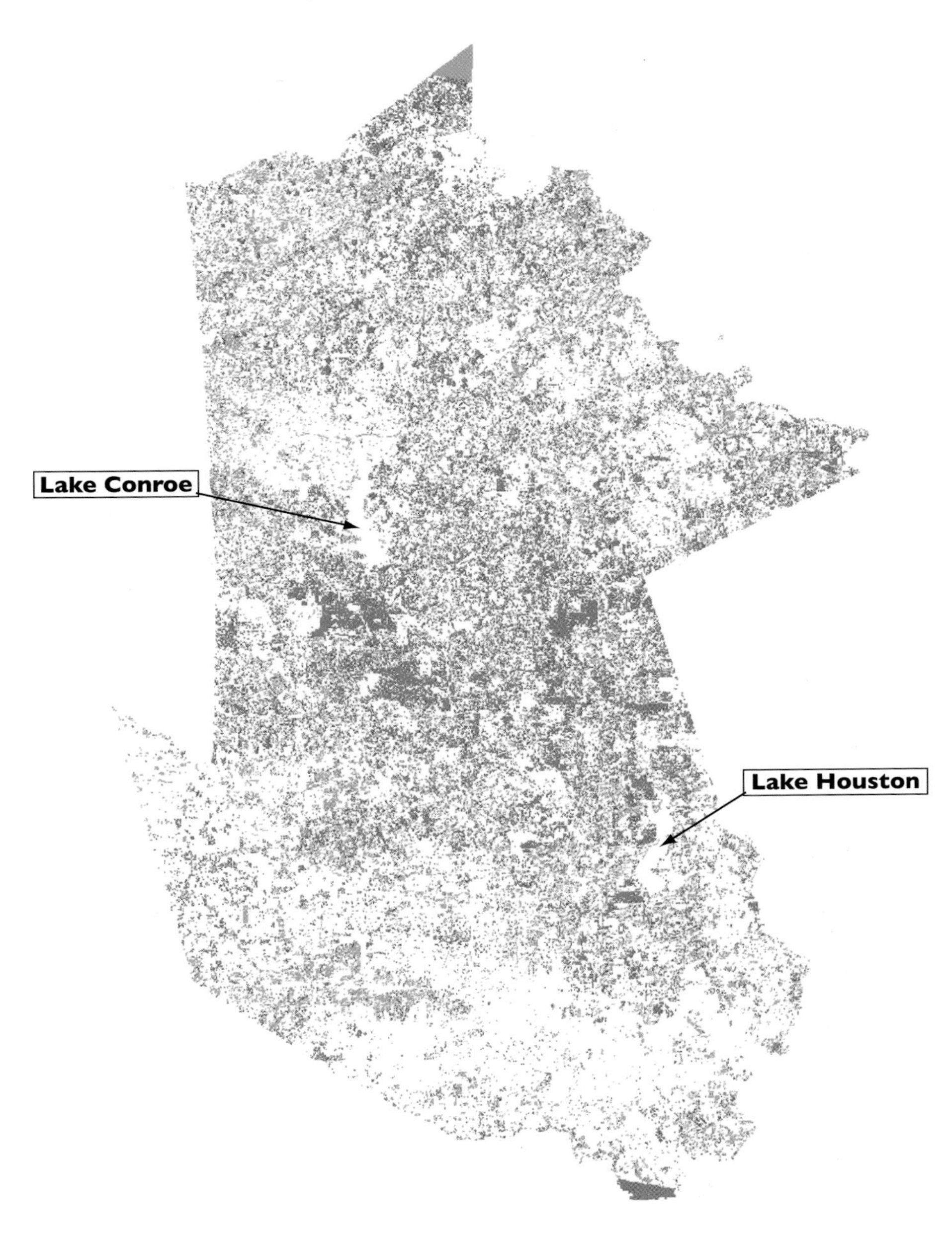

Figure 7.2. Forest change 1979–2000. Red indicates areas that had forest in 1979 that did not contain forest in 2000. Green indicates areas that had forest in 2000 that did not contain forest in 1979.

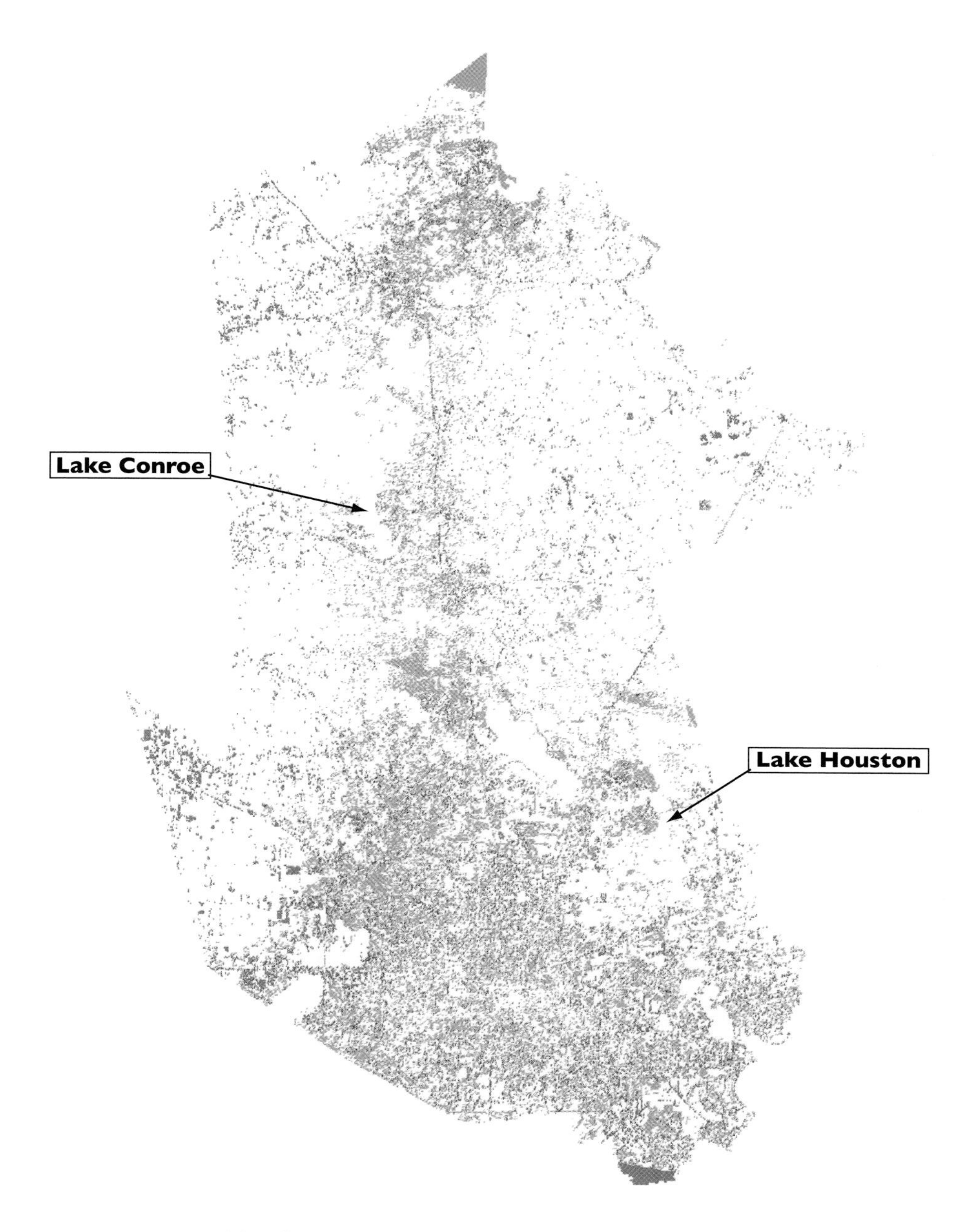

Figure 7.3. Highly reflective surface (HRS) change, 1979–2000. Red indicates areas that had HRS in 1979 that did not contain forest in 2000. Green indicates areas that had HRS in 2000 that did not contain forest in 1979.

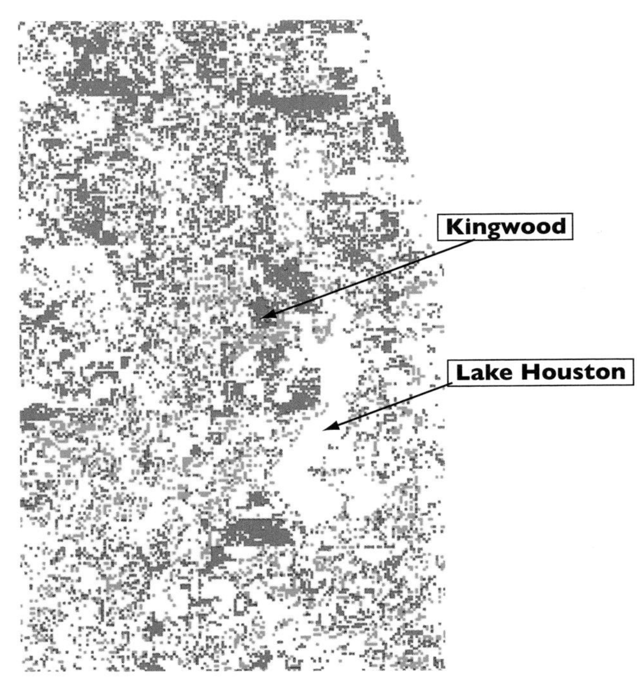

Figure 7.4.
Forest change in the Kingwood–Lake Houston region, 1979–2000. Red indicates areas that had forest in 1979 that did not contain forest in 2000. Green indicates areas that had forest in 2000 that did not contain forest in 1979.

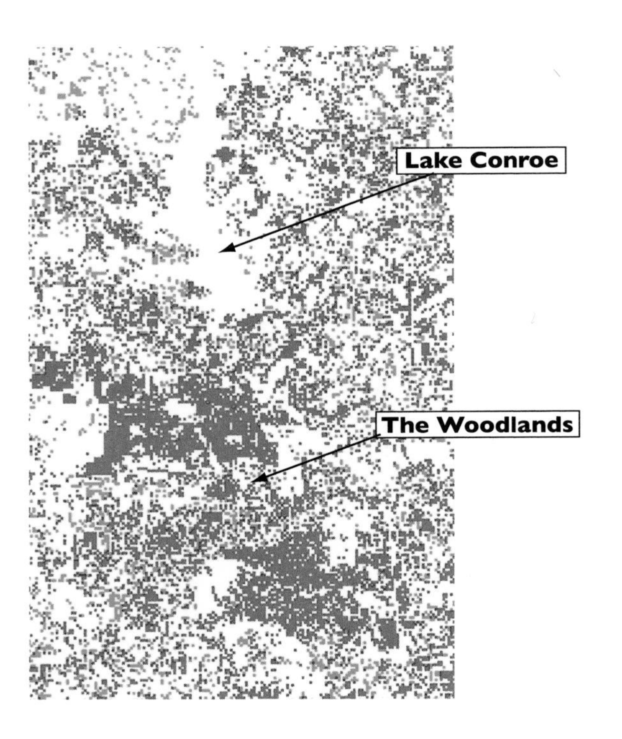

Figure 7.5.
Forest change in the Woodlands–Lake Conroe region, 1979–2000. Red indicates areas that had forest in 1979 that did not contain forest in 2000. Green indicates areas that had forest in 2000 that did not contain forest in 1979.

CHAPTER 8

A Tale of Two Texas Cities

Houston, the Industrial Metropolis, and Galveston, the Island Getaway

WILLIAM C. BARNETT

Less than fifty miles separate Houston and Galveston, but the differences between the cities today are striking. Their histories have always been interconnected, and in the nineteenth century the two cities followed fairly similar paths, but in the twentieth century Houston and Galveston took divergent routes. Houston grew into the nation's fourth-largest urban center—with nearly two million residents in the city proper and about five million in the metropolitan area—and a major center for energy companies and other international corporations. Houston is a leading example of the modern American metropolis, with a complex amalgamation of highways, automobiles, industry, skyscrapers, and suburbs. Present-day Galveston, with less than 60,000 people, pales in comparison to Houston's vast size and wealth.

During the first fifty years of their history, however, Galveston had the upper hand. Houston and Galveston originated as Anglo-American settlements during the 1830s, and they were economic centers in the era of the Republic of Texas and after the United States annexed Texas. But Galveston was the larger and more important of the two cities, because its deepwater port was the primary link between Texas and the world. The Island City, not the Bayou City, was seen as the community with the brighter future because it had a strategic location in an era of maritime trade. Nineteenth-century Galveston was a thriving international port, and most immigrants entering Texas and most agricultural exports passed through the island's harbor. But today it is obvious that the seaport's most prosperous days were in the past, and the city has a new identity. Galveston, once the gateway to Texas, has become a getaway, a place for weekend escapes to the beach from much larger urban areas such as Houston.

Houston's rise and Galveston's decline may appear to be a regional story, but the disparate routes taken by the two cities can be understood as a microcosm of important national developments. Two distinct sets of economic and environmental relationships led to the evolution of these cities. In this view, Galveston was a typical nineteenth-century commercial port, while Houston's industrial metropolis exemplifies twentieth-century urban America. Galveston and ports like Boston and Charleston grew because their natural harbors provided crucial points of connection between land and sea. Houston and twentieth-century industrial centers like Dallas and Denver were much less dependent on their natural landscapes because transportation and industrial advances gave them the power to reshape their physical environments.

It is difficult to travel back to 1880, when Galveston, with 22,000 people, had the highest population of any city in Texas and was one of the nation's 100 largest cities. Nineteenth-century Galveston bore little resemblance to cities that took shape in the nation's dramatic urban-industrial transformations. Its landscape and spatial relationships were defined by ships and the harbor rather than by railroads and highways, and its economy was based on maritime commerce, not industrial production. Before railroads and cars, most goods moved along natural waterways, and most cities were ports located on the sites of natural harbors. Trains and automobiles allowed Americans to build transportation routes across the continent, and these human-made systems were remarkably free from the constraints of geography. City builders were liberated from their reliance on coastlines and rivers, and as Americans gained the ability to build cities on varied landscapes, the environmental impacts of these urban centers increased.

The leading cities in the United States today are fundamentally different from Galveston's nineteenth-century seaport, but much can be learned from examining the city and its sluggish growth, particularly in contrast to Houston's remarkable expansion. Galveston's twentieth-century history was marked by an unsuccessful struggle to adapt to modern America. The community's leaders failed in their efforts to keep pace with the nation's urban-industrial transformations, and were unable to maintain Galveston's position as the leading city in Texas. Houstonians, however, were highly successful in using developments in transportation, technology, and industry, and Houston boomed in the twentieth-century's early decades. Comparing the decline of Galveston's commercial port and the rise of Houston's industrial metropolis provides a case study of the remaking of American economic and environmental relationships.

Industrial cities, suburbs, highways, and cars dominate modern America's topography, but this built environment has quite a short history. The industrial city and the suburb developed in the nineteenth century, and the emergence of automobiles in the early twentieth century led to the reconfiguration of U.S. spatial relationships. The complex system of highways and suburbs did not fully develop until the middle of the twentieth century, but it spread across the nation so rapidly and so completely that many Americans take their urban and suburban landscapes for granted. The sprawling Houston metropolis is a prime example of the nation's modern geography, and the city provides a window into the creation of built landscapes of concrete and steel that look the same in vastly different ecosystems. Meanwhile, Galveston offers a view of an earlier set of environmental relations. The city was built on an island where deep and sheltered water met land, and it has always depended on its relationship with the sea. Galveston's ocean access once made it a prosperous trade center, but now the island's beaches attract tourists. Both the old and new forms of Galveston are visible today, and a palpable sense of history hangs over the city.

A number of visitors have been struck by the lingering presence of Galveston's tragic past. Gary Cartwright's 1991 book on Galveston begins with the sentence, "I never go back to the Island without sensing the ghosts." And Edna Ferber, author of the famous Texas novel *Giant*, wrote in 1963, "The city had a ghostly charm. . . . Here was a remnant of haunted beauty—gray, shrouded, crumbling."[1] Both writers referred to ghosts in explaining the island's sense of loss. In 1900, a terrible hurricane struck Galveston, killing 6,000 people. This massive storm demonstrated that locating a city by the sea has dangers as well as benefits, and showed that natural forces possessed the power to devastate urban areas. This tragedy remains the nation's deadliest natural disaster, and the 1900 hurricane still casts a shadow over the community.[2] In Galveston, more than in most U.S. cities, many layers of the past are visible in the urban landscape, which reveals a complex interweaving of the built and natural environments.

It is easy to picture what the commercial seaport looked like over a century ago, because block after block of nineteenth-century buildings remain standing, and the freeways and skyscrapers that dominate Houston and other metropolitan areas have had little impact in Galveston. It is also possible to envision what this Gulf Coast island looked like before the Anglo-American settlers arrived, because there are large areas of open space, including coastal wetlands and prairies. All it takes is a walk along an empty beach to imagine the island as it was 500 years ago. But the opposite is true in Houston. Rapid economic and territorial expansion has erased most evidence of the city's nineteenth-century origins, and the pervasiveness of human-made structures has made it difficult to detect the natural environments that lie buried beneath the built environment. Houston and Galveston have followed fundamentally different paths in their economic and environmental development, and their urban landscapes expose the key points of divergence. Houston provides a roadmap of the route to becoming an industrial metropolis, while Galveston offers a blueprint for a different type of city.

Galveston, the Commercial Seaport

Since the Spanish era, the harbor on Galveston Island's protected bay side was recognized as the best deepwater anchorage between New Orleans and Mexico, and it was the reason a trading post was established on the island. Oceangoing ships could not go up the shallow rivers to the Texas interior, so they took on cargo in Galveston. Cotton was the most valuable commodity that flowed through Galveston to larger cities in Europe and the United States, and the port became the gateway to the rich agricultural lands of Texas. Ships left the island filled with cotton, hides, and wheat, and returned full of European immigrants bound for Texas and other parts of the West. When Galveston was incorporated in 1839, it was the largest settlement in the Republic of Texas, and it held its lead over other Texas cities for the next fifty years.

Galveston's leaders fully expected to continue as the economic center of Texas, and they hoped to become the great commercial metropolis of the American West. They envisioned their seaport becoming the dominant city west of the Mississippi River. These boosters lived in a maritime age, when goods moved by ship and great cities were seaports. It was their dream that the agricultural production from Texas north to Minnesota and west to the Rocky Mountains would flow through their port. They always began their booster efforts by stating that Galveston's harbor made it the closest point of access to saltwater for the landlocked interior West. In 1849, a group called the Citizens of Texas proposed to Congress that the first transcontinental railroad run from Galveston to San Diego. They noted that this southern route was only 1,250 miles from sea to sea, and asserted that it was the path through the Rockies that nature intended. They wrote, "on this line, the vast chain of mountains which traverses the continent from north to south throughout almost its entire length, seems cut down to a level, as if Nature herself had designed this as the great central line of communication between the two oceans."[3] They claimed to be fulfilling nature's intent, and laid out their great advantage—Galveston was closest to the sea.

Railroads were important to these booster visions, but the critical image was of Galveston as the place where goods from the land-locked western interior would meet oceangoing ships. A second major plan came out in 1866, when Galveston leaders proposed a Great Northern Railroad from Texas north to Minnesota, passing through Oklahoma, Kansas, Nebraska, and Iowa. The boosters argued that they could offer these states a more direct route to the ocean than could the Mississippi River or the often-frozen Great Lakes. They said it was natural for the Midwest's grain to flow through Galveston, stating, "Its transit to any other port is unnatural; and it stands ready . . . to rush like the waters, down into the lap of the sea, at Galveston."[4]

While Galveston's boosters promoted their plans to seize new hinterlands, a settler named Lucy Shaw gave a more critical account of the island. The boosters cited Galveston's advantages, but not its disadvantages. Shaw provided a more balanced view in the richly detailed letters she wrote from 1838 to 1850, because she described the environmental problems of life on a sandy barrier island. The businessmen analyzed Galveston's strategic position on the map relative to other cities, but Shaw assessed the island itself. In 1838, she wrote, "The greatest objection I have to make to this place is its being so low and flat and so destitute of trees." She emphasized the barren prairie and the heat, writing, "You don't know how I long for something like shade. The shade of a tree or a bush. . . . The sun pours down such an intense and blazing heat that it is almost painful to see and really painful to feel."[5] Lucy Shaw, raised in Maine, was in trouble, for this was only April of her first year in Texas.

Shaw never fully adapted to the climate, and twelve years later she wrote, "I never knew anything so dreadful as this weather for the nights are as hot as the day, and you cannot lie down ten minutes without a mosquito net." She summed up her view of the island by writing, "There is nothing natural, familiar, or homeish in the appearance of the country."[6] She also described shortages of water, scarcity of wood, deadly yellow fever epidemics, and terrifying hurricanes. Despite all these problems, settlers came to Galveston in droves. Shaw wrote, "It seems as if people were crazy to get to Texas. . . . It seems as if houses spring up in a night, they are building so fast," and she noted the cause, stating, "the Ambassador from Liverpool came here, the first British vessel that has ever entered a Texian port. She is now loading with cotton to take back to England."[7] Lucy Shaw's Galveston was booming, but it was still a flat sandy island with serious environmental constraints.

But to the city's boosters, cotton was almost a sure thing, making all the problems bearable. They were so confident in their superior location they almost never worried about rival Texas cities. They believed a deeper harbor for larger ships would guarantee the city's continued success. An 1885 book, immodestly titled *Galveston: The Commercial Metropolis and Principal Seaport of the Great Southwest,* declared, "The harbor at this city . . . is the one most susceptible to such improvements as would make it a harbor of the first class . . . and when improved . . . Galveston harbor will have a very undeniable advantage, even over the harbor at New Orleans." The authors focused their attention on New Orleans, and ignored the fact that Houston was becoming a serious threat. Their confidence was striking: "All traffic from the interior, north and west of the north-western Gulf coast, if it would seek the sea for water transportation at the nearest available port, must come to Galveston."[8]

These boosters compared Galveston with Chicago, New Orleans, and St. Louis, and found that every one of these cities fell short of their city's position. They boldly proclaimed, "The progress this city will make will surpass anything of the kind heretofore witnessed in the career of American cities." Galveston's promoters believed that their harbor could not be beat, because the laws of nature and economics favored their location. They declared, "Commerce, in obedience to the universal law of economy, will seek the sea at the nearest available point and will flow back and forth through the port of Galveston as the great current of the ocean sweeps into and out of its harbor."[9]

These predictions may now seem like pie in the sky, but leading businessmen and politicians saw Galveston as a critical piece in the nation's transportation puzzle. Railroad tycoons Collis P. Huntington and Jay Gould paid multiple visits to Galveston, and Presidents Grant and Harrison also came to assess the seaport. In the 1880s and 1890s, Huntington and Gould worked to add the port to their transportation empires, and Huntington helped lobby for federal funds to improve Galveston's harbor so it could become a terminal for his Southern Pacific line. Equally encouraging were 1880s conventions with business leaders from Colorado and Kansas, men who shared the vision of produce from the landlocked West flowing through Galveston instead of through Chicago and New York.[10] The island's dream of being a great metropolis did not come true, but it was taken seriously at the time.

Galveston's Decline

What happened to Galveston? How did the leading city in nineteenth-century Texas, with the best natural harbor between New Orleans and Mexico, fail to become a major twentieth-century city? Many Galvestonians, and some historians, emphasize the prosperity of the 1890s, when the city achieved a deeper harbor, and point to the terrible 1900 storm as the crushing blow to the city's dreams.[11] The hurricane was devastating, but blaming it alone for the city's decline is too simple and ignores several factors. First, the port was quickly rebuilt, and cotton exports peaked in the 1910s. Second, and more important, Galveston had already lost its lead over other Texas cities before the 1900 storm. In 1890, its population fell to third after Dallas and San Antonio, and in 1900 Houston pushed Galveston to fourth place. Rival Texas cities that Galveston had ignored, and interior cities like Chicago, had all decisively passed the island city.

The arguments made by Galveston's boosters provide insight into this failure. First, these men placed too much emphasis on oceangoing trade, and assumed their harbor was a trump card. They were using a centuries-old model in which great cities were seaports. But railroads changed everything, and Galveston's leaders failed to see that their rivals were not just ports like New Orleans, but railroad centers like Houston, Dallas, and San Antonio. Edward King, a Massachusetts journalist, toured the South in 1873, and his comments on post–Civil War Galveston and Houston are revealing. He noted that Galveston "has assumed a commercial importance which promises to make it a large and flourishing city, although it has many rivals." King then wrote, "Galveston has but one railroad exit, the line leading to Houston, where all the railroads of the grand new system will centre. Although the businessmen of Galveston are confident that the cotton crop will all fall into their hands, those of Houston think differently."[12] King's awareness of the impact that railroads had on New England cities allowed him to envision a future that Galveston's leaders did not see.

The city's boosters had a one-sided view of technology—they saw how it could help them but did not envision what it could do for rival cities. This blind spot applied to railroads and harbor improvements. As Galveston gradually deepened its harbor to permit increasingly larger ships to enter, Houston planned a much bolder technological intervention. Viewed by Galveston as an interior city and thus not a threat, Houston engineered a shipping channel and a harbor. The innovation would allow Houston to seize Galveston's dream and become the site where Texas railroads met the sea.

Houston's growth reveals another error in Galveston's map of the way to become a great city. Galveston was again using an outdated formula, improving the port but not developing industry. Cotton, the main export, was processed before coming to the island and was shipped to textile mills in other cities. Edward King of Massachusetts also recognized this deficiency in Galveston's development. He concluded his observations on Galveston with the statement, "Few cities, with a population of twenty-five or thirty thousand are more spirited; though manufacturing, as a solid basis, is, nevertheless, a supreme need."[13] As rivals from Houston to Chicago built industries and added value to agricultural goods, Galveston remained trapped in the southern pattern of New Orleans and Charleston, simply loading raw cotton onto ships. Unfortunately for Galveston, industrialization, not commerce, was the route to dramatic urban growth in this era.

But even if the boosters had realized their errors, converting Galveston from commerce to industry, and from a seaport to a railroad hub, would have been very difficult. As Lucy Shaw revealed, Galveston Island had serious environmental constraints. It did not have the quantities of land, water, timber, coal, or other minerals needed to build an industrial metropolis. And the island was not at the center of a network of railroads but was at the end of the line, dependent on fragile bridges to the mainland. Furthermore, it had a history of hurricanes, and the 1900 storm confirmed the idea that a low-lying island was not a wise place for expensive industrial development.

Galveston's story of decline might appear to be an isolated event, but local causes cannot fully explain this downward spiral, because larger economic forces were at work as well. It is too easy to blame nature and the 1900 storm for the city's economic problems. A broader look at American maritime cities during this era shows that the decline of this Texas seaport was part of a widespread pattern. A variety of commercial ports located in very different regions of the United States shared Galveston's story. These seaports ranged from New England to the Deep South, and from the Gulf of Mexico to the Pacific Coast, and they shared a common fate. Many of them were unable to adapt when sweeping economic changes remade urban America. As the American economy shifted from commerce to industrialization, and from shipping to land transportation, a variety of seaports were pushed to the nation's periphery.

The nation's seaports had been founded on the principle that a region's major city ought to be located at the site of its best harbor, but the idea did not hold true in the twentieth century. This saltwater principle had defined the nation's cities from its start, and New York, Philadelphia, Baltimore, Boston, Charleston, and New Orleans, all seaports, were the leading cities in the early nineteenth century. But the post–Civil War era brought the railroad, and interior cites like Chicago, St. Louis, and Cleveland surpassed coastal cities like Charleston and Boston. Meanwhile, seaports from Providence and Portsmouth in New England to Norfolk and Savannah in the South dropped from the nation's twenty largest cities. Galveston shared their fate, as it was also a loser during the nation's dramatic urban-industrial transformation.

A number of historians have analyzed the cities that were the biggest winners as industrialization reshaped urban America, but too little attention has been paid to the losers. Much can be learned from books like William Cronon's *Nature's Metropolis,* which analyzes Chicago's industrial explosion.[14] But accounts of successful urban giants like Houston and Chicago can make their rapid growth seem inevitable. Failure can be at least as instructive as success, and studying cities that were unable to make the leap to industrial metropolis yields a more complete view of this complex process. The decline of Galveston's port and the rise of Houston's railroad hub reveal the dramatic changes that occurred in the nation's shift from the nineteenth to the twentieth century.

Houston's Rise

In the late nineteenth century, Galveston's leaders focused most of their energies on shipping cotton and deepening their harbor, but Houston's business leaders were active on several fronts. They sought to expand their railroad connections, to build industry, and to improve Buffalo Bayou, the shallow, muddy stream from Houston to Galveston Bay. Houston would eventually achieve all three goals, but it was a lengthy and gradual process. In 1873, when Edward King toured Texas, there were signs that Houston was making progress in building railroads and manufacturing, and King predicted success. He wrote, "Houston is one of the most promising of the Texan towns. It lies fifty miles from Galveston, on Buffalo Bayou, and is now the central point of a complicated and comprehensive railway system." He continued, "This bayou Houston hopes one day to widen and dredge all the way to Galveston," and concluded, "Houston grows daily in commercial importance, and should be made a prominent manufacturing centre."[15]

At the time of King's visit, railroads were driving Houston's economic growth, and industrialization was linked with cotton, cattle, and lumber production.[16] Texas had over 1,000 miles of completed track, Houston had rail links north to St. Louis and south to Galveston, and connections west to San Antonio and east to Louisiana were under construction. King described the city's growing industrial capacity, including railroad repair shops, foundries, cotton mills, meatpacking plants, and sawmills. After reviewing the expansion of Texas railroads, King declared, "A new Texas is springing up, which, in commercial glory and power, will far surpass the old."[17] By 1891, Houston boosters realized railroads were the key to the future, and used the slogan "Hub City"; in 1900, they used the term "the Iron Ribbed City" to communicate this idea.[18]

Even as Houston grew into the railroad center of Texas, the city's leaders sought to improve the port. In 1870, the federal government made Houston a port of entry with a customhouse, and funded a study of plans to dredge a ship channel. That same year, the Buffalo Bayou Ship Channel Company began dredging the bayou, with the initial goal of deepening the waterway from four feet to over six feet. In the 1870s, local boosters, the federal government, and shipping magnate Charles Morgan funded the Buffalo Bayou Ship Channel Company and the Houston Direct Navigation Company.[19] Progress was slow, however, and barges carried freight to Houston, but oceangoing ships could not reach the city. As Houston's rail network grew in the 1880s, dredging was scaled back, but the idea of a ship channel was revived in the 1890s. Congress passed a bill for a twenty-five-feet deep channel in 1896, and work began the next year, but with funding problems and delays. The project's bold goal became realistic in 1902, when Congress allocated $1 million, while also awarding funds to Galveston's storm-damaged port. By 1908, the deepwater route was partially open, and it was fully completed to great celebrations in 1914.[20] Houston's channel was a key turning point in the city's urban rivalry with Galveston, as it took away the island's key advantage. Intriguingly, the project was approved before the 1900 storm, but not guaranteed until after the disaster.

A third event instrumental to the rise of Houston also occurred at the very outset of the twentieth century. In January 1901, the huge Spindletop oil well came in near Beaumont, Texas. This massive oil gusher, located east of Galveston and Houston, ultimately proved to be more important to Houston's economic boom than the ship channel or the Galveston hurricane. While cotton was the commodity that created Galveston's nineteenth-century prosperity and gave Houston its start, petroleum was the natural resource that fueled Houston's remarkable twentieth-century growth. The Spindletop strike was the critical factor in shaping Houston's new economy, as the city would become a national center for oil refineries, pipelines, and petrochemicals. In fact, Spindletop symbolizes the emergence of the system that defined twentieth-century America, because it initiated the nation's modern petroleum economy.

The oil gusher near Beaumont in 1901 was not the nation's first successful oil well, but petroleum's economic role completely changed after Spindletop. Petroleum produced in Pennsylvania in the nineteenth century by Standard Oil was sold for illumination and lubrication, but the unprecedented quantities of Texas petroleum were used as fuel oil. Pennsylvania's kerosene had competed with whale oil and then electric lighting, but oil from Spindletop and later Texas wells began to replace coal as a fuel for locomotion, heat, and industry. Texas railroads and steamships quickly switched from coal to the new and cheaper liquid fuel, and gasoline, previously seen as a by-product in the creation of kerosene, would emerge as the primary fuel for the internal combustion engines used in automobiles. Large quantities of gasoline were manufactured in the enormous oil refineries that were built on the Texas Gulf Coast. This shift from coal to oil and especially gasoline began in the Southwest and then spread across the nation and overseas. In 1911, just a decade after Spindletop, gasoline outsold kerosene for the first time, and Winston Churchill decided to convert the British navy from coal to oil.[21]

The Spindletop gusher initially brought prosperity to Beaumont, a sawmill town, but the petroleum industry that was born at Spindletop soon moved its headquarters to Houston. The first big refineries were near Beaumont, but by the 1920s Houston was the center for manufacturing gasoline. Many corporations that would play leading roles in the twentieth-century petroleum economy were founded at Spindletop, including Gulf Oil, Texaco, Sun Oil, and Mobil's Magnolia Oil. The Texas Company, later called Texaco, moved its offices to Houston in 1908, and Gulf Oil relocated to Houston in 1916. By this period, additional oil discoveries had been made even closer to Houston, in Humble and Goose Creek, the latter of which is between Houston and Galveston. The city of Houston, with its network of railroads and its new shipping channel, was rapidly becoming the focal point of the booming Texas petroleum industry.[22]

Prior to the discovery of huge oil fields in Texas, Standard Oil dominated the business, but Texas newcomers challenged the powerful corporation and permanently changed the industry. At the turn of the twentieth century, Standard Oil controlled over 85 percent of crude oil supplies and over 80 percent of refining capacity, and it also had tight control over oil transport by pipelines and tankers.[23] But Gulf, Texaco, Sun, and other Texas companies developed Spindletop's vast new oil field and progressed from oil drilling to involvement in all stages of the business. These new companies were able to grow because Texas laws banned Standard Oil from entering the region as a direct competitor. Andrew Mellon, a major investor behind the formation of Gulf Oil, declared, "The real way to make a business out of petroleum [was] to develop it from end to end; to get the raw material out of the ground, refine it, manufacture it, distribute it."[24] Gulf Oil, Texaco, Sun Oil, and other Texas companies followed this strategy. They were major oil producers that also became familiar corporate icons to U.S. consumers because they sold gasoline at their many drive-in service stations.

Houston developed a complex transportation network for petroleum products, and varied industries related to petroleum emerged in the first three decades of the new century. The unprecedented quantities of petroleum being produced in Texas could not be transported in barrels on railroads, so oil companies began building tank farms for storage, and pipelines and tanker ships for transport. Pipelines linked Houston with oil fields in Humble, Goose Creek, and Beaumont, and then with Oklahoma, Louisiana, and West Texas fields, and the ship channel became the epicenter of petroleum processing and transport. There were eight refineries along the channel by 1930, and by this date the natural gas industry was also flourishing. The abundant quantities of oil and gas to use as fuel attracted other industries to Houston, especially chemical plants, fertilizer works, cement plants, oil-tool businesses, and steel, iron, and metal works.[25]

Houston's promoters had many successes to celebrate, and their most difficult challenge may have been defining the new industrial metropolis that was taking shape. In 1910, a local bank published a book with an unwieldy title: *Houston, Texas, the Manchester of America, Where 17 Railroads Meet the Sea, the Site of the World's Next Big City, Nearest the Panama Canal.* The slogan "where 17 railroads meet the sea" captured the reason Houston pushed ahead of Galveston, but two years later boosters were using different language. In 1912, they published a bird's-eye view labeled: "Houston—a Modern City." The legend read: "The Workshop of Texas. More Factories, More Wage-earners and Largest Payroll in the Southwest. The Greatest Railroad Center in the South. Oil and Rice Center. Lowest Tax Rate."[26] The 1910 book stressed shipping and railroads, without mentioning petroleum or heavy industry. Just two years later, however, Houston was marketed as an industrial city based on factories and refineries, and the phrase "lowest tax rate" promoted the community's probusiness political climate.

Houston, the Industrial Metropolis

Houston was indeed becoming a new type of American city, an industrial metropolis based on the emerging petroleum economy. Great nineteenth-century industrial cities like Chicago and Pittsburgh developed because of railroads, coal, and steel, but Houston's economic growth was based on petroleum refineries, petrochemicals, and automobiles. In 1899, just prior to the Galveston storm, the Spindletop gusher, and full funding for the ship channel, Houston had about 40,000 residents, and it was a small urban center with promise. Only thirty years later, the population was close to 300,000, and the city was transformed. Houston hosted the Democratic Party's national convention in 1928, and the event symbolized the city's arrival as a major U.S. city. The delegates who came to the 1928 convention did not know it, but in Houston they were seeing the future of urban America.[27]

Houston boomed in the years between 1910 and 1930, and during this period it developed an economic system and a landscape that would come to characterize urban America in the twentieth century. The city's economic focus was shifting from agriculture to industry, and transportation was moving from shipping to railroads and then to automobiles. Agricultural goods and water transportation certainly did not disappear, as Houston held a day of celebration in 1924 when the port's cotton exports reached one million bales. The *Houston Press* headline announced, "Millionth Bale Honored . . . City Kneels to King Cotton," but the fact remained that industry, especially the petroleum industry, was passing agriculture as the foundation of Houston's prosperity.[28]

The rapid evolution of Houston's new economy created both excitement and concern. A 1923 *Houston Chronicle* editorial suggested that the emphasis on petroleum was dangerous and that diversification was needed. It stated, "Texas is leaning on oil, just as she once leaned on cotton. In industry, as well as in agriculture, the one crop idea is playing too prominent a part. . . . Everybody is on the hunt for an oil well, just as formerly, everybody sought a cotton patch."[29] The comparison to cotton was faulty, however, because petroleum created related industries ranging from petrochemicals to automobiles, and these businesses grew together. A more sanguine *Houston Chronicle* editorial in 1923 proved to be quite prophetic, stating, "We are not dreaming when we imagine half a million people in Houston, or even a million. We are not guided by fatuous optimism when we visualize the Ship Channel lined with industries. . . . The raw material is here, the power is here, the demand is here. It is only a question of moving in more people and more money." The editors emphasized the powerful link between industry and oil, asking, "Where is there a better supply of cheap fuel?"[30]

In fact, the essential groundwork for this industrial boom was rapidly being put into place. That same year Paul Brown wrote a series of articles on Texas, which he labeled "an agricultural state lubricated with petroleum." He called the Houston Ship Channel "this miracle of port development" in 1923 and marveled at its network of railroads, oil and natural gas pipelines, refineries, and docks. Brown wrote, "Houston emphasizes the fact that oil pipe lines have been brought to the municipal docks, so that vessels may be bunkered with oil while loading and discharging cargo. She proclaims with honest pride that this is the only port in the world with this facility."[31] Five years later, Houston Natural Gas Company advertisements combined the slogan "Economy and Prosperity Flow in the Path of Natural Gas" with a drawing of industrial facilities along the ship channel. The ads boasted, "with the coming of this cheaper, cleaner, more efficient fuel, new industries are being attracted to the Texas Gulf Coast shores," and this claim was accurate. The natural gas industry began to grow in the 1920s, and came into its own after World War II. The expansion of natural gas was one of the final developments in Houston's petroleum-based economy, and it reveals that city government and public utilities were key elements in this new urban form.[32]

The automobile was also a crucial component in the creation of the nation's twentieth-century urban landscape, and Houston's car culture was evolving in tandem with its petroleum industry. The city was building a network of paved roads, moving from 26 miles of pavement in 1903 to 196 miles by 1915, and these urban streets were being connected with country roads to smaller towns and other cities.[33] Texas roads were built using concrete and petroleum products made in Houston, and beginning in 1933, road building was funded with taxes from gasoline sales.[34] In addition to building roads, Houston was manufacturing motor vehicles, as a Ford Motor Company assembly plant opened in 1919 and a Chevrolet Motor Company plant opened in 1929.[35] A 1930 *Texas Monthly* article depicted the city as the center of automobiles in Texas, reporting, "Houston . . . is the focal point of five State highways . . . providing nine outlets from the city. . . . It may truly be said . . . that 'all roads lead to Houston.'"[36]

The rapid growth in the number of motor vehicles brought serious traffic problems in the 1920s, and traffic became a political issue. By 1930, there were 97,902 automobiles registered in Harris County, almost 100 times as many as the 1,031 cars in 1910.[37] In 1928 and 1929, an elite group called the Forum of Civics published *Civics for Houston,* and the magazine's goal was to convince readers of the need for city planning and zoning. One editorial stated, "Houston was the first city in the United States to install the traffic control light system at street corners. . . . Since that time she has rested on her laurels. . . . Unless we step lively this ghost is going to haunt us for the rest of our days." A 1929 article also depicted Houston as unusually dependent on cars, stating, "Shopping on wheels is a habit so prevalent in Houston as to astonish visitors. . . . Rolling patronage is provided for in Houston as in no other city in the world. . . . In Houston, a much greater percentage of people own and drive their own cars." The author saw Houston grocery stores and restaurants as national innovators in moving to locations with parking on the city outskirts.[38] By the late 1920s, Houston was committed to a transportation system based on cars and was making innovations that would spread to other U.S. cities.

The city planners who published *Civics for Houston* lobbied for street planning, zoning, and mass transportation, but the countervailing ideas of free enterprise and a probusiness stance had more power in Houston. In 1929, the Forum for Civics stopped publishing *Civics for Houston*. The private group concentrated its efforts on newspaper articles and on supporting the 1929 report by the City Planning Commission, which proposed a mass transit system, zoning, and street and highway plans.[39] However, Houstonians committed to private enterprise and opposed to government regulations blocked this detailed planning proposal. An antizoning group called the Houston Property Owners League formed to oppose the plan, and their 1930 protests to the city council were successful. The push for city planning reforms came to an abrupt end, as the two-year study for Houston's future was shelved, and the City Planning Office closed.[40] The urban planning trend that spread across the nation in the 1920s was defeated in Houston, and in that decade the booming city continued to sprawl outward, guided primarily by the dictates of the automobile and the petroleum economy.

Advertisements in *Civics for Houston* in 1928 and 1929 capture some of the tensions in this critical period. Ads for cars and tires ran beside promotions for railroads and public transportation, and celebrations of Houston's growth shared the page with critiques of the sprawling city. Houston Electric Company ads promoted mass transit over cars, asking, "Did you ever try to keep an appointment . . . attempt to park your car and fail to find a place. . . . Exasperating, isn't it? And so unnecessary too, with street cars and buses affording convenient and economical urban transportation." A Houston Gas and Fuel Company promotion titled "What's Happened to Main Street?" showed skyscrapers and cars replacing one-story businesses and horse carriages. The text read, "Along with a lot of other good things of yesterday the Main Street of Emporiums, Soda Palaces and General Stores is fast becoming a mere memory—passed on to make way for something better. In its place is rising a new Main Street dedicated to awesome skyscrapers, glittering movie palaces and stately hotels."[41] Unlike the ad for streetcars, this promotion suggested current trends were inevitable and represented progress.

Other advertisements revealed that wealthy Houstonians were insulating themselves from industrialization's changes. Every issue promoted River Oaks, an upper-class development. One ad read, "With the expansion of the city commercially and industrially, with the encroachment of shops and stores into the very domain of these fine old mansions, most of them must go. . . . What a pity! River Oaks is determined that no such fate shall ever befall any home within her gates." This campaign appealed to elite Houstonians' sense of nostalgia for the past, even as it sold new homes dependent on commuting by car. Unlike the rest of Houston, this wealthy community had zoning to keep out commercial development. The magazine's readers also saw ads for the Galveston Houston Electric Railway selling tickets to Galveston with the slogan, "From the heat of the city to the breezes of the Gulf—Just 75 minutes of cool pleasant riding on the Interurban."[42] Weekend excursions from Houston to the island provided much business for the Interurban, which opened in 1911, before cars were widely available, and before air-conditioning. By the late 1920s, the railway linked two cities that were going in different directions, as Galveston was becoming a nostalgic escape from industrial Houston. Just forty years earlier, Galveston was the larger city, but Houston's phenomenal expansion helped to convert the island from a gateway into a getaway.

Galveston, the Island Getaway

Nineteenth-century Galveston had a central position in the nation's maritime economy, but during the twentieth century the city was unable to maintain its role as an economic gateway. Galveston still had a commercial port, but it did not become a major petroleum center, and it failed to develop a significant industrial base. While Houston was surging forward as an industrial metropolis, Galveston was being pushed to the periphery of the nation's modern economy. The Island City's residents were forced to reinvent their community's identity, and Galveston evolved into a vacation getaway. The island's economy had once been dedicated to commerce, labor, and production, but in the early decades of the twentieth century Galveston was reconfigured as a center for tourism, leisure, and consumption.

Galveston did not become a vacation getaway overnight, and the seaport had always attracted sightseers. In 1843, Matilda Houstoun visited Galveston by yacht, and she described duck hunting, sport fishing, and riding "on a most beautiful beach."[43] When Edward King toured Texas in 1873, he praised the island's beaches, flowers, and climate, writing, "There could hardly be imagined a more delightful water-side resort than Galveston. . . . The little city has built a splendid hotel as a seductive bait for travelers."[44] A decade later, in 1883, the huge Beach Hotel opened. It was not located downtown by the port, but across the island on the Gulf Coast beach, which was a clear sign of tourism's economic role. Nineteenth-century Galveston was a commercial port, first and foremost, but it was also a resort for people who went to the ocean seeking cool breezes, a healthier climate, and opportunities for physical recreation and leisure.[45]

In the twentieth century's first decades, Galveston moved away from being a gateway and embraced its position as a getaway. The Beach Hotel was destroyed by fire before the 1900 storm, but an even larger hotel, the Hotel Galvez, was built in 1911. The Galvez was also located on the beach rather than downtown by the harbor, and it stood behind the protection of the huge seawall built after the 1900 hurricane. This concrete barrier was a response to the storm, and it stretched for miles along the Gulf Coast beachfront. While the seawall was designed to protect the city from storm waves and preserve the commercial seaport, it probably did more to boost the growth of the island's tourist industry. The 1911 opening of the Hotel Galvez on beachfront land protected by the seawall marked the city's new economic direction. Some

Hotel Galvez on the beach in Galveston. Courtesy Houston Metropolitan Research Center, Houston Public Library.

of Galveston's leading families invested in the project, and as the cotton business declined, the city's elites continued to shift money into hotels, tourism, insurance, and banking.[46]

The hotels and other tourist businesses were located on the Gulf Coast instead of on the island's sheltered side, and their growth shifted the city's economic center to the beach. Tourism boomed in the years between 1910 and 1930, and investment was by the seawall, not downtown. Visitors flocked to beachfront restaurants, nightclubs, and casinos. The decline of the port and the business district made it seem necessary to tolerate illegal activities connected to the tourist industry. Local officials largely ignored gambling, Prohibition violations, and prostitution because such activities were boosting the struggling city's economy by bringing visitors to the island. An island that did not modernize along with other cities, Galveston seemed separate from the rest of Texas. It became common to view Galveston as a place apart from mainland Texas, with a different set of rules, and it was a "sin city" long before Las Vegas. This "wide-open" era in Galveston lasted until the Texas Rangers shut down the vice districts in the 1950s.[47]

Many of Galveston's visitors in the years between 1910 and 1920 arrived by train, but by 1930 the automobile was a central part of tourism on the island. The Galveston Houston Electric Railway that ran ads in *Civics for Houston* had a short career. The Interurban rail service began in 1911, the same year the Hotel Galvez opened, but it closed in 1936. During the 1920s and 1930s, automobile use boomed, and buses replaced Houston's streetcars. Weekend visitors reached the island in cars, and a new concrete highway connecting the two cities was finished in 1928.[48] *Texas Monthly* reported in 1930, "On fine Saturdays and Sundays during the Summer months, ten thousand cars daily pass over the section between Houston and Galveston."[49] When the Interurban railway shut down in 1936, plans were already in the works to convert its right-of-way into the multilane highway that would be named the Gulf Freeway.

Galveston gained new highways and hotels in these decades, but much of its urban landscape remained unchanged, and the fact that it was not a modern metropolis like Houston was a key part of the city's appeal. In many ways, the city's downward spiral and its failure to industrialize left a nineteenth-century commercial seaport frozen in time. While Houston boomed into a huge industrial giant, downtown Galveston stagnated, which meant that the city had a manageable scale, a lack of industry, quaint nineteenth-century homes, and few highways. This nostalgic urban landscape was noted in a 1930 discussion of tourism. A Texas Monthly journalist wrote, "Galveston's long miles of smooth beaches and its unequalled facilities for bathing, boating, and fishing, make it preeminently the Summer playground of the Southwest; while in its older sec-

Trolley tracks on Broadway in Galveston. Courtesy Houston Metropolitan Research Center, Houston Public Library.

tions, the romantic suggestion of age give it a charm and an individuality such as are possessed by no other city in the State."[50]

As the decades passed, the appeal of Galveston's nineteenth-century urban landscape grew, and historic preservation became an important part of the city's efforts to attract visitors. Residents worked to bring tourists to the port area by restoring old buildings. The strategy succeeded in picturesque ports like Nantucket, Newport, and Portsmouth in New England and in Key West, New Orleans, and Savannah in the South, and it helped Galveston as well.[51] Historic preservation began in the same year the vice districts were shut down, as the Galveston Historical Foundation saved the Williams-Tucker house from destruction in 1957. Howard Barnstone's 1966 book, *The Galveston That Was,* boosted this preservation movement because its stark photographs captured the faded glory of the city's architecture. Restorations of commercial buildings along the Strand in the old seaport's business district and of a number of Gilded Age mansions followed, with financial support from George Mitchell, a Galveston-born and Houston-based oilman, and others. As a result of these efforts, tourists now enjoy the island's beaches and its charming historic districts.[52]

Present-day Galveston is full of ironies and contradictions. The city lost its fierce rivalry with Houston, but now the economy depends on a stream of visitors from its former rival. Galveston did not just fail to become an industrial city, it failed to such an extent that it is now an escape from the modern metropolis. In the twentieth century, Houston and its suburbs expanded so much that people have come to view Galveston as a part of Houston's metropolitan area. The two cities remain tied together despite the divergent paths they took during the nation's urban-industrial transformations, but their main link is the pavement of the Gulf Freeway, not the waters of Buffalo Bayou.

There is also significant irony in the fact that Galveston's economic decline a century ago and its decades of stagnation created a landscape that appeals to modern tourists. Galveston Island retained its beaches and open spaces, the old seaport and residential districts kept their historic buildings, and highways were not built through the downtown area. Today's tourists enjoy shopping, eating, and drinking on the Strand, because it offers a chance to go back in time to a walking city. The appeal of maritime cities like Galveston, Nantucket, and Key West is rooted in the fact that their urban landscapes are so very different from the cities and suburbs most Americans call home. When visitors walk through these historic seaports, they enjoy urban environments built for people, not for cars and factories. The popularity of these walking cities, and of replicas like New York's South Street Seaport and Baltimore's Harborplace, offers a critique of the nation's industrial cities.

Cities of Consumption

At the start of the twenty-first century, the contrast between Galveston and Houston may appear stronger than ever, but Houston is showing signs of becoming more like Galveston. In the twentieth century, Houston became the fourth-largest city in the United States because of industrial production and the petroleum economy. In that era, Galveston developed an economy based on tourism and historic preservation. Tourism is now the world's largest business, and Galveston has reversed its environmental relationships and become a center of consumer culture. During the twentieth century, Houstonians paid little attention to tourism, historic preservation, and mass transit, but there are recent signs of change. These include hosting the Super Bowl and Major League Baseball's All-Star Game in 2004, the revitalization of Old Market Square, and the construction of light rail. A national shift away from production is occurring as manufacturing moves overseas, but consumption's economic role continues to expand. Tourism is crucial to cities from New York City to Los Angeles, and today's fastest-growing urban centers are not based on factories. Las Vegas offers one vision of postindustrial America, with a booming economy devoted to consumer services and mass consumption, not production.[53]

Economic transformations are reshaping the nation, and cities as diverse as Las Vegas, Orlando, and Galveston are succeeding as centers of consumer culture. At the same time, industrial cities like Houston, Detroit, and Pittsburgh are trying to remake their images. Houston's industrial boom was viewed as progress and rarely questioned until the environmental movement's critiques of the petroleum economy. In 1952, a respected study of the oil industry gushed over Spindletop. The authors wrote, "A new age of human progress was born. . . . It was a crushing blow to all forms of monopoly. . . . Waste and fraud have been eliminated. . . . Spindletop stands as a fountainhead of American progress."[54] Fifty years later, many Americans view the petroleum economy from the opposite perspective, linking it with corporate greed, environmental damage, and urban sprawl. As Houston seeks to improve its image and attract tourists, Galveston may provide insights. When its port declined, the island was pushed to the nation's periphery. The city reinvented itself by changing from an economic gateway into a tourist getaway. Galveston based its revival on consumer culture and, in shifting to mass consumption, moved toward the nation's new economic center.

PART 3 ENVIRONMENTAL ACTIVISM AT THE GRASSROOTS

Sweeping issues like economic development and urbanization can obscure what happens to people in a rapidly growing, expansive place like metropolitan Houston. As a city of the South, Southwest, or Sun Belt, superficial assumptions arise about "red state" insensitivity to environmental reform and the possible lack of assertiveness of environmental protest. The overarching and unrelenting commitment to economic growth also obscures the fact that not everyone always benefits from a booming economy, condones ignoring environmental despoliation, or thinks of nothing more than possessing a ranch-style home and a new car. In some respects, Houston did miss the drama of the rise of the modern environmental movement in the 1960s and hardly led the country in implementing strong environmental legislation. But to suggest that environmental activism bypassed the city is to ignore significant chapters in its recent history. The impacts of industrial production, relentless growth, the vagaries of the weather, and high energy use of all kinds have taken their toll on the Houston metropolis. In some cases, the urban area is an ignominious leader in unwelcome categories—ozone production, Superfund sites, water pollution, flooding, and weather-related disasters. It makes sense, therefore, that since there is a substantial environmental agenda, there is likely to be a fitting environmentalist and grassroots response. The essays in this section clearly demonstrate some of the key responses.

The first two essays discuss NIMBY (Not in My Backyard) issues that uncovered serious environmental racism issues in an African American neighborhood in northeast Houston in the 1970s. Indeed, the resulting protests and lawsuit (the *Bean* case) may be considered among the first—if not the first—environmental justice incident publicly debated in the United States. Robert Bullard's essay is as much a memoir as a scholarly article. Bullard, and his wife, Linda McKeever Bullard, were early activists in the environmental justice movement in Houston, beginning with the controversy in Northwood Manor, a predominantly black community suffering under the placement of a newly proposed landfill. Bullard's well-known Houston waste study "provided the backdrop for one of the earliest empirical studies of environmental racism" in the country. Elizabeth Blum's essay covers the historical ramifications of the *Bean* case—concerning events in Northwood Manor—introduced in Bullard's essay, but provides a larger context for understanding the role of women in the environmental justice movement nationally and in Houston. Taken together, the two essays shed light on the powerful racial implications of urban growth in a modern metropolis.

Teresa Tomkins-Walsh discusses the correlation between rising environmental activism in Houston and issues related to Houston's bayous and watersheds—both distinctive topographic features. As she states, "Riverine and coastal waters both suffered from unplanned and unchecked development that led to an array of environmental problems, including destruction of these natural resources, industrial and nonsource point pollution, and flooding." While the activists engaged in protecting the bayous and combating flooding tended to be "affluent, well educated, and middle class," they shared a concern about urban and industrial environmental problems with their counterparts in Northwood Manor and those toxic protesters discussed in Kimberly Youngblood's piece. Youngblood's essay deals with the Brio Superfund site, located in the suburb of Friendswood, southeast of downtown Houston. It explores the complexities attendant to cleaning up a neighborhood Superfund site—the result of petroleum and chemical wastes. Brio was not only a symbol of the underside of oil and petrochemical production in the Houston area but also an important case demonstrating a significant grassroots response to a local environmental threat.

This group of essays brings environmental problems in the Houston area down to the most personal level, and thus provides perspective on how large societal trends—rampant urban growth and industrial production—affect us all by influencing changes in our physical surroundings.

CHAPTER 9

Dumping on Houston's Black Neighborhoods

ROBERT D. BULLARD

Houston's black population is located in a broad belt that extends from the south-central and southeast portions of the city into northeast and north-central Houston. The city's black population was originally concentrated in a few areas.[1] In 1950, two-thirds of Houston's black population were concentrated in three major segregated neighborhoods—namely, the Fourth Ward, Third Ward, and Fifth Ward. Beginning in the 1960s and accelerating in the 1970s, Houston's black population expanded outward, away from the central city, but continued to be concentrated in these outlying areas, generally in the northeast and southeast quadrants of the city (see figure 9.1).

Houston experienced unparalleled economic expansion and population growth in the 1970s. The city's black community made up nearly half a million residents, or 28 percent of the city's total population. Blacks in Houston remained residentially segregated from the larger community.[2] More than 81 percent of the city's blacks lived in mostly black areas, with major concentrations in the northeast and southeast sections of the city. Over 90 percent of Houston's blacks lived in areas in which they were in the majority in 1970. The 1980 segregation index reveals that 82 percent of the city's blacks continue to live in mostly black areas.

This chapter examines solid-waste disposal in Houston from the 1920s through the late 1970s. Much of the analysis is based on the author's work as an

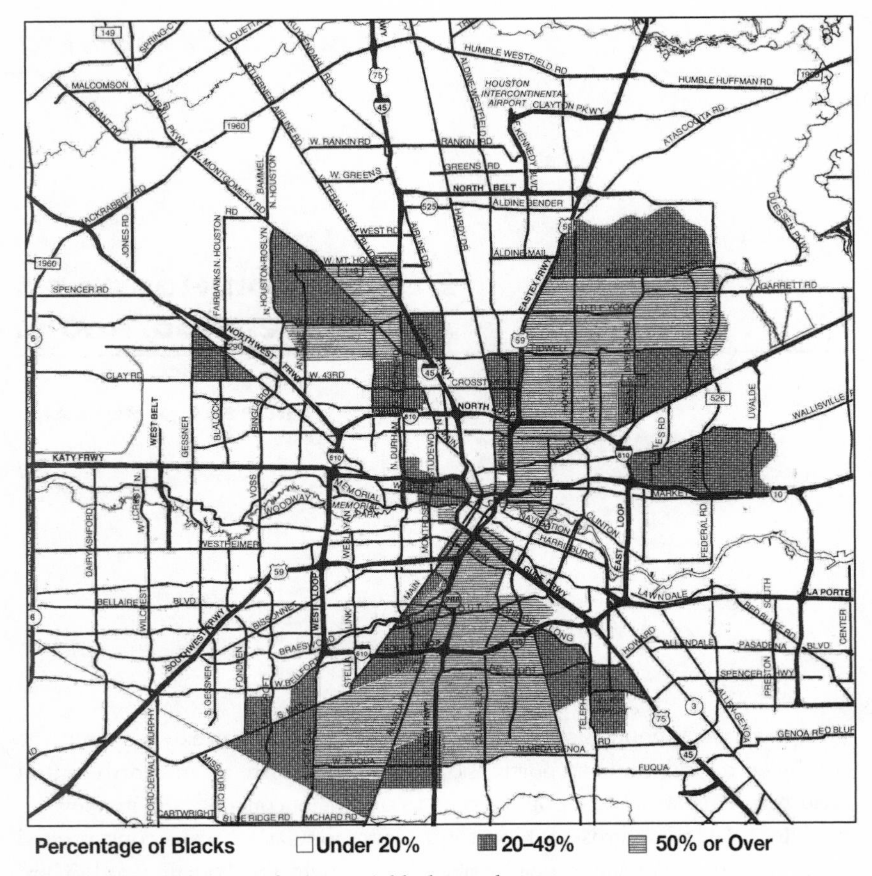

Figure 9.1. Distribution of Houston's black population.

expert witness on an environmental discrimination lawsuit.[3] The central theme of my analysis is that all communities are not created equal when it comes to the siting of locally unwanted land uses (LULUs) such as garbage dumps, landfills, and incinerators. Race and class dynamics, along with political disenfranchisement, interact to place some communities at special environmental and health risks from waste facility siting. If a community happens to be poor, black, and powerless, it receives less environmental protection in the placement of LULUs than an affluent, white, and politically powerful community.[4]

The environmental justice framework is used to uncover the underlying assumptions that may contribute to and produce disparate waste facility siting and unequal protection for minority residents. This framework brings to the surface the ethical and political questions of "who gets what, why, and how much." Waste facility siting is not "rocket science," but is more political science. Quite often it is not hydrology, toxicology, or epidemiology but sociology that carries the greatest weight in siting LULUs.[5]

People of color have known about and have been living with inequitable environmental quality for decades—most without the protection of federal, state, and local governmental agencies.[6] These communities are often victims of land-use decision making that mirrors the power arrangements of the dominant society. Many of the differences in environmental quality between black and white communities result from institutional racism, which influences local land use, enforcement of environmental regulations, industrial facility siting, and where people of color live, work, and play.[7]

The roots of institutional racism are deep and have been difficult to eliminate. Discrimination is a manifestation of institutional racism and causes life to be very different for whites and blacks. Racism has been and continues to be a major part of the American sociopolitical system, and as a result, people of color find themselves at a disadvantage in contemporary society. Environmental racism is real and adversely impacts millions of Americans.[8] What is environmental racism, and how does one recognize it? Environmental racism refers to any policy, practice, or directive that differentially affects or disadvantages (whether intended or unintended) individuals, groups, or communities based on race or color. Environmental racism combines with public policies and industry practices to provide benefits for whites while shifting costs to people of color.[9] Environmental racism is reinforced by governmental, legal, economic, political, and military institutions.[10]

Unofficially "Zoned" for Garbage

Houston has more than 500 neighborhoods. It is also the only major American city without zoning. This no-zoning policy allowed for an erratic land-use pattern. The NIMBY (Not in My Backyard) practice was replaced with the PIBBY (Place in Blacks' Backyard) policy. The all-white, all-male city government and private industry have targeted landfills, incinerators, garbage dumps, and garbage transfer stations for Houston's black neighborhoods. Clearly, white men decided that Houston's garbage dumps were not compatible with the city's white neighborhoods. Having white women on the city council made little difference on Houston's landfill siting in black neighborhoods. This policy changed somewhat when the first African American, Judson Robinson Jr., was elected to the city council in 1971. Robinson had to quell a near-riot over the opening of a landfill in the predominately black Trinity Gardens neighborhood.[11]

Five decades of discriminatory land-use practices lowered property values, accelerated physical deterioration, and increased disinvestment in Houston's black neighborhoods. Moreover, the discriminatory siting of solid-waste facilities stigmatized the black neighborhoods as "dumping grounds" for a host of other unwanted facilities, including salvage yards, recycling operations, and automobile "chop shops."[12]

Houston's no-zoning policy is characterized by "hodge-podge" irrational land-use planning and excessive infrastructure chaos.[13] In the absence of zoning, developers have used renewable deed restrictions as a means of land-use control within subdivisions. Lower-income, minority, and older neighborhoods have had difficulty enforcing and renewing deed restrictions.

Renewable deed restrictions were the only tool many residents had at their disposal to regulate nonresidential uses. However, deed restrictions in many low-income areas were often allowed to lapse because many residents were preoccupied with making a living and may not have had the time, energy, or faith in government to get the needed signatures of neighborhood residents to keep their deed restrictions in force. Moreover, the high occupancy turnover and large renter population in many Houston inner-city neighborhoods further weakened the efficacy of deed restrictions as a protectionist device.

Ineffective land-use regulations have created a nightmare for many of Houston's neighborhoods—especially the ones that are ill equipped to fend off industrial encroachment. From the 1920s through the 1970s, the siting of nonresidential facilities heightened animosities between the black community and the local government. This is especially true in the case of solid-waste disposal siting. It was not until 1978, with *Bean v. Southwestern Waste Management*, that blacks in Houston mounted a frontal legal assault on environmental racism in waste facility siting.[14] A new environmental justice theme emerged around the idea that "since everybody produces garbage, everybody should have to bear the burden of garbage disposal." This principle made its way into the national environmental justice movement and later thwarted subsequent waste facility siting in Houston.

Public officials quickly learned that solid-waste management can become a volatile political issue. Generally, controversy centered around charges that disposal sites were not equitably spread in quadrants of the city. Finding suitable sites for sanitary landfills has become a critical problem, mainly because no one wants to have a waste facility as a neighbor. The burden of having a municipal landfill, incinerator, transfer station, or some other type of waste disposal facility near one's home has not been equally borne by Houston's neighborhoods.

It is clear that *Bean v. Southwestern Waste Management* exposed the racist waste-facility siting practices in Houston. The lawsuit also changed the city's facility siting and solid-waste management practices after 1979. From the 1920s to the late 1970s, black neighborhoods in Houston became the dumping grounds for the city's household garbage.[15] Over the past fifty years, the city used two basic methods to dispose of its solid waste—incineration and landfill. From the

1920s to 1978, a total of seventeen solid-waste facilities were located in Houston's black neighborhoods (figure 9.2). Thirteen solid-waste disposal facilities were operated by the city of Houston from the late 1920s to the mid-1970s. The city operated eight garbage incinerators (five large units and three mini-units), six of which were located in mostly black neighborhoods, one in a Hispanic neighborhood, and one in a mostly white area (see table 9.1).

All five of Houston's large garbage incinerators were located in minority neighborhoods—four black and one Hispanic. All five of the city-owned land-

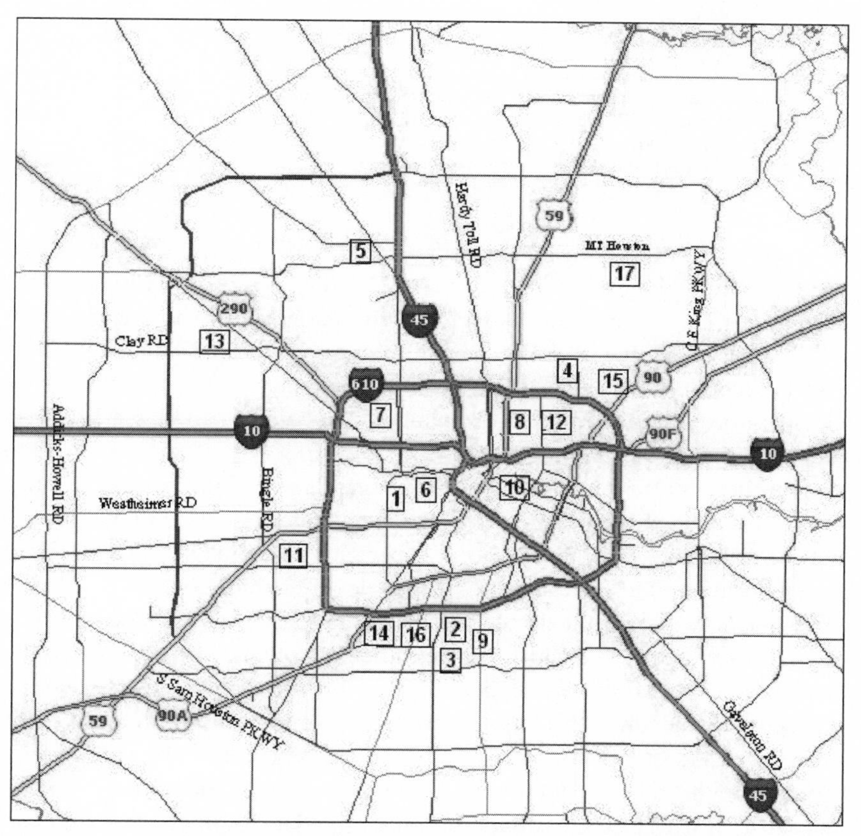

Figure 9.2. Houston municipal solid-waste facilities. (1) Fourth Ward Landfill; (2) Sunnyside Landfill; (3) Reed Road Landfill; (4) Kirkpatrick Landfill; (5) West Donovan Landfill; (6) Fourth Ward Incinerator; (7) West End/Cottage Grove Incinerator; (8) Kelly Street Incinerator; (9) Holmes Road Incinerator; (10) Velasco Incinerator; (11) Westpark Mini-Incinerator; (12) Kelly Street Mini-Incinerator; (13) Northwest Service Center Mini-Incinerator; (14) Holmes Road Landfill (1970); (15) McCarty Road Landfill; (16) Holmes Road Landfill (1978); (17) Whispering Pines Landfill.

Table 9.1. City of Houston Garbage Incinerators and Municipal Landfills

Neighborhood	*Location*	*Incinerator*	*Landfill*	*Target Area**	*Ethnicity of Neighborhood***
Fourth Ward	Southwest	1	1	Yes	Black
Cottage Grove	Northwest	1	—	Yes	Black
Kashmere Gardens	Northeast	2	—	Yes	Black
Sunnyside	Southeast	1	2	Yes	Black
Navigation	Southeast	1	—	Yes	Hispanic
Larchmont	Southwest	1	—	No	White
Carverdale	Northwest	1	—	Yes	Black
Trinity Gardens	Northeast	—	1	Yes	Black
Acres Homes	Northwest	—	1	Yes	Black

* Target areas are designated neighborhoods under Houston's Community Development Black Grant (CDBG) program.

**Ethnicity of neighborhood represents the racial/ethnic group that constitutes a numerical majority in the census tracts that make up the neighborhood.

fills were sited in black neighborhoods. One of the oldest city-owned incinerators was located in Houston's Fourth Ward. This site dates back to the 1920s. Other city-owned incinerators included the Patterson Street site, the Kelly Street site, the Holmes Road site, and the Velasco site located in the mostly Hispanic Second Ward, or "Segundo Barrio." The costs of operating these large incinerators and the problems of pollution generated by these systems were major factors in their closing.

Although blacks composed just over one-fourth of the city's population from the 1920s through the 1970s, all five of the city-owned landfills and six of the eight city-owned incinerators (75 percent) were built in Houston's black neighborhoods. During this same period, eleven of thirteen city-owned landfills and incinerators (84.6 percent) were built in black neighborhoods.

The city contracted with a private company to conduct a "pilot project" of mini-incinerators that were supposed to be more efficient, cost less to operate, and burn cleaner. In 1972, Houston invested $1.9 million in a contractual agreement with Houston Natural Gas Company for these mini-incinerators, which were thought to be "pollution free." Three sets of mini-incinerators were installed in the city. One site was located on Westpark in southwest Houston, one was located on Kelly Street near the North Loop and Easttex Freeway, and another site was located on Sommermeyer Road in northwest Houston.

The Sommermeyer incinerator was located at the Northwest Service Center in the mostly black Carverdale neighborhood. The Carverdale neighborhood was also the host community for the first city-owned garbage transfer station. In 1983, the Houston City Council awarded a contract to Waste Management of Houston and appropriated $1 million to pay the contractor, at the cost of $11.58 per ton of garbage.[16]

The Kelly Street mini-incinerator was sited in the mostly black Kashmere Gardens neighborhood, and the Westpark facility was built across the Southwest Freeway near the mostly white Larchmont neighborhood. Pilot tests of the mini-incinerators found them not to be "pollution free" because they performed with mixed results. The mini-incinerators did not meet the pollution standards of the Houston Air Quality Control Board and were shut down after a short period of operation in the mid-1970s.[17]

The Texas Department of Health (TDH) is the state agency that grants permits for standard sanitary landfills. From 1970 to 1978, the TDH issued four sanitary landfill permits for the disposal of Houston's solid waste (table 9.2). Privately owned sanitary landfills in Houston followed the pattern established by the city. Three of the four privately owned landfill sites are located in black neighborhoods.

Historically, Houston's black neighborhoods were dumped on while receiving less than their fair share of residential services—including garbage collection, water, and sewer services. While some progress has been made, Houston's neglected black neighborhoods are visible signs that race still matters. In a 2002 report on Houston's neglected neighborhoods, *Houston Chronicle* reporter Mike Snider wrote that hasty annexations by the city left a legacy of blight.[18] Even the best efforts of Houston's Super Neighborhood Program, the cornerstone of the administration of Mayor Lee Brown, Houston's first African American mayor, have done little to reverse the decades of systematic neglect. Under the program, the city is divided into eighty-eight "super neighborhoods." These neighborhoods can choose to create resident councils and develop "action plans," which are essentially wish lists, to deal with a range of urban issues, including routine maintenance such as ditch cleaning and as well as sidewalk construction, street repaving, parks and green space, libraries, multipurpose centers, and other capital improvement projects.[19]

Many of these "invisible" black neighborhoods are easily identified by their

Table 9.2. Privately Owned Houston Sanitary Landfills Permitted by the Texas Department of Health, 1970–1978

Site	*Location*	*Year Permitted*	*Neighborhood*	*Neighborhood Ethnicity*
Holmes Road	Southeast	1970	Almeda Plaza	Black
McCarty	Northeast	1971	Chattwood	White
Holmes Road	Southeast	1978	Almeda Plaza	Black
Whispering Pines	Northeast	1978	Northwood Manor	Black

substandard housing, unpaved narrow streets, open ditches, illegal dumps, and lack of sewer and water lines, sidewalks, curbs, and storm drains.[20] When it rains, it floods. *Houston Chronicle* reporter Kristen Mack concluded that a sizable share of the mostly black Acres Homes neighborhoods was as impoverished in 2002 as it was ten years ago.[21] Homoizelle Savoy, a thirty-year resident of Acres Homes, summed up her community's plight: "Thirty years is neglected, isn't it? Ain't no feeling—I know we've been overlooked."[22]

Segregated Black Neighborhoods

Although Houston has more than 500 neighborhoods, nine predominately black neighborhoods shouldered the burden for waste disposal sites from the 1920s through the late 1970s. The nine predominately black Houston neighborhoods with solid-waste sites include the Fourth Ward, West End/Cottage Grove, Kashmere Gardens, Sunnyside, Carverdale, Acres Homes, Almeda Plaza, Trinity Gardens, and Northwood Manor. Although these black neighborhoods are rich in history, they all have received less than their fair share of city services, residential amenities, and environmental protection. From neighborhood-level "snapshots," one can see the impact of Jim Crow segregation in shaping black Houston.[23]

Fourth Ward

Dating back to the 1860s, the Fourth Ward is the oldest black neighborhood in Houston. Former slaves in 1866 moved into the Fourth Ward and founded an area called "Freedmen's Town" (near the current Jefferson Davis Hospital and Allen Parkway Village public housing project). The neighborhood is often referred to as the "Mother Ward," because many black Houstonians can trace their roots to this neighborhood. Development that began in the 1860s continued through the mid-1920s. The mid-1920s saw this neighborhood become the center for black culture in Houston; over 95 percent of Houston's black businesses and black professionals were located in the Fourth Ward.

The Fourth Ward has experienced extreme pressures, as it lies just west of the downtown area. Housing in the neighborhood consists mainly of row houses. The Fourth Ward was the scene of the city's bloodiest race riot (Houston Riot of 1917). The city of Houston operated a garbage dump and garbage incinerator in the neighborhood in the 1920s. In 1936, Harris County's Jefferson Davis Hospital was built on fill land from the dumpsite. The city's solid-waste department currently uses part of the fill property as a maintenance facility.

A large segment of Freedmen's Town was torn down under the city's urban renewal and slum clearance program and during the construction of a white public housing project (San Felipe Courts, later renamed the Allen Parkway Village public housing project) in the early 1940s. Allen Parkway Village was demolished in 1997, and 500 market-rate houses were rebuilt on the site.[24]

West End/Cottage Grove

The predominately black neighborhood of West End/Cottage Grove contains some of the oldest housing in the city, dating back to the 1880s; over one-third of the housing in the area was constructed prior to 1940. In addition, elderly persons comprise a large share of the neighborhood's population. The neighborhood was the site of Houston's Patterson Street incinerator. This incinerator was located in census tract 514, which was over 57 percent black in 1970. This neighborhood has experienced an in-migration of nonblacks in recent years. The 1980 census showed that blacks comprised 42 percent of the tract population. However, blacks remain highly concentrated in the proximate area where the incinerator was located.

Kashmere Gardens

Kashmere Gardens is located just north of Houston's downtown. Part of this predominately black neighborhood includes census tract 207 (the location of Houston's Kelly Street incinerators). Land for this area was subdivided in the 1930s, with over 57 percent of the housing units built in the 1940s and 1950s. Over 70 percent of the housing units in the neighborhood are single-family units. The racial composition of the neighborhood has undergone little change over the years. Blacks comprised 90 percent of the population in 1970 and 89 percent of the population in 1980.

Sunnyside

The Sunnyside neighborhood is located approximately nine miles south of the Houston central business district. Development of the northern part of the neighborhood occurred in the early 1940s along Holmes Road. The 1970 census shows that Sunnyside (census tract 329) was over 97 percent black. This neighborhood, developed on the periphery of the city, remains as a distinctly black area.

The neighborhood contains three waste disposal facilities: the Sunnyside Landfill located on Bellfort Street, the Reed Road Landfill located on Reed Road, and the Holmes Road Incinerator located on Bellfort Street off the South Freeway. The city of Houston landfill that is located on Bellfort Street is adjacent to Sunnyside Park and just one block from Sunnyside Elementary School. The houses in the neighborhood are small, wood frame structures.

A few of the streets in the neighborhood are still gravel roads. Reed Road served as a commercial or business strip for the neighborhood (including small shops, stores, a public library, and so forth). Trash and other debris can be found scattered along Bellfort Street at the entrance to the neighborhood from the South Freeway. In addition, there continues to be illegal dumping at the entrances to the landfill site on Bellfort Street.

Carverdale

Carverdale is located approximately twelve miles west of the downtown area. The neighborhood, which covers about 1.5 square miles, was developed during 1953 and was annexed by Houston in 1955. The 1970 census indicates that Carverdale had a population of 1,664 residents, of which over 77 percent were black.

Carverdale residents continue to be concentrated in the northern portion of census tract 528. Over 75 percent of the neighborhood residents are homeowners. Census tract 528 has experienced a slight increase in population from 2,033 in 1970 to 2,106 in 1980. Carverdale was the site of a city of Houston garbage mini-incinerator pilot project in the early 1970s. The city-owned garbage transfer station is less that one block from the nearest Carverdale residents' homes.

Acres Homes

Acres Homes is located in northwest Houston some nine miles from the central business district. Acres Homes is both rural and suburban in character. Census data reveals that Acres Homes (census tract 525) was 75.6 percent black in 1970 and 78.9 percent black in 1980. The neighborhood developed between 1949 and 1953. Over two-thirds of all housing units in the neighborhood are owner-occupied. Most of the housing in the neighborhood is single-family, wood frame structures on pier and beam foundations.

Some of the older sections in the Acres Homes neighborhood do not have sidewalks or street lighting. The newer sections of Acres Homes were developed after 1970 and include mostly middle-income single-family homes. Two waste disposal sites were located in this neighborhood: the West Donnovan site and the International Disposal Inc. site on Nieman Lane.

Almeda Plaza

Almeda Plaza is located south of Houston's downtown area. The neighborhood lies in the southeastern corner of census tract 332, just south of the Holmes Road Landfill. The neighborhood was over 58 percent black in 1970. Residents who live on the northern streets of the neighborhood (such as Almeda Plaza, Summertime, and Monticello) have an unobstructed view of the Holmes Road Landfill sites. Permits were granted to operate landfills in the neighborhood in 1970 and 1978.

The entire census tract in which Almeda Plaza is located has undergone a dramatic demographic transition, from 26 percent black in 1970 to over 80 percent black in 1980. The neighborhood is primarily single-family residential; there are some apartments located on Almeda Road off Almeda Plaza. The Almeda Plaza Park is a major gathering place for youth activities (it includes a basketball court and children's playground). Open ditches in the area run along Almeda Plaza and west of the Holmes Road Landfill, and standing water in the drainage ditches gives off odors and may serve as breeding grounds for mosquitoes. Illegal dumping continues at the landfill even though the sites are closed.

Trinity Gardens

Developed in the late 1930s, Trinity Gardens is located in northeast Houston. The southeast sector of the neighborhood lies within census tract 216 and covers approximately 2.3 square miles. The neighborhood was 76 percent black in 1970 and over 90 percent black in 1980. There are two public schools in the neighborhood: Kashmere Gardens Senior High School and Houston Gardens Elementary. Trinity Gardens is one of the city's Community Development Target Areas. The Kirkpatrick landfill is located within the Trinity Gardens neighborhood in the general vicinity of the railroad tracks to the north, Hunting Bayou to the south, Homestead Road to the west, and Kirkpatrick Drive to the east.

Northwood Manor

Northwood Manor is a predominately black suburban neighborhood in northeast Houston. Developed largely in the 1970s, the subdivision consists primarily of single-family, middle-income homes. Northwood Manor falls within census tract 224, which had a population of 18,726 in 1980 (over 64.5 percent were black).

The Northwood Manor neighborhood conforms roughly to census tract 224.03 (census tracts subdivided in 1980). The geographic boundaries of 224.03 include East Mount Houston on the north, railroad tracks on the east, East Little York on the South, and Homestead on the west. The total population of tract 224.03 was 8,449 in 1980; blacks comprised over 82.4 percent of the tract's population in 1980.

Residents in the neighborhood are served by the suburban 17,800-pupil North Forest Independent School District (a district where blacks comprised over 80 percent of the student body in 1980). A total of seven North Forest schools are located in the Northwood Manor area.

The 195-acre Whispering Pines landfill is across the street from the North Forest Independent School District bus facilities, Smiley High School Complex,

Service Center, and the Jones-Cowart Stadium and athletic fields. The landfill is only 1,400 feet from the 3,000-student Smiley High School Complex, which did not have air-conditioning in 1982.

As early as 1983, the Whispering Pines landfill mound could be seen rising above the surrounding landscape on East Little York. A more graphic and panoramic view of Northwood Manor's "Mount Trashmore" (the name given to the landfill by area residents) can be seen from modern Jones-Cowart Stadium, an outdoor arena where mostly young black high school students practice and play across the road from the sanitary landfill.

The overall housing in Northwood Manor is well maintained. Over 88 percent of the residents own their homes; nationally, about 46 percent of blacks own their homes.[25] The neighborhood appears largely to be made up of younger families with children, and while driving through the neighborhood, it is not unusual to see children playing in the yards and on the sidewalks.

The neighborhood streets leading to the Whispering Pines landfill (East Little York Road, Homestead Road, and East Houston–Dyersdale Road) are littered with various types of trash and debris (for example, car tires, paper, boxes, mattresses, and so forth). In addition, several schools are located off East Little York Road, a major street that does not have sidewalks. Children who walk to school often walk in the street or along the grassy "trails" along the road. East Little York Road is the major thoroughfare for the solid-waste trucks that dump at the Whispering Pines Landfill.

Black Resistance

Whites singling out black neighborhoods for landfills, garbage dumps, incinerators, and other polluting operations is not new. Prior to 1967, few blacks resisted these unfair practices. Jim Crow and white power ruled the day. On May 16, 1967, black Houstonians picketed the Holmes Road dump in the southeast Sunnyside neighborhood, where an eight-year-old black girl drowned. Not only was environmental racism seen as unfair, but it had actually killed an innocent black child. This hit a nerve and galvanized black resistance to being dumped on.[26]

The landfill protesters and another group protesting racism in Houston schools (African American parents were charging that black students were disciplined more harshly than white students) combined forces, which later fueled the conditions for the 1967 Texas Southern University "riot," the only major civil disturbance that occurred in the city during the turbulent 1960s.[27] The Houston police were called to put down a disturbance on campus involving Texas Southern students, who were part of the protestors, which turned into a deadly confrontation with the police. Gunfire ensued, and a patrolman was killed by a ricocheting bullet. The origin of the shot was never determined. The police cleared some 480 students from a men's dormitory and in the process destroyed several thousand dollars' worth of the students' personal property.[28]

Four years later, in 1971, the first major controversy newly elected City Councilman Judson Robinson Jr., Houston's first black city councilman, had to deal with involved a city-owned dump. Councilman Robinson had to quell a near riot at the Kirkpatrick landfill in the mostly black Trinity Gardens neighborhood. Black protesters were demanding that the city-owned landfill close. The facility was later closed after six months of intense protest demonstrations.

Controversy surrounding landfill siting peaked in the late 1970s with the proposal by Browning-Ferris Industries (a company headquartered in Houston) to build the Whispering Pines sanitary landfill in Houston's mostly black middle-class Northwood Manor neighborhood. As a sociologist at Texas Southern University, a predominately African American institution, I was asked in 1979 by attorney Linda McKeever Bullard (my wife) to conduct a study of the spatial location of all of the municipal solid-waste disposal facilities in Houston. The request was part of a class-action lawsuit (*Bean v. Southwestern Waste Management*) she filed against the city of Houston, the state of Texas, and the locally headquartered Browning-Ferris Industries.

The lawsuit stemmed from a plan to site a sanitary landfill in a suburban, black, middle-income neighborhood of single-family homeowners. *Bean* was the first lawsuit in the United States that charged environmental discrimination in solid-waste facility siting under the Civil Rights Act. The Northwood Manor neighborhood of single-family homes was an unlikely location for a garbage dump—except that over 82 percent of its residents were black.

In order to obtain the history of waste disposal facility siting in Houston, government records (city, county, and state documents) had to be manually retrieved because the files were not computerized in 1979. On-site visits, windshield surveys, and informal interviews—done in a sort of "sociologist as detective" role—were conducted as a reliability check.

After collecting the data for the lawsuit and interviewing citizens from other African American neighborhoods, it became apparent to me and my ten graduate students at Texas Southern University that waste facility siting in Houston was not random. Moreover, this was not a chicken or egg (which came first) problem. Black people did not move in next to the dumps. In all cases, the residential and racial character of the neighborhoods had been established long before the waste facilities invaded the neighborhoods.[29]

The suburban Northwood Manor neighborhood consists primarily of single-family homeowners. No major industrial or commercial firms are found in the area. Many Northwood Manor residents thought they were getting a shopping center or new homes in their subdivision when construction on the Whispering Pines landfill site commenced. When they learned the truth, they began to organize to stop the dump.

Northwood Manor resident Lonnie Anderson learned of the proposed dump at a community meeting to discuss the drainage, sewage, and rat problems already affecting the area. Anderson and other plaintiffs in the lawsuit collected more than 2,000 signatures on a petition opposed to the dump. Judge Gabrielle McDonald said the petition indicated strong sentiment against the dump, but she ruled it inadmissible.[30]

It is ironic that some of the residents who were fighting the construction of the waste facility had moved to Northwood Manor in an effort to escape landfills in their former Houston neighborhoods. Patricia Reaux recounts her experience with landfills: "About seven years ago, I lived near another dump on Hirsch Road. . . . There were rats so big you had to use a gun to kill them. The smell was awful, and the water sometimes was not drinkable."[31]

Local residents formed the Northeast Community Action Group (NECAG)—a spin-off organization from the local neighborhood civic association—to halt the construction of the facility. They later filed a lawsuit in federal court to stop the siting of the landfill in their neighborhood. The residents and their black attorney, Linda McKeever Bullard, charged the TDH and the private disposal company (Browning-Ferris Industries) with racial discrimination in the selection of the Whispering Pines landfill site.[32] Residents were upset because the proposed site was not only near their homes but also near their schools.

Neither Smiley High School nor the other fifteen North Forest schools are equipped with air-conditioning—not an insignificant point in the hot and humid Houston climate. Windows are usually left open while school is in session. Seven North Forest Schools are found in Northwood Manor and contiguous neighborhoods. Testifying in court, Toley Hart, North Forest superintendent, expressed his concerns about the Whispering Pines landfill: "I can't see any reason other than racial ones for putting landfills in our area. . . . I worry about how this will affect the attitudes of our schools' young people, whose parents have worked hard to have a nice community and good homes. To have something like this happen is bound to change them."[33]

Superintendent Hart, who is white, also testified that the district successfully opposed another attempt to site a garbage dump in the area in the 1960s (when the area was 90 percent white), and the Harris County Commissioners Court, which then licensed all landfill outside the Houston city limits, denied the application after strong protests from the area residents. Hart explains, "About 25 years ago, this area was about 90 percent white, and there were no landfills here. . . . It was not until the early '60s that the first one [dump] was put on Settegast Road in what was then the only minority pocket in the district."[34] The demographics of Northwood Manor also shifted from majority white in 1970 to majority black in 1980.

The 1979 lawsuit finally went to trial in 1984. The lawsuit was first argued in front of U.S. District Judge Gabrielle McDonald—the only African American female judge in Texas at the time. The residents and their attorney were seeking a temporary injunction to block the opening of the 195-acre landfill. In December 1979, Judge McDonald denied the temporary injunction.

In her written order denying the temporary injunction, Judge McDonald said she "might very well have denied this permit" by the TDH licensing the landfill site. "It simply does not make sense to put a solid waste site so close to a high school, particularly one with no air conditioning. . . . Nor does it make sense to put the landfill so close to a residential neighborhood. But I am not TDH. For all I know, TDH may regularly approve of solid waste sites located near schools and residential areas, as illogical as that may seem."[35] Judge McDonald also described the landfill siting both "unfortunate and insensitive," but she thought the plaintiffs had failed to prove that the permit was motivated by "purposeful racial discrimination."[36]

The case was later transferred to U.S. Federal District Judge John Singleton's court in 1984. Judge Singleton repeatedly referred to the black plaintiffs as "nigras" and their neighborhoods as "nigra areas." Most people from the South know that the word "nigra" is code for "nigger." The judge's clerk, a white female, had to keep reminding Judge Singleton that blacks in his court were offended by him referring to them as "nigras." The judge stopped referring to the plaintiffs as "nigras," but ruled against the black residents, and the landfill was built.

Conclusion

The NECAG plaintiffs lost their legal battle in federal district court. However, *Bean v. Southwestern Waste Management* did produce some changes in the way waste facility siting was dealt with in Houston. The Houston City Council, acting under intense political pressure from local residents, passed a resolution in 1980 condemning the siting of the landfill in the Northwood Manor community and prohibited city-owned trucks from carrying Houston solid waste to the controversial Whispering Pines landfill.[37] The city also provided the Northwood Manor residents a much-needed park. The only problem is that East Little York Park was built next to the landfill.

The TDH updated its requirements of landfill permit applicants to include detailed land-use, economic, and sociodemographic data on areas where it proposed to site standard sanitary landfills. Black Houstonians sent a clear signal to the TDH, city government, and private waste disposal companies that they would resist any future proposals to site waste facilities in their neighborhoods. The landfill question appears to have galvanized and politicized a part of the Houston community, the black community, that for years had been inactive on environmental issues.

Black Houstonians soon realized that they could not rely on city leaders for environmental justice and protection. They turned to the courts and direct action. Although the 1979 *Bean* lawsuit did not prevail in court, it remains a watershed case because it was the first time in U.S. history that Americans challenged environmental racism using civil rights law and is cited in every environmental racism/environmental justice and Title VI case filed in the United States. It was also a turning point whereby African Americans decided to end their communities' traditional role as the "dumping grounds" for other people's garbage. However, it took until 1990 for the second environmental discrimination civil rights lawsuit to be filed in the United States.

The Houston waste study provided the backdrop for one of the earliest empirical studies of environmental racism. It laid the foundation for disparate impact methodology and designing environmental justice impact assessment; it provided a case study for *Dumping in Dixie,* the first environmental justice book; and it infused the environmental justice framework, concepts, and principles into the college curriculum.

Over the past two decades, "environmental justice," "environmental racism," and "environmental equity" have become common terms. Out of the small and seemingly isolated environmental struggles in Houston and other communities across the country emerged a potent grassroots environmental justice movement. For millions, this movement became a unifying theme across race, class, gender, age, and geographic lines.

Millions of Americans ranging from constitutional scholars to lay grassroots activists now recognize that environmental discrimination is unfair, unethical, and immoral. They also recognize that environmental justice is a legitimate area of inquiry inside and outside of the academy. Environmental justice is also a civil rights and human rights issue. This was not always the case. Two decades or so ago, few academicians, government bureaucrats, environmentalists, or civil rights or human rights leaders knew anything about or understood the racial dynamics involved in environmental decision making.

Today, we see groups such as the National Association for the Advancement of Colored People (NAACP), the NAACP Legal Defense and Education Fund, Earthjustice Legal Defense Fund, Lawyers Committee for Civil Rights Under the Law, International Human Rights Law Group, Center for Constitutional Rights, National Lawyers Guild's Sugar Law Center, American Civil

Liberties Union, Legal Aid Society—and the list goes on—teaming up on environmental justice and health issues that differentially affect poor people and people of color.

Environmental racism and environmental justice panels have become "hot" topics at conferences sponsored by law schools, bar associations, public health groups, scientific societies, social science meetings, and even government workshops. Environmental justice leaders have also had a profound impact on public policy, industry practices, national conferences, private foundation funding, and academic research. Environmental justice courses and curricula can be found at nearly every university in the country.

Half a dozen environmental justice centers and legal clinics have sprung up across the nation—four of these centers are located at historically black colleges and universities: Environmental Justice Resource Center (Clark Atlanta University–Atlanta), Deep South Center on Environmental Justice (Xavier University of Louisiana–New Orleans), Thurgood Marshall Environmental Justice Legal Clinic (Texas Southern University–Houston), and Environmental Justice and Equity Institute (Florida A&M University–Tallahassee).

The environmental justice movement has continued to make its mark in the twenty-first century. From New York to California, environmental justice activists are challenging environmental racism, international toxics trade, economic blackmail, corporate welfare, and human rights violations. Groups are demanding a clean, safe, just, healthy, and sustainable environment for all. They see this as not only the right thing to do but also the moral and just path to ensure our survival.

CHAPTER 10

The Gunfighters of Northwood Manor

How History Debunks Myths of the Environmental Justice Movement

ELIZABETH D. BLUM

> Women's history scholarship does not simply add women to the pictures we already have of the past, like painting additional figures into the spaces of an already completed canvas. It requires repainting the earlier pictures, because some of what was previously on the canvas was inaccurate and more of it misleading.
> —Linda Gordon, "U.S. Women's History"

On October 26, 1979, residents of Northwood Manor, a primarily African American neighborhood in northeast Houston, filed a lawsuit that alleged civil rights violations in the siting of a nearby landfill.[1] The first legal attack of its kind, the case proved valuable in establishing awareness of the environmental disparities faced by minorities several years before Benjamin Chavis Jr., then executive director of the United Church of Christ Commission for Racial Justice, coined the term "environmental racism."[2] African American women played a pivotal role in this suit, from both the legal and grassroots sides, defining issues and solving problems.

In the almost thirty years since the *Bean v. Southwestern Waste Management* case, the environmental justice movement has become a potent and visible, if not always successful, force. Myriad groups exist to spread knowledge and expertise about the issues, communities have sprung up in opposition to various inequities in the distribution of waste, and President Bill Clinton issued an executive order requiring federal agencies to consider environmental justice when making decisions. This national-level attention developed as the result of efforts by several very vocal leaders, including Robert Bullard, director of the Environmental Justice Resource Center. Bullard and others, beginning in the early 1980s, emphasized the originality and even revolutionary character of the "new" environmental racism movement: leaders asserted that their environmental activism stressed urban, human concerns, and included minority populations, which other environmental activism lacked.

This definition and the myths surrounding this national movement greatly marginalize or ignore the substantial level of activism by African American women over the course of the twentieth century. A more detailed examination of African American women's activism across the twentieth century fells these environmental justice myths. The examples of women's activism behind the *Bean* case, as well as additional case studies of activism by the National Association of Colored Women (NACW) in the 1910s and 1950s, serve as valuable reminders that, as Linda Gordon noted, women's stories alter historical pictures completely.

NACW Environmental Justice Activism during the Progressive Era

During the Progressive Era, reformers began copious numbers of crusades against the growing problems of industrialization and urbanization. Both African American and white women played important roles in these campaigns, although in an era of pervasive racism, the two rarely worked together. For African American women, the NACW became a potent and empowering vehicle for reform.[3]

Beginning with its formation in 1895, the NACW actively campaigned for many of the same urban environmental issues as white female Progressive reformers.[4] The NACW operated as an umbrella organization for local African American women's groups across the country, under the nationwide motto of "lifting as we climb." Although educated, highly literate, middle-class African American women formed the leadership corps at the national and local levels, geographic dispersal of NACW clubs was far less limited. Reports in the national newsletter flowed in from clubs in Alabama, South Carolina, Tennessee, Georgia, New York, Rhode Island, Indiana, and California.[5]

The NACW strongly promoted an agenda of remolding stereotypes through activism. In one of its first actions in 1895, the organization publicly decried a letter from John Jacks, president of the Missouri Press Association. Since "the man has not only slandered women of negro [sic] extraction but the mothers of American morality, on a question that not only involves the good repute of the present generation, but generations to come," the NACW passed a resolution condemning his views.[6] In addition, Mary Church Terrell, in 1901, wrote furiously in reaction to "The American Negro," penned by Hannibal Thomas. "In order to prove the utter worthlessness and total depravity of colored girls," Terrell wrote, "[the author states that] under the best educational influences they are not susceptible to improvement."[7] Terrell then reiterated the group members' commitment to remolding negative stereotypical images through their activism. The NACW saw all of its projects, including urban environmental activities, as directly tied to the fight against racism.

In 1912, as a part of its activism, the NACW sponsored a "Swat That Fly" campaign. African American women, through reports of doctors, had become convinced of the danger many insects posed in spreading disease. At the national level, Dr. Mary F. Waring, national chairwoman of the Health and Hygiene Department, contributed numerous articles on the dangers of flies, mosquitoes, and fleas in spreading disease and increasing the death rate among African Americans. The year after the first "Swat That Fly" campaign, Waring advocated that women start with the "clean up" of individual homes. Waring listed various methods to reduce the number of flies, specifically with regard to garbage. She urged women to "provide covers for the garbage can; all waste, decayed food should be destroyed. When you know that the female fly lays about 150 eggs at a time, even though her life is short, see what a multiplicity of flies will be produced if conditions are favorable. Therefore, clean up; clean your basement, your yard, your gutter, and your garbage can."[8] By 1917, Waring urged group projects on the women, rather than simply engaging in individual efforts. She recommended that women's clubs to "go into the backways and the tenement houses, into the alleys and basements, into the places unfrequented by garbage wagons and street cleaners and help teach and preach 'cleanliness.'" Group activity, she reminded her readers, should be supplemented by good individual habits and cooperation. She exhorted members to "clean your backyards, and hoist your screened windows at night . . . [and organize boys clubs to] pick up all the waste paper, fold it and sell it; you will spade up waste places and plant seed." By setting this example, Waring believed, entire neighborhoods would follow suit.[9]

To impress her audience with her seriousness and their responsibilities, Waring justified her campaign for clean-up days with maternalistic reasons. "Flies cause so many of the intestinal diseases that carry off so many young children during the summer. If we *clean up* more there will be less white coffins to buy. The work of prevention of disease is the work of everyone. The care of children is properly the work of club women."[10] Other NACW members concurred with Waring on the responsibility of women outside the home. This maternalistic rhetoric by African American women paralleled in many ways language and logic used by white women dealing with public health issues, especially municipal housekeeping, at the same time. One NACW article stated that because of the health responsibilities of women in their homes, "it is necessary then that housekeeping be extended to encompass the city. The health and welfare of the public need such vigilance as only one of domestic tending and training can give."[11]

Not all NACW members agreed with this justification, indicating a lack of unity among the African American community even within the organization. One woman quoted extensively an article by Lizzie Holmes, which argued that women should spend no more time on childrearing than men. "Let the woman live for herself, not for unborn children," Holmes stated. She continued: "Woman has been considered too much as woman, and not enough as a human being. . . . To arrive at her best, she simply needs consideration as a fellow member of society."[12] Some women of the NACW decried the emphasis on activism as solely a function of women's roles as wives and mothers.

NACW Environmental Justice Activism during the 1950s

During the 1950s, the NACW continued with its Progressive Era philosophy of "lifting as we climb," hoping to present a positive image of African Americans to improve the conditions and position of African Americans in general. The NACW women also continued their environmental activism, often engaging in tactics similar to those used during the Progressive Era. In 1955, for example, in the midst of the Montgomery bus boycott and the eruption of the national civil rights movement, NACW women launched a nationwide campaign to improve African American neighborhoods. Financed in conjunction with the Sears-Roebuck Foundation, the Community Project Contest involved a competition between clubs in the NACW's five regions for the best improvement of an urban neighborhood.[13] In justifying the organization's involvement in the project, President Irene Gaines stated first and foremost the NACW's concerns "about the problems of children."[14] Her justifications included worries about "the general health and well being of underprivileged citizens caused very often by poor living standards involving bad housing, ill health, malnutrition, lacks [sic] in education, employment and recreation."[15] In prompting the local clubs to enter the contest, Gaines wrote that "we see all about us the ugly spread of housing decay and erosion. Houses are weather beaten, dilapidated and decadent. Clusters of such houses are breeding grounds for crime, delinquency and demoralized lives." In addition to this moral reasoning, Gaines again emphasized a maternal justification for neighborhood improvement. "The greatest prize of all in this contest—a prize that every participating Club can win—is the better neighborhood you can develop for yourselves and your children."[16] Gaines saw environmental work as one way not only to improve children's lives but also the image and status of African Americans across the country.

Project sponsor Sears-Roebuck intended the first year of the contest to be a pilot program for the African American women. In each of the NACW's five regions, the organization chose a club to represent efforts in the contest. The five groups chosen included one from Washington, D.C., from the northeast region; California from the southwest region; Alabama from the southeast region; Washington State from the northwest region; and Illinois from the central region. If successful, Sears-Roebuck indicated, the company would "extend it [the project] to include the participation of our club women in every state" where the organization existed.[17]

A brochure sent to the various regional offices described the prizes, examples of various types of projects that could be undertaken, and judging criteria for the project. The brochure mentioned numerous examples of possible projects, each of which had environmental overtones, pushing for improved local areas through clean-up crusades.[18] One of the suggested projects was the renovation of the outside or inside of homes. Local women's clubs could form committees to organize the residents to plan for "major and minor repairs; for replacement of windows, doors, porch railings, roofs, floors, ceilings and all outside renovations." They could also "organize and work with the housewives of selected areas in interior paint-up, clean-up and home decorations." Groups could hold neighborhood clinics on how to complete household renovations or improvements, on zoning requirements, or on the types of plants to use in landscaping projects. Backyards could be cleaned "on a block basis to remove trash, and accumulations of discarded furniture, ashes, and old paper." Women might enlist youth groups through anti-litter campaigns, clearing vacant lots for a playground, or providing garbage cans to residents.[19]

After numerous difficulties and obstacles, final judging of the project took place at the group's national headquarters in New York between May 6 and 12, 1957, with the winner announced by Sears-Roebuck representative Harry Osgood at a dinner on June 1, 1957. Each of the winning projects revealed a deep concern among middle- and lower-class African American women with local environmental issues. Tuscaloosa, Alabama, took that state's first prize. Clubwomen there began a process of improving sanitary conditions in their city through clean-up projects and through enlisting additional city services. They removed "unsanitary and disease-breeding hog pens" from their neighborhood, replacing them with "attractive vegetable and flower gardens." Other efforts at the individual level "in home improvement and general community beautification" included improvement of walks and driveways and landscaping and tree-planting projects. The work also involved cooperation with city officials to install sewers, instead of the open drainage ditches still in use, and to make streets more usable.[20]

The Longview, Washington, group won both the Washington State prize and the national prize for the entire contest. Victoria Freeman, president of the Longview Woman's Study Club, initiated, organized, and energized the widespread campaign in her neighborhood. Longview had long suffered from neglected or nonexistent services from the white urban government, including "dusty, muddy streets, trash-filled alleys, overgrown vacant lots and other eyesores and inconveniences." One woman lamented that "it is so dusty here you can't even sit out on your porch in the evening." The women initially began an "intensive Clean Up Campaign" to remove garbage from the neighborhood alleys, but control of the street dust quickly rose to the top of the women's list of needed improvements. Freeman led the Study Club before the city council to petition for "local Improvement District" status for road and sidewalk paving. Mayor H. R. Nichols refused the request for concrete streets because of the low tax value of the homes in the area, but he and the city engineer suggested that a "light bituminous surface (two shot oil) could be applied to the streets for $16 a lot if the City Council would approve such a street." When residents requested the improvements for Seventh, Eighth, and Ninth streets, along with Delaware and Douglas streets, the city applied the price tag of $1,845. Undaunted, the women raised the money through producing and selling turkey and ham dinners. Although most of these fund-raisers were held at private residences, including Freeman's, restaurants also began to order from the women, who eventually served more than 1,000 of the meals. The Longview project spurred other improvements, with the addition of concrete sidewalks and funding for the club to begin work on "installation of drainage facilities" in the area.[21]

Although Sears-Roebuck failed to renew the Community Project Contest for any subsequent year, the NACW proudly touted the accomplishments of the clubwomen in each of the five regions. National publications especially recognized the Longview club for its contributions. The 1959 national brochure presented Longview as an example of what only a few women could do with drive and determination. "A single club of 10 women in Longview, Washington[,] changed the complexion of a whole town, winning prizes totaling $6,500.00."[22] Each of these examples from the NACW demonstrates that urban environmental reforms remained a vital component of African American women's activism from the early twentieth century through the 1950s.

African American Women's Activism during the *Bean* Case

African American women's activism certainly continued in the latter half of the twentieth century and outside the organizational boundaries of the NACW. In 1979, a local campaign to stop the construction of a landfill in Houston demonstrates many of the key themes seen in earlier twentieth-century activism by African American women. Although their presence is often neglected in traditional accounts of the case, the neighborhood women used the environmental struggle to continue to battle long-embedded racism inherent in American society.

By the 1980s, Houston emerged, along with many other southern cities, with intensely segregated residential areas. Most African Americans lived in the northeast and southwest areas of the nation's then fifth-largest city. This trend had been visible for some time. In 1950, sociologist Henry Bullock noted that "Houston's Negro population is very tightly concentrated in a few areas. Although the population has responded to the suburban movement like all other urban populations, tradition has prevented basic changes in the geography of Negro settlement." This concentration existed mostly in two areas during the 1950s, the Fifth Ward (northeast) and the Third Ward (south-central). As the white population expanded geographically over the next thirty years, so did Houston's African American population, but the overall pattern of segregation persisted.[23]

Segregation manifested itself not only in relegating minorities to certain areas within the city but also in city services being allocated preferentially to wealthier, usually white, neighborhoods. The all–African American community of Riceville in southwest Houston demonstrates the neglect paid to some African American areas. Robert Bullard noted that "although the annexation by the city occurred in the sixties, Riceville still did not have many city services as late as 1982. Public water facilities did not serve the Riceville area, city sanitary sewers were not provided, and storm drainage and runoff water flowed alongside roadside ditches to open drainage ditches. Even now [1987], the streets in the neighborhood are gravel-topped roads riddled with potholes."[24]

In the late 1970s, events unfolded in Houston that revealed some of the environmental dilemmas associated with a racially segregated city. During the summer of 1977, attorney Paul Philbin attempted to benefit two of his clients with mutual interests. He suggested to William Walker that a site existed that might be appropriate for a landfill in northeast Houston at the intersection of East Houston and Dyersdale roads.[25] It was also close to railroad tracks and a major thoroughfare, Little York Road, which led directly to U.S. Highway 59. Significantly, Smiley High School, its athletic facility, and its administrative offices, as well as the residential neighborhood of Northwood Manor, also sat near the proposed location.[26] In October 1977, to take advantage of the property available, Walker organized Southwestern Waste Management Corporation to purchase the land.[27] Soon after, when his health began to fail, Walker negotiated with Browning-Ferris Industries (BFI) to manage the project.[28] With the land purchased, initial agreements in place, and state permits issued, BFI began construction at the site in March 1979.[29]

In accordance with state laws, after conducting numerous geologic and engineering tests on the property, as well as contacting a landscape architect, Walker applied for a Type I landfill permit with the Texas Department of Health (TDH) in February 1978. (Type I landfills receive only municipal solid waste, not hazardous, or "Class 1"–type, industrial waste.) After noticing the public hearing in various local newspapers and the *Houston Chronicle*, TDH hearings examiner John Richards II set the permit hearing date for April 18,

1978. Because of "intense" public interest, the TDH moved the hearings closer to the neighborhood. After four days of testimony and despite community opposition, Richards recommended the permit be approved. The commissioner of health, upon Richards's evaluation, issued permit number 1193 to Southwestern Waste on July 7, 1978. Since no party filed a motion for rehearing within fifteen days after the commissioner issued the permit, it became final on July 22, 1978. This chain of events set in motion years of contentious litigation and citizen upheaval over the site, as African Americans mobilized against what they perceived as undue environmental burdens against the African American neighborhood.

Many residents initially believed the construction company planned a shopping center in the neighborhood.[30] Once they realized that a landfill, and not a shopping center, was under construction, however, they immediately mobilized for opposition. Later testimony revealed that "when plaintiffs learned that a land fill [sic] site was under construction they organized, circulated petitions signed by hundreds of persons and raised funds to oppose this construction."[31] Jearaldine Adkins, a member of a local civic group called the Association of Community Organizations for Reform Now (ACORN), initiated the community's involvement in the fall of 1979. She began handing out leaflets, going door to door, and speaking to local organizations to inform them that the nearby construction effort actually involved a landfill. Through these informational efforts, Adkins activated numerous area networks, including the local civic club and local churches, which spread the word throughout the community.[32]

After being contacted by Adkins, Margaret Bean (now Margaret Lair) immediately began to distribute leaflets and go door to door as well. She noted that "a lot of people still didn't know about this dump, even after this thing was getting started."[33] Others aided in these preliminary efforts. Patricia Reaux "started calling and walking the streets, telling as many neighbors as I could that we were getting a dump in our area and then we started getting petitions together to see what we could do."[34] Louise Black also assisted with the telephone network, calling people to relay information about the landfill status. She "started knocking on the doors and letting everyone know it was going to be a dump and getting them involved and telling them to come on, help us, see what we could do."[35]

Through these early efforts, and later organized involvement, the triumvirate of Bean, Black, and Reaux became known across the area as the "Gunfighters." As Reaux explains it, the name stuck "because they knew that if we were going to fight, we would fight to the end." The three initiated a tag-team approach to their door-to-door and media activities. "We would keep our notes and make a point," Reaux remembered. "And if Margaret hit them with some, and I didn't think that she hit a good enough job on it, then I was going to come back and get them with it. Or if Louise thought that I missed something, she was going to get it."[36] Becoming close friends through the activism, the Gunfighters formed the leadership core of activities against the landfill.

The refusal to accept the landfill emerged from various reasons, including maternalism. Northwood Manor women also maintained a well-developed commitment to the aesthetics of their neighborhood, participating in frequent clean-up days through the local civic club well before the establishment of the landfill. Organized on an as-needed basis, Reaux and others targeted a certain location and mobilized volunteers to assist in picking up trash and debris. Reaux also provided refreshments for the task, bringing sodas and food for the workers, who often began their clean-ups in the early morning on a weekend. Accompanied by as many as a hundred residents, Northeast Manor clean-ups removed garbage along with copious amounts of old tires and abandoned junk. They coordinated their efforts with the city, which provided large trash receptacles for the crews. Some residents, of course, were more cooperative than others in the efforts. Reaux remembered some neighbors sitting on their porches watching as the clean-up crews emptied their yards of unwanted junk. In addition to the garbage pickup, residents also participated in house painting and other improvement efforts. Reaux believed the appearance of the entrance to the neighborhood was especially important, and sometimes mowed the grass there herself.[37] Through these clean-up activities, which many African American women engaged in throughout the twentieth century, the women of Northwood Manor demonstrated a firm commitment to the aesthetics of their neighborhood and a willingness to become personally involved in improving and maintaining its appearance.

The landfill's proximity to the local high school exacerbated many of the women's concerns, causing them to worry over the health and safety of their children. Margaret Bean articulated the concerns of many when she stated that the neighborhood "didn't have sidewalks out here. I was concerned because the kids were walking home and . . . these big old Browning-Ferris trucks [would be] going up and down the street. I thought some of these kids might get hit by these trucks."[38] Mildred Douglas remembered her primary concern as "just overall safety of the children and the family environment, because over a long period of time, I knew that that would cause some type of health problem."[39] The Northwood Manor women's initial concerns and justification for activism centered around the health and welfare of their families. Although maternalism became more muted after the development of the legal case, at least publicly, women continued to use it as justification for their actions.

Many of the women included strong elements of racism and classism in their justification for opposing the landfill. Specifically, the residents believed that the landfill had been sited in their area because the neighborhood was predominantly African American. Patricia Reaux, for example, stated succinctly that she thought she "had been discriminated against as a resident of the Northwood Manor area."[40] Dorothy Grimes included both racism and maternalism in her reasoning, stating that "I felt as though I was being discriminated against [in the siting of the landfill] and that it was endangering my children's life, as far as healthwise . . . they are being unjust to me and my family."[41] Grimes added, "I had an obligation to my children to defend them, not to go through the turmoil of the dump."[42] This belief began soon after the discovery of the nature of the site, stemming from research initiated by the residents themselves.[43]

Their belief in racist motivations for the landfill siting also stemmed from direct knowledge of an earlier attempt in 1970 at siting a landfill in the neighborhood. In 1970, the population had been approximately 30 percent African American and 70 percent white; by 1980, huge demographic shifts had occurred, changing the composition of the neighborhood to approximately 83 percent African American and 18 percent white.[44] Grimes stated that she believed the siting involved racial discrimination because "several years before then [the siting of the landfill] when the area was predominantly white, it was mentioned about a landfill being put out there. . . . I felt as though since it was predominantly African American now they think it would be easier to push it."[45] Reaux also remembered the earlier attempt to place a landfill near the neighborhood. She echoed Grimes's statements, saying, "When the area was predominantly white, we did not get it. Now that I am a homeowner I find I have a landfill in the very same area where it was once denied."[46] Reaux's opposition to the dump hinged on her experience living near a dump as a child and led her to connect racial discrimination with the siting. She stated, "The reason why I am opposed to this particular landfill is because I feel that the reason for it being placed in the center of my current community is because I'm African American, because our neighbors have always been victimized by some unpleasant something and the dump seems to be something that's either following me or I'm following it. But I feel that it was placed there because of the fact that it is a black community and it's just another form of downgrading our communities."[47]

With the area networks spreading the word of the almost-completed landfill, in late 1979 the local civic clubs formed the Northeast Community Action Group (NECAG) as an umbrella organization to fight the landfill.[48] In addition to continuing the initial work of notifying the area's residents about the proposed landfill and spreading the word through the local churches, the NECAG also called various city agencies, which advised them to initiate legal proceedings.[49] Armed with this advice, the group hired Linda McKeever Bullard, a young African American attorney in the area.

McKeever Bullard capitalized on the residents' early research and feelings of discrimination, enlisting their assistance in research efforts. Reaux remembered that McKeever Bullard "made us do a lot of research. Because she couldn't do it all, nor did we have the money. So then we had to go to the city planning department, and start looking up to see where all the landfills were. We went to the library and did some research on existing versus when some had closed. You know, there were times when all of us had an assignment to do."[50]

This research reinforced their beliefs that racism had played a part in the initial siting decision. As they examined maps of landfill locations, the residents noted "every one of them were [*sic*] in an African American area. So, that clearly defined racist. You know, if you could find one in a predominantly white area, then you could say, okay, that rule didn't fit. But it fit. Every one."[51]

Residents began to engage in protest-type activities in addition to legal actions. One day the group organized a protest at the landfill site.[52] After meeting at Louise Black's home, core members of the NECAG began a carpool to collect residents. At the appointed time, about 100 people gathered to walk from the neighborhood to BFI's landfill entrance.[53] Carrying signs of protest and singing hymns, the demonstrators refused to allow any of the dump trucks to enter or leave the facility, forming a human chain across the entrance roadway. The group quickly decided that the men would be the ones arrested if need be, so that the women of each family would be available to post bail. As a truck finally approached the landfill, Judge Al Green, a prominent African American activist and precinct judge, took a stand in the vehicle's path, backed by the rest of the residents. Seeing its trucks unable to pass, BFI called the police to disperse the residents. The officers, known by the residents, attempted to resolve the issue peacefully, without arrests. By the end of the day, the residents eventually dispersed. Police made no arrests, but the protest successfully prevented BFI's business for a portion of the day.[54]

Hiring a lawyer, of course, also entailed paying the lawyer. A significant portion of the NECAG's activities involved fund-raising to pay for McKeever Bullard's legal efforts. The Gunfighters controlled the moneymaking efforts, as well as the purse strings of the organization. Charles Bean, vice president of the organization, noted that "Charles Stradit [NECAG's president] and myself, we didn't pay much attention to the money, because we had great confidence in the women. There was no need to watch none of those women. We knew that they were doing right."[55] The women spread fund-raising over a wide array of activities, including collecting money on street corners from cars every weekend; holding car washes, garage sales, bake sales, fish fries, and banquets; and selling dinners. Overall, Reaux and Bean estimated that they raised over $60,000 from the community, which went directly to McKeever Bullard.[56]

Because the residents had a late start on the process, McKeever Bullard rushed to file a suit before the scheduled opening of the facility on November 1, 1979.[57] She initially requested a temporary injunction to stop the facility from opening, to be followed by a permanent injunction later. From the outset, McKeever Bullard, led by the residents, framed her argument in terms of discrimination against the African American neighborhood. She specifically brought action under the Fourteenth Amendment to the Constitution, "which prohibits the deprivation of property without due process of law or the denial to any person equal protection of the laws; and the laws of the United States"; as well as the sections in the U.S. Code forbidding discrimination on the basis of race.[58] She set up her argument, which would be refined over the rest of the case, in the original complaint:

> The decision . . . to grant Southwestern's permit to operate the aforementioned solid waste facility in the geographical area in question is based on considerations of race, class and socio-economic factors, the effect of which are [*sic*] to unconstitutionally discriminate against plaintiffs and the class they represent. The geographical area in which plaintiffs reside is already presently unduly burdened with solid waste facility [*sic*] in this area will have the effect of permanently depriving plaintiffs and the class they represent of their property by accelerating the depreciation of their property.[59]

McKeever Bullard's other elements of the case included that the facility would become a nuisance to the neighborhood and would not be operated in accordance with regulations. She never incorporated environmental claims in the case to any specific degree, however, choosing instead to focus almost exclusively on the claims of racial discrimination. Perhaps this stemmed in part from the timing of the suit, brought only a week before the opening of the facility, when she thought residents could not yet claim environmental harm. In addition, claiming environmental harm may have also damaged the legal status of the case. If environmental issues became the focus of the case, the U.S. District Court may have returned McKeever Bullard directly to the TDH for remedies. Instead, U.S. District Court Judge Gabrielle McDonald found that she had jurisdiction, as the TDH had no remedies or methods of hearing cases based on claims of racial discrimination. Judge McDonald noted that the "Solid Waste Disposal Act permits revocation of a permit on rehearing only for 'reasons pertaining to public health, air or water pollution, land use, or violation of this Act or any other applicable laws or regulations controlling this disposal of solid wastes,'" but that was not what the plaintiffs had claimed, and it would be "useless to refer them back to TDH."[60]

McKeever Bullard realized she could prove her case of racism in the siting through statistics indicating that, in Houston, African American neighborhoods disproportionately held the burden of landfills.[61] To meet this requirement, she decided to hire her husband, Robert Bullard, as an expert. Bullard, a sociologist who completed his Ph.D. only three years earlier at Iowa State University, began to gather data to assist his wife's case. At the time his wife filed the case, Bullard served on various urban advisory committees on housing and crime issues. He had also accepted a visiting professorship at Rice University for the spring semester of 1980. Texas Southern University later hired him to a permanent position as associate professor of sociology in 1980. In his role as director of research at the university's Urban Resources Center, Bullard had four publications through Texas Southern University Press and numerous presentations and lectures to his credit, mainly on housing issues relating to African Americans in Houston.[62]

Little of the eleven-day November hearing on the injunction or the case itself and subsequent legal proceedings moved favorably for the plaintiffs.[63] Mc-Keever Bullard had few resources to compete with a well-heeled corporation and its capable defense attorney, Sim Lake of the prestigious law firm of Fulbright Jaworski. At the preliminary injunction hearing, presided over by Judge McDonald, Bullard presented his evidence to prove a historic tendency to discriminate by race with environmental hazards in Houston. Defense lawyers, however, successfully poked large holes in his methodology and statistics, often providing evidence themselves to refute Bullard's findings directly. Expressing doubts about Bullard's evidence, Judge McDonald refused to grant the injunction to the plaintiffs, since they had, in her opinion, failed to establish the intent to discriminate.[64]

While the preparation for the case proceeded, the NECAG began to encounter internal problems of its own. In the fall of 1980, hoping to defuse the local opposition, BFI hired a consulting firm, Image Transition, to help promote the landfill in the neighborhood. Charles Moore, head of the company, decided to work surreptitiously within the community group to defeat opposition. In September 1980, as a part of its ongoing fund-raising, the NECAG organized a banquet. Brought in by trusted member Judge Green, Moore began to attend meetings shortly before this, even offering to sell tickets to the banquet. Charles Bean expressed excitement over Moore's involvement since he had been told that Moore "was the one that went to Iran when they had those hostages there."[65] Excited about the offer for help from the prominent African American man, and unaware of his affiliations, the group gave him $500 worth of tickets to sell. They never saw the money or Moore at a meeting again.[66] Residents believed that Moore planted "a bunch of seeds" of doubt about the group's activities.[67]

Moore focused more effort on African American community leaders outside the group. He developed lists of influential African Americans in the area, keeping BFI informed of their position on the landfill and his efforts to influence them. One woman with whom Moore encountered some success was

Emma Horn, an African American woman who had run against Ernest McGowan for his city council seat. Horn, the local school board president, served as president of the Northeast Houston Community Uplift Project and supported clean-up campaigns for her area. Moore had received information that Horn was "very strong, very vocal in just about every event on the northside."[68] After talking with Moore, Horn agreed to support the landfill. She attended several meetings of the NECAG, attempting to persuade members that BFI offered opportunities for increased tax revenues for the area.[69] As a reward, Moore spoke with BFI, mentioning that they should support Horn's efforts for a clean-up project since "this approach is timely and most positive."[70] Moore's and Horn's actions vividly demonstrate the lack of unity in the African American community over environmental issues.

In July 1984, almost five years after the original filing of the case, *Bean v. Southwestern Waste Management* began before Judge John Singleton, who replaced McDonald on the bench. Bullard, with more time to prepare and gather evidence, adjusted his strategy to deliver more powerful statements during the case. The additional effort failed to impress the judge. Less diplomatic than McDonald, Singleton struck out at Bullard's methods and research in harsh language. He noted that Bullard's research "is not reliable" and "gives a distorted view of the facts."[71] Like McDonald, Singleton's opinion again resulted in a loss for the plaintiffs. Still undeterred, McKeever Bullard filed an appeal with the Fifth Circuit. The district court, however, affirmed the lower court's decision in January 1986. In a brief notice, the appeals court documented their reasoning succinctly. They agreed with Singleton that "Dr. Bullard's methodology was subjective, imprecise, and sometimes inaccurate or misleading."[72] Ultimately, the case petered out, with the plaintiffs losing and the landfill opening. It continued to operate into the twenty-first century, well beyond its conceived life span.

Debunking the Myths of the Environmental Justice Movement

The impact of the *Bean* case lies not in its result. The case itself failed, of course. For Robert Bullard, however, and numerous others, *Bean* yielded a new cause to which they energetically applied themselves. Over the next few years, the movement gained in prominence, drawing support from other scholars and activists. During the 1980s, activists in the environmental justice movement developed definitions of their movement in strong opposition to their perceptions of the "mainstream" environmental movement. Environmental justice advocates present a view of the environmental movement as elitist, completely dominated throughout American history by wealthy, educated white men who center their concerns on animals and landscapes rather than on human problems. For example, one activist in Alabama colorfully stated, "if it does not hoot in the night, or swim upstream, environmentalists are not interested."[73] Sociologist Dorceta Taylor described the early environmental movement as dominated by whites and by issues of "natural resource conservation, and wilderness and wildlife preservation." Carl Anthony, Earth Island Institute president, stated that "with its focus on wilderness, the traditional environmental movement on the one hand pretends there were no indigenous people in the North American plains and forests. On the other, it distances itself from the cities, denying that they are part of the environment."[74] Generally, environmental justice advocates see the Big Ten groups as the entire environmental movement.[75] Called the "mainstream" environmental movement by Robert Gottlieb, conservation and wildlife concerns dominated these groups, which used essentially legalistic, lobbying tactics to push their agendas at the national level.[76]

Bullard, among others, led the charge claiming differences from the mainstream movement. Dorceta Taylor concurs with Bullard, going so far as to state that the environmental justice movement "represents a revolution within the history of U.S. environmentalism."[77] According to activists, the "new" movement stresses human and urban-centered concerns, focuses on marginalized populations like minorities and the working class or poor, and values the participation of women within their organizations for the first time.

Martin Melosi, among others, developed a critique of the historical myths espoused by environmental justice advocates. Incorporating much of the literature and background of environmental history, Melosi notes that the environmental movement has historically demonstrated interest in urban, health, and human-centered issues. Although the most well-known Progressive Era environmental activists included John Muir, Gifford Pinchot, and Theodore Roosevelt, "lesser known activists struggled with the blight of pollution, health hazards and the physical degradation of cities." These early urban environmental reformers consisted of professionals, including engineers, doctors, and sanitarians, as well as citizens' groups often dominated by middle-class women. Melosi notes that some urban reform, like "the effort to build more urban parks and to plan for city growth[,] were reminiscent of concerns of the conservationists and preservationists." However, to most of the urban reformers, "fundamentally the environmental problems that mattered the most were those affecting human beings." Melosi's discussion of the early urban-centered environmentalism effectively refutes the environmental justice movement's contention that environmentalism is only concerned with parks and endangered animals.[78]

Building upon the critique by Melosi (and others), the three case studies presented of African American women's activism also refute many of the claims of the environmental justice activists. Comparisons between these case studies reveal a long-standing history of women's involvement as well as complexity and depth to the environmental justice movement.

From the early twentieth century, African American women played an important role in reform efforts directed at environmental problems within the cities. Each of these groups of women worked actively to protect and improve their urban environment. In the years between 1910 and 1920, the NACW directed a "Swat That Fly" campaign to reduce disease; in the 1950s, it moved to a national community improvement project to clean up neighborhoods; and in the late 1970s, the women of Northwood Manor fought against a local landfill, actions that spawned a prominent national movement.

The activists of the modern environmental justice movement also repeatedly claim that the concerns they voice represent new issues within the environmental movement. Bullard states that "despite some exceptions, the national [environmental] groups have failed to sufficiently make the connection between key environmental and social justice issues." Taylor concurs, noting that the environmental justice movement is "more ideologically inclusive than more traditional ecology groups. It integrates both social and ecological concerns much more readily and pays particular attention to questions of distributive justice, community empowerment, and democratic accountability."[79] Judi Bari, an outspoken forest preservation activist in the radical Earth First! group, states that environmentalism failed to reach out to working-class people. The problem, she concludes, is "that the environmental movement tends to separate itself from the general social-justice movement in this country. Environmentalists ghettoize themselves among these privileged people and try to limit themselves to wilderness issues that only privileged people can worry about." She continues, "I think traditional environmentalism leaves no room for workers; they're declared immoral because they work in the factory."[80]

However, again as seen through the earlier examples, African American women consistently incorporated social justice issues, like fighting poverty and racism, into their environmental activism. The NACW and the Gunfighters saw their environmental struggles as linked to the struggle against racism. Both also certainly saw the poverty of African American communities as something directly affecting their struggle. In addition, in all three struggles described above, each group of women saw duty to their children and families as a pivotal justification for their activism. For them, the social justice issue of improved health, especially for children, played a vital role in becoming active and continuing the struggle.

Many environmental justice case studies also present a prevailing, simplistic view of the community struggles. In these stereotypical snapshots, a minority community bands together, united in the fight against the big, monied corporation. Using the media and word-of-mouth, the community reveals the evil corporation as a racist, classist entity. Outcomes in the typical story remain uncertain. Some communities triumph against oppression, while the companies outgun and outlast others. Few accounts reveal the complexity behind environmental justice struggles, including the struggles within communities over these issues.[81]

The cases presented here fail to follow the stereotypical format of clear-cut unity within the African American community, as often portrayed by activists. At Northwood Manor, for example, the Gunfighters faced opposition from local leader Emma Horn, who supported the landfill, and Charles Moore, who worked secretly on behalf of the landfill to assist Southwestern Waste. Approximately fifty years earlier, the NACW bickered over the use of maternalism versus feminism to achieve goals. Through these case studies, a more layered picture of struggles for environmental justice emerges. These complicated stories piece together images of fragmented African American communities, torn apart through racism over the ideas of community benefits and degradation.

Despite claims by modern environmental justice advocates, African American women's urban, social justice–centered activism is hardly new or revolutionary. Certainly, the *Bean* case remains an important precedent that heightened awareness and empowered activists for a long-term struggle. However, in claiming their post-1980 activism was a break with prior tradition, environmental justice advocates marginalize previous efforts by African American women and contribute to their invisibility in the historical record. To conclude, thorough knowledge of the environmental movement and a historical evaluation of African American women's involvement in environmental activism serve as valuable correctives to the historical fallacies presented and maintained in the debate over environmental justice.

CHAPTER 11

"To Combine Many and Varied Forces"

The Hope of Houston's Environmental Activism, 1923–1999

TERESA TOMKINS-WALSH

Most environmental activism begins at the local level. John Muir's Sierra Club formed in 1892 to explore the California wilderness, but he mounted a national campaign to protest construction of the Hetch Hetchy dam in California. Muir's journalistic talent, connections in Washington, D.C., and shared wilderness outings with President Theodore Roosevelt coalesced to into a powerful national promotion for California's wilderness and the nation's natural resources, an effort that added significantly to Sierra Club's membership rolls. Over time, the Sierra Club formed local chapters and campaigned for wilderness protection and other environmental issues across the country. As the oldest environmental organization in the country, the Sierra Club began as a small group devoted to wilderness recreation and developed into a highly visible national organization.

As Samuel Hays has noted, however, most environmental groups remain local, and many disappear without the expanded support and increased visibility that sustain national organizations.[1] One reason groups remain local is the singularity of certain environmental features. Marjorie Stone Douglas brought national attention to the problems of the Florida Everglades. Although the problems besetting the Everglades were symbolic of certain kinds of environmental mismanagement, the Everglades remained a distinctive feature of Florida, even as the area was designated a national park.[2] Both the Sierra Club and the Everglades campaigns demonstrate how personal charisma or powerful symbolism can attract national audiences to local environmental problems.[3]

Houston's bayous and watersheds are its distinctive topographic feature. Riverine and coastal waters both suffered from unplanned and unchecked development that led to an array of environmental problems, including destruction of these natural resources, industrial and nonsource point pollution, and flooding.[4] Environmental activism in Houston and Harris County often centered on these watersheds through direct preservation efforts and flood management projects, improvement plans for recreation and beautification purposes, and efforts to boost Houston's image as an attractive venue for new businesses, family life, and nature recreation. In the early 2000s, local officials proposed that Congress approve a study of Buffalo Bayou as a potential National Heritage Area—to acknowledge its historical, recreational, scenic, and cultural value to the nation—but neither legislative recognition nor environmental campaigns to preserve Houston's bayous have yet excited the national imagination.[5]

This chapter explores the dominant themes, involved individuals, and organizations that permeate the history of environmental activism in Houston. Endeavors to manage environmental concerns in Houston—or environmental activism there—must be understood not so much as a movement but more as the efforts of individuals whose passion, means, and leadership have inspired others for a moment or for a lifetime. The history of Houston's environmental activism is the story of the legacy of personal influence, the story of coalition building and rebuilding, and the story of a small core community of individuals who worked to achieve environmental and policy improvements without effecting structural change. Many of these environmental activists were women. Most of those involved were private citizens, but some were public officials who initiated new policies and collaborated with the environmental activists. Environmental activism, both wilderness and urban, evolved from the 1920s forward, as activists broadened their scope of interests in the 1950s and 1960s to address citizens' rights issues and extended their reach in the 1990s to embrace sustainable development programs.[6]

By the 1920s, urbanization in Houston brought the common kinds of environmental problems arising in growing cities across the country: degraded air, noise, traffic, loss of open space, and broad-spectrum pollution.[7] Houston's problems in the 1920s included a city that had grown too fast from the well-organized grid plotted on Buffalo Bayou to a patchwork of streets with little evidence of design.[8] Increased automobile traffic required wider streets, and drives were built along the bayous. The 1920s brought construction of multistoried buildings; the downtown area increased by the addition of at least thirty buildings constructed by Jesse Jones alone.[9] As steel and concrete stirred pride and wonder in Houstonians, many yearned for the quiet of nature, a desire satisfied by the development of bucolic suburban retreats and wilderness recreation outside the city.

A famous Texan, Will Hogg, advocated comprehensive planning for all of Houston, but he developed the River Oaks suburb as a model of private planning.[10] Hogg selected 1,100 acres of high elevation, luxuriant foliage, and natural drainage contiguous with Memorial Park on one side, Buffalo Bayou on the other, and the River Oaks Country Club on the third.[11] When Country Club Estates Corporation finished the first house in 1925, Hogg promoted environmental protection by mounting a marketing campaign that promised enduring natural amenities. At the entrance to River Oaks, large columns "physically and psychologically remove[d] the entrant from the blemishes of urban life." Carefully managed street traffic and underground utilities preserved the ambience of urban wilds, and advertising touted River Oaks as a wildlife sanctuary, protected from the smell and noise of excessive automobile traffic. The homeowners association paid gardeners "to help owners maintain the commitment to natural beauty." Promises to prospective homebuyers included environmental beauty and security. Because the cost of homes was beyond the reach of average buyers, River Oaks acquired a certain exclusiveness, but deed restrictions also ensured that neither Jewish nor nonwhite homebuyers would reside in River Oaks.[12]

Salubrious private space was a priority for developers and elite Houstonians, but public officials and philanthropists looked to Houston's public spaces to create small islands of respite from city life for those with fewer means. From its beginnings, Houston lacked the appropriate ratio of park space to population.[13] In 1898, Sam Brashear campaigned for mayor with a plan for improved infrastructure mainly for city services but also for a "city-owned system of parks and greenspaces." He won against incumbent H. Baldwin Rice, who supported minimal government. Brashear moved among the "commercial elite" but viewed government as a means to social reform. Having enjoyed recreation in private parks himself, Brashear wanted to offer that experience to all the citizens of Houston. The new mayor paid $23,000 for a sixteen-acre site located just south of Buffalo Bayou at the northern edge of downtown. The city developed the space into a park that opened in 1899, with a ceremony for 1,000 attendees who listened to Brashear announce that he intended to provide a park in every ward of Houston. Brashear ran for reelection in 1900 and won against a challenger who attacked his Progressive government, but he resigned on January 21, 1901, presumably because he resented the criticism.[14]

Mayor H. Baldwin Rice created the first Board of Park Commissioners in 1910, with one lawyer and two real estate investors.[15] In 1912, the commissioners retained landscape architect Arthur Comey to develop a plan for a park system. Comey suggested a major park across from the campus of Rice Institute, noted in the plan as Pines Park.[16] In 1914, George Hermann, one of the first park commissioners, willed 285 acres to the city for Hermann Park on the southern edge of downtown in the area where Comey had proposed Pines Park. Hermann, a prominent Houstonian who built a fortune from cattle and real estate businesses, donated land for a number of Houston landmarks, including the Museum of Fine Arts. Hermann Park opened in 1920, becoming the site for the Houston Zoo and the city's first racially integrated public golf course in 1922.[17] In 1923, the Women's Club of Houston celebrated the passing of a woman's right to vote by offering the city a decorative commemorative statue. They raised $75,000 to commission the statue of Sam Houston, which still graces the entrance to the park.[18]

If Hermann Park is emblematic of Houston's public amenities inside the 610 Loop, Memorial Park symbolizes upscale suburban living, although it also rests just inside the West Loop. In 1923, Camp Logan Hospital, the last vestige of the army camp, closed, and *Houston Chronicle* columnist Iona B. Benda memorialized the camp's significance.[19] During the war, soldiers and their wives found housing with Catherine Emmott, whose two sons also served. Emmott wrote a letter to Benda, suggesting that the camp was "sacred ground" and should become a memorial.[20] Emmott asked the Houston community to submit ideas for the camp's use. Benda called for a public meeting, which resulted in Emmott's appointment as committee chair, and Emmott began a campaign of persuasion, talking, meeting, and impressing anyone who would listen.[21] The Hogg brothers listened.[22] Planning to expand their upscale River Oaks subdivision, Mike and Will Hogg possessed 875 acres, which they sold at cost to the city in 1924. Afterward, they purchased another parcel of acreage and sold it to the city as well. In May 1924, the city took title to the Camp Logan acreage and established Memorial Park.[23] Opposition to the park included protests from those who believed it was too large a space to remain undeveloped.[24]

From the beginning, the plan was to keep Memorial Park a wild sanctuary for plant and animal species. Over time, a golf course was added, as were hiking trails and picnic areas, but to a large extent and with great vigilance, the original plan has been honored. Ima Hogg monitored the park for fifty years to prevent inappropriate development.[25] When Hogg became too frail, she assigned oversight to a committee of three: Terry Hershey, Sadie Gwin Blackburn, and Frank Smith.[26] Their informal watchdog role eventually coalesced into the Memorial Park Conservancy, which evaluates proposals for park activities and advises the city concerning park issues.

All three of these early public parks were significant in Houston's history, and the stories of their acquisition by the city represent the avenues by which Houston has achieved urban improvements: foresight of a remarkable public official, personal philanthropy, and a community coalition inspired by a dedicated individual. While the establishment of city parks was essential to improved quality of life in Houston, another impulse toward environmental activism arose among those who enjoyed leaving the city to experience wilderness.

Joseph Heiser was Houston's first wilderness activist and organizer.[27] Labeled "Houston's John Muir," Heiser was born in the Bayou City. After serving in World War I, he returned in 1920 to begin a forty-year career as an accountant for Texaco.[28] Heiser's father was Houston's first superintendent of parks under Mayor Sam Brashear.[29] To start the Outdoor Nature Club (ONC), Heiser ran an open invitation in the *Houston Chronicle* for those readers who wanted to strengthen bonds among nature lovers, to study local flora and fauna, to cooperate with comparable regional and national organizations as well as with sportsmen's clubs to create support for meaningful legislation, to work with local initiatives aimed at civic improvement through beautification, and finally to broadcast the natural assets of Houston.[30]

Key to Heiser's early mission was the "enjoyment" of outdoor life, but club members shared social and recreational activities along with an active scientific curiosity. Birds and their habitats interested the members of the organization, but they also formed study groups for botany, photography, conchology, and other pursuits. A serious conservationist and preservationist, Heiser brought national attention to his campaign to preserve holly trees by planting, rather than cutting, during the Christmas ritual. He successfully promoted the mockingbird as the Texas state bird, and he, along with the ONC, campaigned to protect the roseate spoonbill.[31] To promote the organization's education mission, Heiser presented a nature program at Will Hogg's Forum of Civics. In addition, he produced a newsletter from 1924 to 1927 and served on the conservation committee.

Urban activism addressing beautification issues arose at the same time.[32] A small group of women organized the Garden Club of Houston in 1924 to promote "horticulture, conservation, and civic beautification."[33] Within a year, membership had increased to forty women, and activities included a flower show and participation in Heiser's campaign to save holly and yaupon trees. The club worked to preserve wildflowers and undertook maintenance of the grounds of the art museum under the direction of a landscape architect. Private gardens were displayed to the public to raise funds for club activities. Like members of the ONC, Garden Club members studied the scientific basis of their avocation, such as soil conditions and horticulture, and they shared that information with newcomers and the community.[34]

Will Hogg, who sold the city the valuable land for Memorial Park, was the eldest son of a former Texas governor, graduated law school, inherited his father's wealth, and immersed himself in Houston's civic affairs. In 1927, Mayor Oscar Holcombe appointed Hogg to lead the City Planning Commission.

As planning commissioner, Hogg promoted economic growth and aesthetic improvements, insisting that they were connected. Although the commission vigorously addressed Houston's problems, Hogg soon realized that the commission was powerless and penniless and unable to implement its recommendations. To circumvent the city government's lack of support and attention, Hogg used his own wealth to achieve major objectives.[35]

Both the civic arts movement and Progressive Era ideology influenced Will Hogg.[36] While garden clubs preoccupied themselves with horticultural inquiry and dressing private and public gardens, Hogg promoted comprehensive city beautification and city planning by distributing trees and plants.[37] In fact, he believed that beautification was a means to convey to both government officials and citizens the benefits of city planning. Hogg wanted to make the city pleasant and useful through beautification and scientific planning and believed that citizens should care for their own homes and streets as well as their city and the region. Although Hogg was a reformer, he was also a businessman; he supported gradual change within the system and always supported Houston's continued economic growth and business ethic.[38]

When Will Hogg died in 1930, his passing deprived the city of a strong voice in favor of city planning and his coalition building. Of particular relevance to future urban environmental activism was Hogg's Forum of Civics, established to address civic issues and to coordinate the activities of an array of environmental, civic, and beautification groups to effect more efficiency in Houston's urban management.[39] Will Hogg's Forum of Civics did not survive him, but it was a prototype for community coalitions that would emerge again in the 1970s and the late 1990s.[40]

Interwar environmental activism is not as well documented as earlier efforts or the activism that emerged after World War II.[41] In Houston, the ONC and the garden clubs continued recreation and beautification efforts during the interwar years. During the Great Depression, ONC members met regularly in members' homes.[42] They organized field trips to area bayous, the San Jacinto River area, the Highland Park Reservoir, and many other local areas. The ONC initiated a letter-writing campaign to support establishment of Everglades National Park during the 1940s. The group affiliated with the Texas Wildlife Federation, and its art committee sponsored an exhibit at the State Convention of Garden Clubs. Also in the 1930s, the ONC began honoring their deceased members by planting magnolia trees along Buffalo Bayou in collaboration with the Junior Chamber of Commerce's tree-planting program. Most significant, during the 1930s the ONC established a wildlife sanctuary on Vingt-et-un Islands in Trinity Bay, where members had discovered a colony of nesting spoonbills, the largest in the United States. Along with the National Audubon Society, ONC members, including Joseph Heiser, proposed to lease the island from the state of Texas and hire a warden.[43] By 1931, the ONC had successfully established the sanctuary, banded the roseate spoonbills, and adopted the spoonbill as their organizational symbol.[44]

Houston's history provides an interesting example of the confluence of environmental and urban history. Environmental historians study nature, landscape, and wilderness recreation and protection. Urban environmental historians explore the essential role of cities in the environment, including the impact of culture on environmental change.[45] Urban environmental historians find in early urban reform efforts the genesis of later environmental activism, while others suggest that the modern environmental activism was a distinct post–World War II movement. Stephen Fox placed the roots of modern environmental activism in the wilderness ethic of John Muir and the conservation agenda of Gifford Pinchot implemented during in the Progressive Era.[46] Fox defined the "radical amateur" as a catalyst for environmental activism across the middle twentieth century and into the post–World War II environmental movement. He argued that the unpaid energies and passions of radical amateurs—white, middle-class, affluent (independently sustained without the need to work for wages)—connected the reformers of the early and mid-twentieth century to activists in the 1960s and 1970s, beginning with efforts that usually began as NIMBY (Not in My Backyard) campaigns among philanthropic elites who zealously worked for community improvements.[47]

Agreeing with continuity across the twentieth century, Robert Gottlieb has argued that there was more than one environmental movement, and that modern environmental movements emerged from nineteenth-century traditions of activism among reformers, professionals, and radicals. Gottlieb defined "alternative" or local groups as critical of the environmental mainstream groups, local rather than national, interested in citizen empowerment, and involved in direct action. He further asserted a "direct lineage" between the public health and municipal housekeeping crusades of the Progressive Era and the environmental justice campaigns of the 1970s and 1980s.[48]

Also advocating continuity in the history of environmental activism, Martin Melosi traced the roots of the modern environmental movement to the Progressive Era, and he identified two groups that were involved in urban environmental reform: professional experts working from technical and scientific principles, and the "civic/aesthetic" group, which included citizen volunteers (largely women) and environmental pressure groups. Relevant to the Houston case is Melosi's observation that early environmental reformers hoped to improve the city but did not advocate alteration of the basic economic structure.[49]

In contrast to historians arguing for continuity, Samuel Hays recognized a significant cleft between pre–World War I conservation and municipal housekeeping and post–World War II activism. Hays argued that early-twentieth-century conservation emerged from the planning of a leadership committed to the powers of science, technology, and government to develop and distribute resources efficiently. The movement's leaders were more interested in efficiency than in the proliferation of the democratic process. Conversely, according to Hays, post–World War II environmental activism emerged from concerns cultivated within the middle class and communicated to a resistant and often hostile leadership. Environmentalism, according to Hays, was consumer oriented, broad based, and a direct product of post–World War II affluence.[50]

Evidence in the Houston case supports both views of the history of environmental activism, because threads of wilderness protection and urban environmental activism were entwined from the beginning, but the environmental activism that emerged after World War II did acquire new dimensions. By the late 1950s, pressing new issues invigorated Houston's environmental activism.[51] Citizens who had enjoyed the wilderness recreational activities with the ONC refocused their attention to issues that required public participation in the legislative process, new methods of citizens' outreach, and protection of public access to natural resources.

In 1959, members of the ONC became aware of a recent Texas Supreme Court decision. From *Luttes et al. v. the State of Texas,* ownership of Texas beaches became an issue for a few alert conservationists in the ONC.[52] The facts of the case were that the Shell Oil Company held an oil and gas lease from J. W. Luttes, who held title to Potrero de Buena Vista, land granted from the Mexican government and confirmed by the Texas legislature in 1852. Contesting this claim was the state of Texas, which argued that the land grant in Laguna Madre along the Texas Gulf Coast was public land and unavailable for oil exploration. Specifically, land primarily under water belonged to the state, and the primarily dry land above belonged by grant to Luttes and his lessee, the Shell Oil Company. At issue was the boundary line, Luttes arguing that the line had shifted during the 100 years since the original grant; hence the "mudflats" originally excluded from Luttes's land were now his private property. Central to the dispute was the procedure for calculating the shoreline. Conflict arose from understandings of Spanish law, *Las Siete Partidas;* Spanish-English translations of the original land grants; and the difference in measurements of mean high water level and mean high tide level.[53] In the 1958 decision, the Texas Supreme Court ruled in favor of private ownership, and "overnight, fences appeared on Texas beaches from Brownsville to the Sabine River, denying access to the public."[54]

Armand Yramategui, an energetic and charismatic ONC member since 1951, realized that public access to Texas beaches was at stake and that the court ruling represented a successful ploy by an oil company to secure oil-drilling rights along the treasured Gulf Coast shoreline. One night in 1959, Yramategui announced the significance of the case at an ONC meeting and proposed that the club fight the ruling.[55] Yramategui's contribution to Houston's environmental community would inspire activists and shape Houston long after his death, but that night in 1959, the ONC, under the leadership of L. A. M. Barnette, a Humble Oil Company engineer, declined to fight the ruling.[56] Ten ONC members, including Yramategui and Sarah and Army Emmott, decided to organize Texas Beaches Unlimited (TBU).[57]

Sarah Emmott was the daughter-in-law of Catherine Emmott, who worked so vigorously to preserve Memorial Park.[58] Born in the Panama Canal Zone where her father worked, Sarah moved to Houston in 1913. She received a degree from Houston Junior College, where she studied scientific research methods and worked for her professors. Sarah married Army Emmott in 1949 and began sharing his interest in the outdoors. They were both members of the ONC, and members of countless other local and national organizations. Eventually, they would become prominent figures in the Houston environmental community and start an environmental library in the attic of Emmott Bookbinders, and they donated money to establish a central library for environmental resources in Texas.[59]

Conducting research with Anella Dexter, Sarah Emmott determined that public access to Texas beaches was a legally upheld residual from Spanish law. They worked in libraries, copied notes by hand, and typed them laboriously at home.[60] All Texas legislators received letters from the Emmotts and the Dexters, and TBU activists sent chain letters to inform the public of the issues at stake.[61] Texas representative Robert Eckhardt of Houston responded to the argument that beaches were a public resource to be preserved for the future.[62] Facing tremendous opposition and supported by some legislative technicalities, Eckhardt's Texas Open Beaches Law passed on July 16, 1959; it was the first such law passed in the United States. The constitutionality of the law was tested, and Sarah Emmott's and Anella Dexter's research won the case, after which Sarah Emmott was named an honorary assistant attorney general.[63]

A short-lived but successful TBU broadened the tactical reach of Houston's environmental activists in a number of important respects. TBU illustrated the competence and determination of individuals fighting to secure public access to natural resources and to protect citizens' rights through political activism.[64] It expanded the attention of some ONC members from wilderness recreation and protection to include pressing environmental concerns and citizens' empowerment, but many ONC members chose to remain active only in the social outings and nature recreation that were characteristic of the club. Joseph

Heiser, the club's founder, advocated political action, and friction arose among the club's members. Some resigned from the club, while others chose to engage in political activism and wilderness recreation.

Recreation was an important part of the outdoor experience, and Houston's scenic waterways offered plentiful opportunities for recreational activities.[65] Besides recreation, however, Houston's complex of seven major bayous brought the potential for flooding in the city and county, a potential exacerbated by unchecked development in and around the bayou watersheds. By the mid-1960s, homeowners in the Memorial Park subdivision and a flood-management project on Buffalo Bayou collided to produce a campaign for citizens' rights and environmental protection. Two major floods along Buffalo Bayou, one in 1929 and one in 1935, led to the formation of the Harris County Flood Control District (HCFCD) in 1937.[66] To protect Buffalo Bayou, the U.S. Army Corps of Engineers constructed Addicks and Barker reservoirs and began a program of straightening and cementing bayous in collaboration with the HCFCD.[67]

During the mid-1960s, a group of homeowners in Houston's Memorial subdivision formed the Buffalo Bayou Preservation Association (BBPA) to protect the natural beauty of their neighborhood bayou. In 1966, Terry Hershey and a number of other individuals observed an area along the bayou near Chimney Rock that was ravaged by fallen trees and bulldozed undergrowth. They soon learned that the HCFCD was rerouting Buffalo Bayou without public notification. Outraged by the condition of the bayou and the county's failure to proffer public notification, Hershey, a Memorial resident herself, joined BBPA and quickly became its most visible and energetic activist.[68] Houston developer and oilman George Mitchell lived on Buffalo Bayou and was a founding member and president of the organization.[69] Hershey's first acts involved community coalition building, but before long she was traveling to Washington, D.C., to testify before the House Appropriations Subcommittee by invitation of young congressman George H. W. Bush. Her testimony led to a halt of the work on Buffalo Bayou and a request by Congressman Bush that the U.S. Army Corps of Engineers restudy the project.[70]

Although the BBPA began as a NIMBY organization, by 1969 members realized that their concerns for Buffalo Bayou applied to all of Houston's and Harris County's watersheds, so the organization expanded its scope and amended its name and became the Bayou Preservation Association (BPA). As a NIABY (Not in Anybody's Backyard) group, the BPA emerged as an organization devoted to watershed oversight and information dissemination. Promoting community education and participation in watershed-management decisions was a major focus of BPA activities. During the 1970s, the BPA orchestrated the formation of the Harris County Flood Control Citizens' Advisory Task Force, a community collaboration of engineers, developers, and interested citizens, and the BPA sponsored a herculean effort to propel Houston and Harris County into the federal flood-insurance program in 1973.[71]

Terry Hershey's role in this first citizen protest against structural intervention along Buffalo Bayou was fortuitous for Houston's environmental community. Born in Fort Worth, she moved to Houston in the 1950s to marry Jake Hershey, an independent businessman, sailor, and philanthropist. When Terry Hershey recognized the danger to her neighborhood and understood the broader scope of environmental problems in Houston, she remembered her mother's lessons on the responsibilities of women to civic activism and embarked on a personal mission to eradicate Houston's problems.[72] Hershey was and remains a central figure in Houston's environmental activism. She studied the problems, sought input from other communities, organized groups, created coalitions, headed protests, and offered financial support to myriad efforts directed toward improvement of quality-of-life problems in Houston.[73] Her first environmental campaign was successful because she, and the community activists she aroused, stepped into uncharted territory by challenging local government in Harris County and by taking their argument to Washington. Congressman Bush gambled considerable local political capital to suggest a restudy of the Buffalo Bayou project, but his courage heartened local activists as much as it antagonized local government.[74]

The fight to save Buffalo Bayou was significant on several fronts. First, it represented a coordinated demand from citizens' groups to be informed of and included in government projects.[75] Second, it uncovered a small subterranean network of disparate groups and individuals in Houston who responded when Terry Hershey organized a coalition. Last, it illustrated the force and inspiration of talented individuals who could coordinate groups into powerful, if temporary, coalitions. Sarah Emmott, along with many other volunteers, lent her voice, energy, and money to the "Fight to Save Buffalo Bayou." Inspired by a television news segment featuring Terry Hershey, Hana Ginzbarg came forward for the first time and offered her services.[76] Ginzbarg joined the BBPA and set up a table in Memorial Park to collect 4,000 signatures supporting a restudy of Buffalo Bayou flood management plans, a foreshadowing of the incredible energy and tenacity that Ginzbarg would apply to the preservation of Armand Bayou.[77]

Environmental concerns including urban beautification, wilderness and open space protection, flood management, and citizens' rights, plus acute need for improved municipal services and pressure from the federal government, led to the formation of regional councils of government across the nation. By the mid-1960s, it became apparent even to most laissez-faire hardliners that the Houston region would suffer without regional planning and interagency cooperation. Although the Houston-Galveston Area Council started from local initiative in 1966, President Lyndon B. Johnson's administration withheld funding from regions where such planning bodies did not emerge, thereby "encouraging" the formation of regional government planning councils across the country.[78]

In December 1965, 100 officials from a seven-county area met to discuss formation of a Houston Regional Planning Commission. It was an idea whose time had come, but still a minority protested. Houston's mayor Louie Welch claimed that such an organization was being forced upon the region by the federal government. Others thought such an organization was moving regional politics too far to the left, but protests were too weak and too few to stop the momentum. In August 1966, the newly named Houston-Galveston Area Council (HGAC) established its bylaws, although its beginnings were rough and ineffective because of mistrust and disagreement. Building a sense of community among constituents, however, was one of the goals for these councils of government established across the country. Through its careful management of local government entities, the HGAC demonstrated that regional solidarity and promotion of the good of the region were in the best interest of each local government.[79]

Federal persuasion inspired the formation of the HGAC, but Houston's environmental activists welcomed the establishment of local chapters of national organizations in Texas and the Houston area. In 1965, the Lone Star (Texas) chapter of the Sierra Club formed, and the Houston group was created a few years later. Houston group members were prominent in the fight against the 1969 Texas Water Plan and devoted considerable energy to wilderness protection.[80] The Houston group supported opening the Addicks-Barker Reservoir area for recreation and wildlife protection. During the 1970s, Houston Sierra Club members participated in Texas Air Control Board meetings and supported areawide planning advocated by the HGAC. Members enjoyed recreational outings, such as a visit to ONC's Little Thicket Nature Sanctuary in 1974 and local biking events. Participation in the Houston Sierra Club provided members with opportunities for political activism and wilderness recreation, and the club often participated in local coalitions.[81]

An autonomous Houston chapter of the National Audubon Society was incorporated in 1969, when it began sponsoring local Christmas bird counts. Houston's ONC had collaborated with the National Audubon Society since the 1920s, but local environmental activists wanted their own chapter so that they could add their energy to local environmental issues.[82] Well-recognized for its birding, Houston's Audubon Society is one of the nation's most active chapters and was a central actor on the city's environmental stage. It participated in the Harris County Flood Control Citizens' Advisory Task Force and encouraged local participation in the Federal Flood Insurance Program, among other efforts.[83]

Post–World War II prosperity brought seemingly boundless business and population development to Houston. By the mid-1960s, Houston was a city of remarkable growth and employment opportunities, but the city was degraded by poor air and water quality, troubled by recurrent flooding, and deficient in park and recreation space. Concerned citizens welcomed formation of the HGAC, but lacking complete confidence in government officials, activists promoted local government oversight and community advocacy organizations dedicated to the improvement of Houston's quality of life.[84] Citizens Who Care (CWC) was a group of women (and initially one man) who decided in July 1968 that Houston's problems required community activism. In November 1968, the CWC organized a luncheon, with Jake Hershey's help, for a group of Houston's elite men, who subsequently organized the Houston Area Forum, raised an impressive sum of money, hired a director, and printed promotional brochures. A short time later, the Houston Area Forum folded into the Chamber of Commerce and later disbanded altogether. Disappointed with these results, the women of the CWC redirected their energies. Having realized that there were a number of small groups dedicated to improving Houston's quality of life, the women of the CWC decided that Houston needed a coordinating organization to facilitate environmental improvement efforts.[85]

Influenced and trained by groups in other cities, such as the San Francisco Planning and Urban Renewal Association (SPUR) and the Citizens Action Program (CAP) of Chicago, the founders of the CWC formed the Citizens' Environmental Coalition (CEC) formally in June 1970, but their informal activities had begun as early as 1969. Throughout the spring of 1970, members of the CWC collaborated with other groups to sponsor a community education program on air quality, in preparation for June 1970 hearings by the Texas Air Control Board. Their efforts persuaded 1,500 citizens to attend the air quality hearing, a success that encouraged the formal establishment of the CEC. In June 1970, members of the CEC met to organize, with articles of incorporation, bylaws, and a second important project on their agenda. The CEC provided participating groups and the Houston community with networking, resources, and information. By August 1970, the CEC had twenty-seven group members.[86]

Having succeeded with its Air Quality Information Campaign, the CEC's second major undertaking carried out over the last half of 1970 was formation of a Land Use Planning Subcommittee sponsored by the Sam Houston Resource Conservation and Development Area.[87] The subcommittee planned to study existing legislation first, define current and projected problems, and finally make recommendations for future legislation and policy. Represented on the subcommittee were a number of professional organizations and volunteer groups. Cochair Terry Hershey assigned study areas: the Sierra Club assumed responsibility for reporting on the National Land Use Policy of 1970, the Junior Bar Association helped with interpretation of existing bills and policies, the HGAC reported on the other councils of government in Texas and elsewhere, Planned Parenthood studied population trends, the Audubon Society researched activities of other Audubon chapters regarding land-use issues, and so forth.[88] As the subcommittee's work continued into the fall of 1970, Hershey helped set up a meeting to include the "real estate and development community." Having received acceptances to five of eight invitations the day before the meeting, the plan went forward. However, only one invitee, a representative the Houston Board of Realtors, attended.[89] Neither the development community nor local governments were ready in the 1970s for the recommendations of the CEC's Land Use Subcommittee, which listed eight points, including support for a National Land Use Policy Act, a proposal to amend the Water Resources Planning Act, acceleration of soil surveys, and strengthening soil and water conservation districts.[90]

The CEC's Subcommittee on Land Use was an important step in Houston's environmental activism because it pushed Will Hogg's idea of the Forum of Civics to a new level by coordinating environmental, political, and professional groups as well as government agencies to influence land-use legislation. Little progress resulted from the CEC's subcommittee, but the CEC was a success as the communication nexus for Houston's environmental community. Over time, it offered telephone answering services, an information hotline, and recommendations for expert speakers, among other services.[91] During the late 1980s, the CEC sponsored a radio program on KPFT, Radio Pacifica, titled *Talk of the Earth*. By the 1990s, the CEC had established a Web site and began sending a weekly newsletter by e-mail. In 1998, the CEC opened the Houston Environmental Conference Center, which offers office space to member organizations plus meeting rooms and a conference room that can seat over 200 participants.

Starting in 1970, the same year the CEC studied land-use problems, Hana Ginzbarg spearheaded the campaign to save Armand Bayou, originally named Middle Bayou but renamed in 1970 to memorialize Armand Yramategui, who joined the ONC in the 1950s, just so he could learn the names of the trees growing on his newly purchased tract of land. Of Basque and Mexican parentage, Yramategui spoke fluent Spanish, graduated with an engineering degree from Rice Institute in 1947, embarked on a self-taught earth sciences study program, and became a respected and loved member of Houston's conservation groups.[92] In 1965, Yramategui became curator of the Burke Baker Planetarium, a position the suited him well because his interests ranged "from the smallest earthbound creature to the most distant star."[93] In January 1970, three young men approached Yramategui on a Houston street to offer assistance with his flat tire; then they robbed and shot him. Houstonians and Texans mourned Yramategui's senseless death and recommitted themselves to environmental activism.[94]

Hana Ginzbarg's role was singular and instrumental in the Armand Bayou preservation campaign. An immigrant from Czechoslovakia, she had completed a master's degree and begun doctoral work when she met and married Arthur Ginzbarg. They moved to Houston in 1949, where she taught chemistry at several of Houston's private schools. Both Hana and Arthur Ginzbarg enjoyed outdoor recreation along bayou pathways and noticed bulldozed trees along several portions of Buffalo Bayou.[95] Already a member of the Houston Sierra Club and the Nature Conservancy, Hana Ginzbarg saw Terry Hershey on television about this time and joined the fight to save Buffalo Bayou.[96] As a member of the Bayou Preservation Association, which nominated representatives for each of Houston's major watersheds, Ginzbarg assumed the role of representative for Armand Bayou after Yramategui's death. She devoted herself to the campaign to preserve Armand Bayou as "a small urban wilderness reserve" and protect it from impending residential development. After Yramategui's death, preservation of Armand Bayou monopolized Ginzbarg's life for several years. She sent letters, made phone calls, photographed the bayou, attended political meetings, wrote press releases and newsletter columns, and generally performed like a whirling dervish for whom no task was too insignificant to merit her careful attention.[97]

Armand Bayou presented a significantly different challenge from Buffalo Bayou. Friendswood Development Company owned the land surrounding Armand Bayou.[98] In 1964, when a member of the ONC had proposed a nature sanctuary, company executives were less than enthusiastic. After extended negotiations, the company refused to donate the land but agreed to sell if preservationists could meet the price, so raising the required money was a centerpiece of the campaign. By 1969, the city of Pasadena held municipal jurisdiction over the area after rebuffing Houston's annexation advances. Pasadena, a largely industrial and working-class area, embraced a park project that would bring the area recreational and environmental ambience. Money, land, and development were the triggers in the Armand Bayou preservation campaign, rather than local government action.

As preserved by the campaign, Armand Bayou includes 2,500 acres of the natural wetlands, forest, prairie, and marsh habitats once abundant in the Houston-Galveston area. It offers habitat for 370 species of birds, mammals, reptiles, and amphibians, and it is an effective natural flood management system. As developed, the site contains hiking trails and a farm, and exhibits, field trips, Scout programs, and birding opportunities are available to visitors.[99]

When The Park People emerged in 1979, sixty-six years after Arthur Comey cited Houston's urgent need for park space, Houston still faced a critical lack of parks and open space. A National Urban Recreation study conducted in 1977 found Houston and Harris County deficient (104th in the nation) in park space.[100] The following year, Terry Hershey and Glenda Barrett organized a group dedicated to improving parks and open space in Houston to demonstrate that there was sufficient interest in the city to support a field office for the Bureau of Outdoor Recreation headed by Cris Delaporte, an undersecretary of the Department of the Interior.[101] The original group name, Citizen Open Space Task Force, was changed to avoid the implications of the acronym "COST."[102]

The Park People's original mission was advocacy, its goal to assist other groups to acquire and maintain parkland. That mission included raising Harris County residents' awareness of the importance of urban parks, communicating a park ethic, and preserving and augmenting park space. Over the years, The Park People has been an active organization whose projects include Jesse H. Jones Park, drainage and irrigation for the San Jacinto Battleground, landscaping for project row houses in the Third Ward, and wildflower planting programs. Another recent accomplishment was the Greenway Trails Map for Houston–Harris County, which shows 600 miles of actual and proposed greenway trails, parks greater than five acres in area, and parking areas for trails and parks.[103] Along with the CEC, The Park People has succeeded in building issue-specific coalitions with the city, the county, and other community organizations.

Houston's environmental activism continued through the 1980s and early 1990s, evolving with increasing frequency toward the concept of community collaboration.[104] By 1990, scholars and journalists began to compare environmental activism in the 1970s with activism in the 1990s, as the twentieth Earth Day celebrations got under way. Historically, in Houston, efforts at community-wide coalitions were sporadic, but the late 1990s brought an acceleration and proliferation of coalition building. A systematic and long-term approach toward urban planning for development and environmental amenities inspired these new coalitions, which drew as well from the legacy of past coalitions.[105]

After a preliminary period of organization, the Gulf Coast Institute emerged as a committee of the CEC. Then the city of Houston's Sustainable Development Committee transferred to the Gulf Coast Institute in 1999 and became its steering committee. A first meeting drew sixty attendees to hear discussions of smart growth.[106] Another conference on smart growth in October 1999 attracted 200 attendees.[107] David Crossley, active in the Houston community for twenty years and a former president of the CEC, headed the Gulf Coast Institute. Crossley found his inspiration in *Beyond the Limits,* a book exploring the concept of sustainable development.[108] Another inspiration was David Crockett of the Chattanooga Institute, an organization dedicated to establishing integrated social goals and "the involvement of all citizens in community process."[109] From another conference entitled "Connecting the Visions," Crossley elicited basic priorities across all segments of the community by reviewing and consolidating planning initiatives proposed over the last twenty years.[110] The Gulf Coast Institute spawned related coalitions—such as Blueprint Houston, Livable Houston Initiative, and Imagine Houston—organized to build community support and disseminate information through reports on issues such as transportation, air quality, and economic development.[111]

Like Will Hogg, Crossley traveled the country to observe community process and sustainability ventures. Hogg searched the country for planning schemes palatable to the Houston power elite, sponsored community lectures, and funded the Forum of Civics journal. Like Terry Hershey, Crossley adopted plans from successful organizations in other cities. The CEC produced a monthly calendar and a directory of groups. Hoping to attract a responsive public to information from scientists and technical experts, Terry Hershey and the CEC sponsored luncheons and community events to present position papers on environmental issues. Despite the new emphasis on broad initiatives such as smart growth and sustainable development, coalition building in Houston is not a late-twentieth-century innovation. New coalitions in Houston build not only from the Progressive ideas of Will Hogg in the 1920s and the communication infrastructure begun by the CEC in the 1970s but also from advances in communication technology and the professionalization of environmental activists, many of whom are schooled and credentialed in environmental science, law, and policy. While past coalitions experienced only modest success, each laid the groundwork for future coalition building, and groups formed to address single issues sometimes merged as coalitions after repeated attempts at organizing, each cycle bringing broader participation and success.[112]

Houston's environmental activism exhibits distinctive characteristics reflecting the city's southern history and political culture. While historians of the modern environmental movement often suggest that the New Left, the women's movement, and the civil rights movement were precursors of the post–World War II environmental activism, the Houston case confounds that pattern.[113] First, Houston was not receptive to influences from the New Left, more visible in national organizations and in other regions. One explanation was the fanatical Red Scare that swept the city in the 1950s and lingered through the 1960s and 1970s.[114] Although challenges of the 1960s enraged local government officials as well as the development community and elicited accusations of leftist politics, Houston's environmentalists engaged in reform activism that employed only moderate tactics and contested government practices within an established governmental framework.[115] From its beginnings, environmental activism in Houston supported private enterprise and a progrowth economy, and it transcended political parties, with both Democrats and Republicans devoting themselves to environmental causes.

Second, among the women who were leaders of environmental activism from the 1950s and beyond, most were middle and upper class, and middle-aged. Many belonged to the Junior League and the League of Women Voters, and some supported the National Organization for Women (NOW) financially, but many absorbed inspiration from the first wave of the women's movement through the loyalties and activism of their mothers.[116] For many, activism was a family tradition, although ambient permission may have emanated from the second wave of the women's movement to motivate these environmental activists, as it did other women across the social spectrum.[117] Although these women activists created, inspired, organized, and donated to environmental efforts, during the 1950s and 1960s most women did not assume top leadership titles in the organizations described in this chapter.[118]

Third, Houston experienced a comparatively low-key civil rights movement, and desegregation resulted from a relatively quiet business deal between the business elite, local media, university leaders, and civil rights protestors.[119] Through much of the city's history, Houston's neighborhoods were segregated, and deed restrictions ensured separation, although the ethnicity of certain neighborhoods shifted over time.[120] Historically, neighborhood minority groups organized to address issues similar to those confronted by more affluent, middle-class citizens, but minority activists frequently found themselves preoccupied with the maintenance of basic municipal services or employment issues.[121] When NIMBY issues led to activism, activists in one subdivision would not necessarily be inclined to combat problems in another.[122] Environmental justice campaigns sometimes attracted activists across the social spectrum. (Several of those Houston campaigns are included in other chapters in this volume.) More recent efforts to build community-wide coalitions eventually may narrow this divide.

Estrangement from mainstream national groups is another generalization applied to local environmental activism.[123] In contrast, Houston activists maintained close connections with national organizations and movements. Will Hogg corresponded with city planners and architects around the country; Joseph Heiser maintained close ties with the National Audubon Society and others; and Terry Hershey sat on any number of national boards and served on the Citizens' Advisory Committee on Environmental Quality.[124] David Crossley drew inspiration from initiatives surging across the country, and Houston environmentalists generally sought opportunities to establish local chapters of national organizations and as well as maintain national memberships.

Houston's activists do fit one of the stereotypes: they tend to be affluent, well educated, and middle class.[125] Often considered a shortcoming of environmental activism, that demographic could be more a strength than a weakness when viewed through the lens of "environmental citizenship."[126] Drawing from the traditions of citizens' rights and responsibilities, environmental citizenship suggests that privileged groups who extracted greater benefit from the environment historically bear greater responsibility for environmental improvement.[127] On the other hand, voluntary associations among citizens in coalitions can lead to collaboration among disparate groups and individuals. According to Dave Horton, embracing the concept of environmental citizenship can create the cultural and political space to discuss differential risks experienced by individuals and segments of the community and changes in individual behaviors that can lead to sustainability for the entire community. Houston's environmental activism, tempered by southern tradition and business politics, started toward that possible future, however slowly, when Will Hogg proposed a coalition of citizens' organizations in 1924.

CHAPTER 12

Voices of Discord

The Effects of a Grassroots Environmental Movement at the Brio Superfund Site

KIMBERLY A. YOUNGBLOOD

The Brio Superfund site, located in Friendswood, Texas, about eighteen miles southeast of downtown Houston, became a significant case study in environmentalism and the Superfund process. Grassroots environmental movements alter and affect communities, perspectives, and even the cleanup of hazardous waste sites. The grassroots effort at the Brio site changed the way many Texans viewed their neighborhoods and responded to environmental issues while simultaneously influencing the Superfund process designed to protect citizens and clean up the site.

Congress responded to the public pressure and environmental activism of the late 1970s with the passage of the Superfund law in 1980, which sought to protect the public's health and the environment from hazardous chemicals. Superfund, or the Comprehensive Environmental Response, Compensation and Liability Act (CERCLA), granted the federal government the authority to respond to direct or threatened releases of hazardous substances that endanger public health and the environment. The law ensured that polluters were held responsible for damages and cleanup while authorizing the Environmental Protection Agency (EPA) to locate potential responsible parties (PRPs).[1]

The Creation of a Superfund Site

By the 1940s, Houston's vast oil-refining complex, especially along the Houston Ship Channel, was reaping the benefits of expanded oil production in World War II. In turn, the war played an integral role in the growth of the petrochemical industry in Houston. When natural products such as rubber and silk were no longer available for tires and parachutes, companies like Dow, Monsanto, Humble, and Sinclair came up with solutions to the problem.[2] In fact, the development of synthetic rubber was one of the petroleum industry's most significant contributions.[3] As the oil and petrochemical industry rapidly grew in Texas, Houston's prosperity was measured in part by its relentless expansion outward from its core. Houston's suburban sprawl, with multiple nodes that included homes, shopping centers, industrial parks, and schools, characterized a "boomtown."[4]

The Brio site in Friendswood, encompassing about fifty-eight acres of land, was a fairly desolate, rural piece of pastureland far removed from the booming city of Houston when Ralph Lowe bought it in the 1950s. He built a processing plant to recycle oil and turned chemical company waste into various profitable products on the site. Once reprocessing the waste was complete, he disposed of the leftover by-products, which contained such toxic chemicals as vinyl chloride, fluorine, styrene, and benzene.[5] A great majority of the chemicals and chemical compounds utilized are now known to be hazardous to public health and to the environment. In fact, low-level exposure to vinyl chloride or benzene can cause minor afflictions such as skin irritation or edema, but longer exposure can cause more severe illnesses, ranging from cancer to reproductive problems or even internal bleeding and other disorders.[6]

The petrochemical industry's innovative product development may have fulfilled new consumer-driven demands, but the synthetic organic chemicals invented in the late 1930s and 1940s also posed a new conundrum—how to dispose of the extremely toxic waste or by-products effectively. Individuals like Lowe were only offering to recycle the wastes, not properly dispose of them. In addition, prior to 1976 no regulation existed that monitored their disposal.

Congressman Robert Eckhardt of Texas chaired a committee shortly before the passage of Superfund that obtained information on chemical waste disposal by 1,605 chemical plants since 1950. The report concluded that these plants had disposed of 762 million tons of toxic waste in landfills, pits, lagoons, or ponds, disposal techniques that have contributed to the hazardous waste problem of today.[7] Inappropriate disposal methods were only half the problem, since no one contemplated the possibility of future leaks from the dump areas. "These industries created wastes which half a century later continue to tax society's cleanup capabilities."[8] These problematic chemicals are nonbiodegradable, and their resistance to natural biological processes allows them to become persistent hazards in the air, water, and land, infecting all life.[9]

The Brio case exemplifies the nature of this problem. During its twenty-five years in operation, the plant experienced various contamination incidents and numerous transfers of ownership. Although all the facilities stored chemical by-products or toxic waste, each corporation attempted various manufacturing processes to fabricate diverse products from the waste.[10] The site contained a range of facilities along with waste storage pits and tanks that held the toxic waste. Lowe Chemical Company constructed most of the storage pits during the period from 1964 to 1970, when the site experienced tremendous growth and added a new reclamation operation.[11] From its inception as the Hard-Lowe Company in 1957 until its final conversion as an oil refinery, Brio Refinery, Inc., in the mid-1970s, the site attracted attention from various city and state agencies, such as the Texas Department of Water Resources (TDWR), the Texas Air Control Board (TACB), and the Harris County Pollution Control Department (HCPCD). These agencies gathered environmental information on the many mishaps and spills at the site over the years. State agency records documented environmental and management problems dating from 1962.[12] Much of their data detailed a plant with haphazard management practices with respect to highly toxic chemicals and a wanton disregard for the surrounding environment.

After several years when various owners defaulted on loans or went out of business, the site reverted to Ralph Lowe. Although many of the incidents of leaks, spills, and improperly stored waste resulted in fines and permit violations, the site remained open. These environmental blunders concerned the TDWR, since the spills flowed directly into the main drainage source for the watershed in that part of South Houston via a small stream called Mud Gully.[13]

All surface runoffs from the site drained into Mud Gully and the upper confined groundwater flow zone, the Numerous Sand Channel Zone (NSCZ), and sampling revealed that numerous chemicals from Brio had heavily contaminated this zone. The real dilemma stemmed from the fact that groundwater was a nonrenewable resource in this area and was impossible to purify once it was contaminated.[14] Water contamination and the many violations at the Brio site attested to the seriousness of this situation. Some of the violations included association with a fish kill in Clear Creek in 1968, a sodium sulfide spill of 1,000 gallons in Mud Gully and Clear Creek in 1969, and several violations in 1970 of copper sulfate, styrene tar, and other toxic chemical leaks into Mud Gully and Clear Creek.[15]

The catalog of permit violations and spills reads like a dirty laundry list of stains that continued until the site's closure, including a description of several saturated ground areas with a black, oily substance and puddles of styrene tars.[16] Even more disturbing was the lack of operation and process records at the site. Ralph Lowe and the subsequent owners were negligent in their record-keeping practices, which affected the site dramatically during the cleanup process. The lack of records made it difficult to determine which entity accepted what chemicals and what operations each entity utilized. This problem intensified when the time came to decide who was responsible for the waste and to determine the best method to clean it up.[17] Finally, when Brio's last owners declared bankruptcy in 1982, the TDWR immediately contacted the EPA and suggested that the site qualified for placement on the National Priorities List (NPL) as the first step of the Superfund law.[18]

In 1984, as part of the NPL investigation to see whether the Brio site qualified to be placed on the list for Superfund cleanup, the EPA tested the groundwater and air at the site. The groundwater tests revealed that copper, vinyl chloride, fluorine, styrene, and ethylbenzene contaminated the groundwater; the air inspections discovered toluene, benzene, and other aromatic chemicals at the site.[19] Ultimately, the EPA's findings confirmed that the site held some very volatile chemicals that posed a significant possibility of a public health risk. In 1984, the EPA nominated the site for the NPL, and in 1985 the EPA designated Brio as one of the worst Superfund sites in America.

Frequently, the EPA's investigation of a hazardous waste site reveals many parties who are potentially accountable for cleanup at the site.[20] In the Brio case, the EPA discovered 100 companies responsible for the waste.[21] However, many were no longer in business. Under the law, the identified companies still in business had to complete the cleanup. Through a negotiation process accepted under the law, smaller responsible parties can cash out to larger ones, leaving a group called the Task Force. The Brio Site Task Force (BSTF) consisted of a few petrochemical and oil companies. Of these, two were most recognizable, Monsanto and Amoco. As the cleanup procedure began, the situation became more complicated. Attention turned to public health hazards as a major concern because the once-remote area near Brio was becoming a rapidly growing suburb.

A Grassroots Movement Emerges

During the early 1980s, urban sprawl saw the development of a residential subdivision adjacent to the northwest corner of the Brio site, called Southbend. In 1976, there were only three houses in the area, but soon Southbend was home to hundreds of families taking advantage of affordable suburban living conditions in a quiet and isolated community close to Houston.[22] Residents never really contemplated what had been going on at Brio or what kind of companies operated there.

The EPA's decision to add the site to the NPL list changed the suburban serenity. In 1985, when Brio's designation as a Superfund waste site was made public, there were approximately 660 homes in the Southbend community. Thirty-one of these houses and the new Weber Elementary School—located on the northwest border of Brio—were closest in proximity to the Superfund site. Weber Elementary was a mere 1,200 feet away.[23] While residents pondered how such a threat could be in their backyard, emotions of fear and anger motivated their actions. The media, the government, and the residents all took part, but most important, activist women played a significant role in the events that followed.

Clearly, fear dominated this grassroots movement, influencing each person's perception of the risk involved by living close to the site. The women were not experienced activists or environmentalists. They were housewives and mothers whose dread of the possibility of toxic contamination and the illnesses it could cause motivated them to take a stand and provided them with an undiscovered powerful voice of activism. Samuel Hays describes this type of fear as something that usually lies dormant until a key event occurs, exposing a new and more determined shift in values that compels individuals to protect their families.[24]

One of the most infamous American environmental catastrophes occurred at Love Canal in New York State, the result of a serious incident of toxic-waste disposal by Hooker Chemical Company, made public in 1978. The event prompted a precedent-setting grassroots environmental response that created the means for women to become environmental activists. Although women's involvement in the environmental movement existed long before Love Canal, a new social movement began, with women acquiring a prominent role. Women like Lois Gibbs, a Love Canal resident, became major players in the antitoxics movement because their maternal instincts and innate ability to lead and organize assisted them in spearheading community-based environmental movements.[25] Love Canal was an important inspiration for women associated with the protest against Brio.

Media attention is an important aspect of an environmental movement's success, because it is more than just the daily or nightly news that alters the public's perception of toxic risk. The images and words that reporters employ in these news excerpts instill an indescribable fear in Americans. "One of the most powerful protest images that emerged out of the Love Canal protests was a Mother's Day Die-In, focusing on children's exposure to hazardous wastes and issues of reproductive health."[26] The significant advantage for the environmental movement at the Brio toxic waste site was that one of their staunchest supporters and leaders was a member of the media. This advantage compelled significant modifications in the remediation policy at the Brio site, and neither the EPA nor the PRPs could have foreseen the outcome.

Grassroots environmental movements rely on the media to keep the site and their activities in the public eye. As Hays states, "For most people, knowledge about environmental affairs is shaped by the media, which have a strong bias toward sensational events and personal drama."[27] In fact, by obtaining national media attention, the Love Canal movement proved that the media is a powerful tool in the grassroots environmental movement. This type of environmental awareness can encourage public awareness and even induce action from different segments of a community. In 1978, Love Canal became national news when television networks such as ABC and CBS and national newspapers relayed stories about the site to the public. "The surge of media attention had made hazardous waste the nation's Most Important Environmental Problem."[28]

The Brio environmental movement demonstrates how a similar grassroots environmental movement utilized the media and claimed a decisive role in the resolution of the cleanup remedy provided by Superfund and instituted by the EPA. From the beginning, Marie Flickinger's reports and articles shaped many people's ideas of risk from the site while gaining public attention for Brio. Flickinger, a resident of nearby Friendswood and co-owner of the *Southbelt–Ellington Leader,* a free local weekly newspaper, became the spokesperson and organizer of the Brio environmental movement. She believed that the Brio case and its final remediation affected a broad community beyond the Southbend subdivision, and she resolved to keep the public informed of the happenings and decisions made regarding the site.

The *Houston Post* and the *Houston Chronicle,* along with the local news stations, also sustained the media attention Brio received by regularly updating the public about the latest developments at the site. In this way, the controversy intensified instead of taking a backseat to other local news. The media and community attention induced the EPA and the responsible parties—namely, Monsanto and Amoco—to deal with the remediation process in a way that they may not have contemplated initially. The media also helped energize the Brio environmental movement, providing the case with national publicity from time to time.

The Superfund process involves the public, the government, and the PRPs. Sustaining an open line of communication with the public is part of the EPA public relation plan at any Superfund site. The EPA regularly releases information about the site and the remediation process in EPA community meetings, thereby ensuring public participation in the process. The Southbend activists not only maintained a presence at the EPA community meetings, but they also attended the Brio Site Task Force meetings and mayor's community leaders meetings, which were held to inform the entire community of the activities and progress at Brio. Most significant, Flickinger assisted Southbend residents in conducting their own community meetings, called the South Belt Town Meetings, and reported on them. These meetings would often attract as many as 500 people. Invited speakers such as the mayor of Friendswood or ex-EPA officials were asked to talk about the dangers and recent events at the Brio site. This enabled Southbend activists to keep Brio in the public eye, ensuring further media reports on the Superfund process at the site.

Perceptions of Public Health

After Brio's designation as a Superfund site, claims surfaced in the late 1980s and early 1990s that illnesses that plagued various residents stemmed from the site and toxic exposure. These ailments included everything from bloody noses and rashes to birth defects and cancer. Yet all the soil, air, and water tests conducted by the EPA and the BSTF indicated no threat of toxic exposure to the 660 homes in the Southbend community. Remarkably, quite a number of residents approved of the EPA's findings and felt safe in Southbend. This break in the ranks did not affect how Brio activists felt, and they set out to prove that there was an inherent risk living close to the site. Although most of Southbend's residents were convinced that toxic contamination caused many of their ailments, comprehensive health studies were necessary to prove cause and effect.

A Southbend resident, Cheryl Finley, conducted an informal citizen survey in 1990 regarding birth defects in the area, and she entreated the Agency for Toxic Substance and Disease Registry (ATSDR) to assist her since no government agency had conducted a health survey for the subdivision. Because the EPA had determined that no risk existed, it was difficult to convince authorities that outside surveys were necessary. Finley's survey revealed that the national average for birth defects at the time was 2.7 percent of all live births, but the average for Southbend residents was 35 percent of live births. The survey included 120 pregnant women in the neighborhood between 1981 and 1990. The pregnancies produced twenty-four miscarriages, two stillborn babies, one baby born with a fetal defect that required therapeutic abortion, and twenty-one babies with birth defects.[29] Examples of Southbend's disturbing birth defects included a child born with its intestines outside the body, another born without a brain, and two children born with holes in their hearts.[30]

Cheryl Finley's survey obviously raised a red flag. The ATSDR, which assisted with the survey, became the first government agency to perform a health study at the Brio site. In 1994, its study determined that residents who lived near the Brio site were prone to respiratory ailments, skin rashes, ulcers, and other medical problems and that their children endured frequent vomiting, drowsiness, and severe headaches. However, the ATSDR was not willing to confirm or link the birth defects to the contaminants at the site.[31]

There is another side to the story. Some Southbend residents did not believe that there was any significant contamination from the Brio site and quickly pointed out the number of healthy children born in the community, while insisting that the number of birth defects and health problems experienced by their fellow residents was completely within the norm. These residents, like Carol Womack and her husband, Jimmy, had faith in the EPA's assurances that Brio posed no health threat to their community. In an interview with the *Houston Chronicle,* Carol Womack stated that more than 100 Southbend residents did not believe there was any significant contamination from the Brio site and that the claims of contamination and risk were exaggerated. Carol and Jimmy Womack believed that the developer, Farm and Home Savings, and the many attorneys engaged in the case were to blame for the litigation because Farm and Home convinced a group of litigants to join it in a suit against Monsanto. Residents who testified would receive $85,000 for aiding Farm and Home in its lawsuit against Monsanto.[32]

The birth defects and the ailments that the residents of Southbend endured became points of contention that neither side could agree upon because no potentially liable party wanted to admit the risk or harm to the public health. Furthermore, there were different points of view on the danger to the public health posed by Brio. The fact that the public health agency had already stated there was no risk or danger of contamination influenced the agency's perspective and the statements it made. For instance, when Cheryl Finley took the results of her birth defect survey to the ATSDR offices in Atlanta, one official declared that the new information would cause a severe dilemma since the agency earlier had stated that there was no health hazard from the Brio site. He said that the agency would be reluctant to admit its error and revise the necessary statistics with the new information.[33] Still, the Brio protesters were not going to back down.

Activism Affects the Superfund Process

For better efficiency, the EPA maintains regionally based offices throughout the country along with its office in Washington, D.C. The Dallas regional EPA office initially took charge of the process of remediation at Brio, but the Washington and Dallas offices differed on the best way to proceed. While the Region 6 office in Dallas concerned itself with working out an agreeable arrangement with the PRPs, the Washington office dealt with the political pressure from the growing activist and community groups. By 1993, the publicity and the friction between the community and the regional Dallas EPA office became so acute that the Washington EPA office got involved. The conflict lay between the discrepancies of opinions that the Dallas office and the Brio environmental movement held regarding the issue of public health risk. The media elevated this tension by the many stories about Brio.

The situation came to a head in October 1993, after a much-publicized vinyl chloride release that occurred at the site in August heightened public concern about the site. Afterward, the air analysis revealed that chloride levels were above the federal permissible levels. In fact, the analysis measured vinyl chloride levels at 34,000 parts per billion (ppb), while the federal permissible level is 1,000 ppb every eight hours. Tina Forrester, a toxicologist with the Agency for Toxic Substances and Disease Registry in Atlanta, recommended evacuation of the Southbend subdivision, but her recommendation came too late.[34] Apparently, the contractor in charge of cleaning up the Brio site, Chemical Waste Management, did not forward to authorities for almost a month the air sample results taken on that day. The time delay between the actual leak and the stop-work directive is unexplainable.[35] This episode significantly affected the Washington EPA office's actions and became the impetus for the residents' activism in the remediation resolution.

The Washington EPA office reviewed the situation and mandated thirteen new recommendations for cleanup procedures as well as further tests and studies to determine the public health risk.[36] The additional testing revealed a high level of contaminants in Mud Gully and Clear Creek, the two water sources that ran next to and through the Brio site. The contaminants from Mud Gully forced the Texas Department of Health to ban recreational activities in Clear Creek and to post signs declaring the fish uneatable. Fish-tissue samples taken from Clear Creek revealed dichloroethane and trichloroethane, suspected cancer-causing chemicals.[37] Furthermore, additional studies in March 1994 linked these contaminants to their source—Brio.[38]

It was abundantly clear that media recognition and public outcry prompted Washington's involvement at Brio and influenced its decisions. This type of public involvement yielded the results that the environmental group at Brio needed, and the protesters quickly realized that they could sway the EPA's decisions and the actions of the Brio Site Task Force as well. The Washington EPA office stated that after the Brio investigation, it wanted to "illustrate a new desire in Washington to better involve and respect community interest in all Superfund cleanups nationwide."[39]

For the Sake of the Children

The activists decided to expand their efforts. Surveys and meetings were fine, but it was time to address other issues of safety since the group was adept at using the media to gain public attention. The major area of concern for mothers was their children's health. Southbend children attended Weber Elementary School in the neighborhood. Many children who attended Weber, even nonresidents of Southbend, experienced myriad ailments, including frequent bloody noses and headaches. It was just a matter of time before the mothers of the Brio environmental movement confronted the Clear Creek Independent School District (CCISD) with these issues, and the school district ultimately experienced their wrath. Many mothers from Southbend and the surrounding area were convinced that the proximity of the school caused many of the students' illnesses.

Beginning in 1990, as a form of protest, some of these mothers began keeping their children home from school. They petitioned the school district to allow them to enroll their children in a different school. To ensure the safety and health of the student body, the CCISD entreated the Texas Department of Health and other various agencies to perform air, water, and soil contamination studies. The school district and its board wanted assurance that the school's location was not a factor in the health claims. When all the test results returned negative, the district resolved to keep Weber Elementary open. In response, Southbend mothers and activists picketed the school in an effort to force the school board to shut it down. This action did not always get the desired effect, however, since picketers often frightened many remaining students by yelling over bullhorns that students who continued to attend Weber might die from toxic exposure.[40] Finally, the neighborhood protests, negative media, and recent cleanup procedures at the Brio site forced the CCISD to hold an emergency meeting on March 25, 1992. The consensus at the school board meeting was that it was better to be safe than sorry. Besides, cleanup of the Brio site had begun, and the new studies did show differing contamination risks. The board decided to close Weber Elementary permanently while transferring the children to schools in different locations.[41] The powerful voice of activism had won a victory.

Changes in the Superfund Process

While the public has a voice in the Superfund process, the final responsibility rests squarely on the shoulders of the EPA. In the case of Brio, community input and the environmental movement influenced the EPA's conclusion. The EPA and the Brio Site Task Force decided that incineration would be the best form of remediation, but the Brio environmental movement, with the media's help, vociferously opposed that approach. The safety of incineration became the focal point of every newspaper and news story regarding the remediation process at Brio. Since this was about the same time that the Washington EPA office became involved, the activists won this battle, and the parties involved abandoned the incineration process. As a member of the Brio Site Task Force stated at the time, "We were moving ahead with the incineration remedy, and then all of a sudden . . . Marie Flickinger and the community people suddenly realized that they didn't want to see any incineration in their neighborhood."[42] On September 14, 1989, the *South Belt–Ellington Leader* reported that the Municipal Utility District Thirteen board of directors voted to oppose the consent decree for incineration. They stated that, in their opinion, any airborne particles from the incineration would still be hazardous to someone's health.[43]

The Superfund law gives the EPA the final power to decide the remediation at any Superfund site. However, the EPA has an obligation to be attentive to the public reaction to any remediation choice. As a government agency, it recognized that its actions could have political ramifications. Therefore, the EPA must listen to concerns from the public and also consider the voice of the media. Marie Flickinger stated the resolution of public activism to a *Houston Chronicle* reporter: "Based largely on the continued outcry from the community and intervention of the ombudsman, the incineration remedy was abandoned last year and the process to choose a new remedy began."[44] The $20 million the BSTF invested on the incinerator was lost when the EPA changed this remedial decision.

Even though many hazardous waste disposal experts agreed that incineration was the safest and most effective method of disposal, further debate on the subject was futile. Eventually, all the parties concerned supported a new method of remediation, with a revised Brio consent decree in 1997. The Brio site plan recommended a fifty-five-foot-deep slurry wall encompassing the eighty-nine acres around it, and a gas collection cap placed over the site.[45] The cover system consisted of compacted clay, a liner, and a gas-collection system. On July 2, 1997, the parties signed the amended Record of Decision (ROD), and the federal district court judge approved it on March 5, 1999. The 6,000 linear feet of slurry wall made of bentonite "exceeded the required permeability by a factor of 10," and its depth was forty to fifty feet surrounding the Brio Superfund site.[46]

Activism and the Media

Similar to Lois Gibbs at Love Canal, Marie Flickinger became one of the first identifiable women activists in the Brio environmental movement. Flickinger was the first reporter to investigate and print the happenings at the Brio site. Even Paul Schrader, the Friendswood mayor at the time, stated to a *Dallas Morning News* reporter in 1990 that Flickinger and her paper had been "the primary vehicle through which people have gained information and I think they have been inspirational in providing leadership in the issue."[47] The Dallas EPA representative, Lou Barinka, in the same article described Flickinger's involvement quite differently. He told a Southbend resident, "It's been a war. Either you believe her, or you believe us." Flickinger made sure that her paper, the *South Belt–Ellington Leader,* reported on everything that happened at Brio, including community meetings, legal testimony, and picketing to document cover-ups. Her main motivation was to elevate the public's awareness of the site and activities at the site while maintaining its presence in the news.

Clearly, the antitoxics movement got a boost from women activists who put their housewife duties on hold and utilized their organizational abilities to unite a new grassroots environmental movement. These women, motivated by their maternal instincts and fear of toxic contamination, combined with a mistrust of big industry and government, strove for what they believed to be the basic rights of public health and security.[48] In that respect, Flickinger emulated Gibbs's example, but she was able to push the envelope further because she was the owner and editor of a local newspaper.

As revealed earlier, the final accepted remediation shifted from incineration to containment. The Washington EPA office responded to the public outcry against the regional Dallas office and even ordered new soil and toxic chemical testing and upgrading air monitoring. The informal citizen survey by Cheryl Finley prompted the Agency for Toxic Substance and Disease Registry to perform the first outside health study, and the CCISD eventually felt compelled to close Weber Elementary School. Additionally, the Brio environmental movement modified the decisions in the lawsuits for the case, and it especially determined the actions of one of the major responsible parties, Monsanto, in the case of *Slaughter v. Monsanto.*

Lawsuits and Judgments

Of all the PRPs, Monsanto became the most recognized for its role at the Brio site. A profitable Fortune 500 company, Monsanto also utilized the Brio site for a major portion of the wastes from its chemical plants in southeast Texas. Southbend residents filed a lawsuit against Monsanto in "the largest trial in Harris County history,"[49] claiming that the many illnesses and the decreased property values residents experienced stemmed from the proximity of the toxic waste site to their subdivision. Additionally, in 1987 a group of 178 homeowners filed a civil suit against other companies involved in the development of the subdivision. These companies—Pulte, Ryland, Perry Homes, and others—built and sold homes in Southbend and, according to the residents, did not warn prospective buyers of the neighboring toxic waste site and its dangers.[50]

The lawsuit *Slaughter v. Monsanto,* filed by 1,700 Southbend residents, ended differently than most people imagined. The jury found that Monsanto had no liability for the health and environmental problems of the Southbend residents. However, Monsanto decided to settle with the residents who maintained they had injuries resulting from living next to Brio. In 1992, Monsanto gave them $39 million but declared that the company had done no wrong. The company spokesperson said that paying the residents would cost less than the appeals process and further litigation.[51] Additionally, the lawsuits against the home builders ended with the jury deciding that the Farm and Home Savings Association, the mortgage company, should assume all the home loans for Southbend residents. The home and land developers received the strongest message from the trial's judgment. Developers need to ascertain information about the land on which they proposed to build homes and take responsibility for contamination found on the property.[52] It was revealed in the 1987 lawsuit that Farm and Home knew since 1979 that the property was adjacent to a toxic waste site, but the home builders had no knowledge until 1984, when Brio was added to the NPL. The jury in the case found Farm and Home guilty of fraud, and eventually the company assumed the mortgages of all of the homes in the subdivision as part of the final settlement.[53]

The total settlement amount the Southbend residents received from Monsanto, the Farm and Home Savings Association, and the other chemical companies in the BSTF was $207 million.[54] This settlement is extremely significant because it is one of the largest in environmental history, but no one got rich, and the residents did not receive the admission of guilt they sought. In the end, the man who bought the Brio site and began the operations there, Ralph Lowe, paid the government $400,000 and signed over security interest in thirty acres of land valued to sell for $600,000 to the government. A $1 million settlement was a small price to pay compared to the estimated cleanup costs of around $40 million to $60 million for Brio.[55] Yet the oil and chemical companies did not pay as dearly as it may seem, since Monsanto held insurance policies to cover much of the cleanup costs. It is true that the BSTF spent almost $20 million on building an incinerator that was never used. However, Monsanto's insurance company covered the cost of the incinerator as well.[56] In the end, Monsanto did not bear as much of Brio's cleanup costs as one would have thought.

The Final Analysis

The similarities between the Brio site and Love Canal are quite remarkable. Individual perceptions and emotions not only influenced the Brio environmental movement, but the general feeling from the media reports encouraged the belief that no one would want to live near Brio. As Michael Brown, the journalist instrumental in revealing the Love Canal story, wrote, "So did the emotional trauma weave its way slowly, inexorably, shifting from house to house."[57] The fear of Southbend residents spread through the neighborhood, affording the opportunity for women to profit from this sanguine time and assert their role as environmental advocates. Marie Flickinger capitalized on the power of the media since she co-owned a local neighborhood newspaper, and the Brio movement ascertained expeditiously that agency, powerful interaction with the community, and the inciting voice of the media were potent tools that altered the government's response to an environmental catastrophe. The environmental movement affected many aspects of the Brio case, from the remediation plan to the settlement of the lawsuits.

Eventually, the Southbend neighborhood of 660 homes, including Weber Elementary School, was demolished. Undoubtedly, the proactive stance from the Brio environmental movement, the power of the media, and the participants' newfound agency influenced the execution and the result of the Superfund process, as well as the final resolutions surrounding the case. Yet residents who did not partake in the various lawsuits thought that litigation was the major cause of the problem. It came down to a conflict between those who believed a risk existed and those who did not. The residents who did not believe there was a risk resented the plummeting house values and negative media coverage of a community they had called home. These factors left them with a feeling of shame for where they lived. The residents who adamantly believed that their health was imperiled believed that no amount of money could offer them assurances of good health in the future for their children or themselves.[58] Even after they moved away, questions of contamination still haunted these individuals.

Brio illuminates the technical and political aspects of an environmental dilemma while demonstrating the dynamics that affect the Superfund process. Many people participate in the Superfund process, and the public should maintain its involvement in defining health problems from hazardous waste sites since there are thousands of Superfund sites in the United States that fail to contain their wastes securely and properly.[59] Some of the Brio site still stands today, along with a huge vacant area covered with grass, weeds, and crumbling concrete, as a reminder of how much influence and power a grassroots environmental movement can have on the legal and government system.

During the 1970s, environmentalists made great strides in instituting communication networks and inspiring environmental policies that affected their communities immensely.[60] Hopefully, future environmentalists will employ these archetypes of community involvement and outrage as a way to understand the long-term political and social effects of a grassroots environmental movement as well as a representation of public activism.

NOTES

Introduction

1. William Cronon, *Nature's Metropolis: Chicago and the Great West* (New York: W. W. Norton, 1991); Kathleen Brosnan, *Uniting Mountain and Plain: Cities, Law, and Environmental Change along the Front Range* (Albuquerque: University of New Mexico Press, 2002).

2. There is as yet no comprehensive economic history of Houston. For a general political history, see David G. McComb, *Houston: The Bayou City* (Austin: University of Texas Press, 1969). See also Marilyn McAdams Sibley, *The Port of Houston: A History* (Austin: University of Texas Press, 1968); and Lynn M. Alperin, *Custodians of the Coast: History of the United States Army Engineers at Galveston* (Galveston, Tex.: U.S. Army Corps of Engineers, 1977).

3. A pioneering work of energy, growth, and environment in a region is James C. Williams, *Energy and the Making of Modern California* (Akron, Ohio: University of Akron Press, 1997). An excellent account of the early politics of oil-led development in California is Paul Sabin, *Crude Politics: The California Oil Market, 1900–1940* (Berkeley: University of California Press, 2004).

4. For a collection of essays about energy transitions, see Lewis J. Perelman, August W. Giebelhaus, and Michael D. Yokell, eds., *Energy Transitions: Long-Term Perspectives* (Boulder, Colo.: Westview Press, 1981). For an essay on Houston, see Joseph A. Pratt, "The Ascent of Oil: The Transition from Coal to Oil in Early Twentieth-Century America," in Perelman et al., *Energy Transitions*, 9–34.

5. Martin V. Melosi, "Community and the Growth of Houston," in *Effluent America: Cities, Industry, Energy, and the Environment*, ed. Martin V. Melosi (Pittsburgh: University of Pittsburgh Press, 2001), 194–95.

6. See Joe R. Feagin, *Free Enterprise City: Houston in Political and Economic Perspective* (New Brunswick, Conn.: Rutgers University Press, 1988).

7. On the national level, see Robert Engler, *The Politics of Oil, Private Power and Democratic Directions* (New York: Macmillan, 1961). For Houston, see Feagin, *Free Enterprise City*. For Texas, see George Green, *The Establishment in Texas Politics: The Primitive Years, 1938–1957* (Westport, Conn.: Greenwood Press, 1979). For Houston, see also Joseph A.

Pratt, "8F and Many More: Business and Civic Leadership in Modern Houston," *Houston Review of History and Culture* 2, no. 1 (2004): 2–7, 31–44.

8. Short discussions of many oil-producing regions are found in Augustine A. Ikein, *The Impact of Oil on a Developing Country: The Case of Nigeria* (New York: Praeger, 1990). See also Tony Hodges, *Angola: Anatomy of an Oil State* (Bloomington: Indiana University Press, 2001).

9. I would like to acknowledge the efforts of Steven MacDonald in identifying and collecting research material for this section of the book.

10. W. L. Fisher, J. H. McGowen, L. F. Brown Jr., and C. G. Groat, *Environmental Geologic Atlas of the Texas Coastal Zone—Galveston-Houston Area* (Austin: Bureau of Economic Geology, University of Texas at Austin, 1972), 1.

11. David G. McComb, *Galveston: A History* (Austin: University of Texas Press, 1986), 6, 7–8. See also Houston Geological Society, *Geology of Houston and Vicinity, Texas* (Houston: Houston Geological Society, 1961), 3, 7; Robert R. Lankford and John J. W. Roger, comps., *Holocene Geology of the Galveston Bay Area* (Houston: Geological Society, 1969), vii, 1; and Fisher et al., *Environmental Geologic Atlas*, 7.

12. Fisher et al., *Environmental Geologic Atlas*, 7; Jim Lester and Lisa Gonzalez, eds., *Ebb and Flow: Galveston Bay Characterization Highlights* (Galveston, Tex.: Galveston Bay Estuary Program, 2001), 12; Joseph L. Clark and Elton M. Scott, *The Texas Gulf Coast: Its History and Development*, Vol. 2 (New York: Lewis Historical, 1955), 14–16; Houston Geological Society, *Geology of Houston*, 3.

13. Fisher et al., *Environmental Geologic Atlas*, 7.

14. Ibid.; G. L. Fugate, "Development of Houston's Water Supply," *Journal of the American Water Works Association* 33 (October 1941): 1769–70.

15. Planning and Development Department, *Public Utilities Profile for Houston, Texas* (Summer 1994), III-15; Fugate, "Houston's Water Supply," 1769–70. The geologic formations from which Houston obtains groundwater supplies are upper Miocene, Pliocene, and Pleistocene in origin. See Nicholas A. Rose, "Ground Water and Relations of Geology to Its Occurrence in Houston District, Texas," *Bulletin of the American Association of Petroleum Geologists* 27 (August 1943): 1081.

16. See "Houston," in *Twentieth Century Cities*, part 4 of Association of American Geographers, *Contemporary Metropolitan America*, ed. John S. Adams (Cambridge, Mass.: Ballinger, 1976), 109, 121–24; Houston Chamber of Commerce, *Houston Facts '82* (Houston: Houston Chamber of Commerce, 1983).

17. U.S. Environmental Protection Agency (EPA), *Heat Island Effect: Houston's Urban Fabric*, http://www.epa.gov/heatisland/pilot/houst_urbanfabric.html; U.S. EPA, *Heat Island Effect: Houston*, http://www.epa.gov/heatisland/pilot/houston.html.

18. U.S. EPA, *Heat Island Effect: Houston;* World Travels, Houston Climate and Weather, http://www.wortltravels.com/Cities/Texas/Houston/Climate.

19. Fisher et al., *Environmental Geologic Atlas*, 1.

20. Ibid., 1, 7.

21. Espey, Huston & Associates, prep., *Archival Research: Houston-Galveston Navigation Channels, Texas Project—Galveston, Harris, Liberty and Chambers Counties, Texas*, April 1993, 8.

22. McComb, *Galveston*, 121–49.

23. David Roth, "Texas Hurricane History," National Weather Service, Lake Charles, La., 2004, http://www.srh.noaa.gov/lch/research/txhur.php.

24. Espey, Huston & Associates, *Archival Research*, 10.

25. Eugene Jaworski, "Geographic Analysis of Shoreline Recession, Coastal East Texas," College Station, Texas A&M University, Environmental Quality Note 3, June 1971, 1–13.

26. Fisher et al., *Environmental Geologic Atlas,* 15, 20; Robert R. Stickney, *Estuarine Ecology of the Southeastern United States and Gulf of Mexico* (College Station: Texas A&M University Press, 1984), 247–80.

27. Lester and Gonzalez, *Ebb and Flow,* 9–11.

Part 1. Energy and Environment

1. Brian Black, *Petrolia: The Landscape of America's First Oil Boom* (Baltimore: Johns Hopkins University Press, 2000), 10, 7.

2. David E. Nye, *Consuming Power: A Social History of American Energies* (Cambridge, Mass.: MIT Press, 2001), 208.

Chapter 1. A Mixed Blessing

1. See Williams, *Energy and the Making of Modern California;* and Sabin, *Crude Politics.*

2. See McComb, *Houston: The Bayou City.* See also Sibley, *Port of Houston*; and Alperin, *Custodians of the Coast.*

3. Frederick Law Olmsted, *A Journey through Texas, or, a Saddle Trip on the Southwestern Frontier* (New York: Dix, Edwards, 1857), 366. For an even earlier reaction to the region by an anonymous traveler, see Andrew Forest Muir, ed., *Texas in 1837* (Austin: University of Texas Press, 1958), 3–41.

4. For a useful overview of energy trends before the energy crises of the 1970s, see Sam Schurr and Bruce Netschert, *Energy in the American Economy, 1850–1975: An Economic Study of Its History and Prospects* (Baltimore: Johns Hopkins University Press, 1960).

5. George Rogers Taylor, *The Transportation Revolution, 1815–1860* (New York: Harper and Row, 1968). For a view of changes in Texas cities during these years, see Kenneth Wheeler, *To Wear a City's Crown: The Beginnings of Urban Growth in Texas, 1836–1861* (Cambridge, Mass.: Harvard University Press, 1968).

6. Andrew Forest Muir, "Railroads Come to Houston, 1857–1861," *Southwestern Historical Quarterly* 46 (July 1960): 42–63; S. G. Reed, *A History of Texas Railroads* (Houston: St. Clair, 1941); Earl Fornell, *The Galveston Era: The Texas Crescent on the Eve of Secession* (Austin: University of Texas Press, 1961).

7. Scott Stabler, "Free Men Come to Houston: Blacks during Reconstruction," *Houston Review of History and Culture* 3, no. 1 (2005): 40–43, 73–76; James M. Smallwood, *Time of Hope, Time of Despair* (Port Washington, N.Y.: Kennikat Press, 1981).

8. For the Southern Pacific, see Kenneth J. Lipartito and Joseph A. Pratt, *Baker and Botts in the Development of Modern Houston* (Austin: University of Texas Press, 1991), 32–46.

9. For a collection of essays about energy transitions, see Perelman et al., *Energy Transitions.* For an essay on Houston, see Pratt, "Ascent of Oil."

10. For a discussion of the use of coal by the railroads, see Pratt, "Ascent of Oil."

11. For the best source on economic developments in Houston before and after the turn of the twentieth century, see Harold L. Platt, *City Building in the New South: The Growth of Public Services in Houston, Texas, 1830–1910* (Philadelphia: Temple University Press, 1983). For Texas as a whole, see John Spratt, *The Road to Spindletop: Change in Texas, 1875–1901* (Dallas: SMU Press, 1955).

12. Judith Walker Linsley, Ellen Walker Rienstra, and Jo Ann Stiles, *Giant Under the Hill: A History of the Spindletop Discovery* (Austin: Texas State Historical Association, 2002);

James A. Clark and Michel T. Halbouty, *Spindletop* (New York: Random House, 1952); Diana Davids Olien and Roger Olien, *Oil in Texas: The Gusher Age, 1895–1945* (Austin: University of Texas Press, 2002); Carl Coke Rister, *Oil! Titan of the Southwest* (Norman: University of Oklahoma Press, 1949).

13. Walter Rundell Jr., *Early Texas Oil: A Photographic History, 1866–1936* (College Station: Texas A&M University Press, 1977), 81–92, 119–34. See also Henrietta Larson and Kenneth W. Porter, *History of the Humble Oil and Refining Company: A Study in Industrial Growth* (New York: Harper and Row, 1959).

14. Brian Black, *Petrolia: The Landscape of America's First Oil Boom* (Baltimore: Johns Hopkins University Press, 2000).

15. For striking photographs of the pollution, waste, and danger to workers in the early development of the Spindletop field, see Rundell, *Early Texas Oil,* 40–47. For descriptions of the development of the field, see John O. King, *Joseph Stephen Cullinan: A Study of Leadership in the Texas Petroleum Industry, 1897–1937* (Nashville, Tenn.: Vanderbilt University Press, 1970).

16. Pratt, *Growth of a Refining Region,* 13–88.

17. Joseph A. Pratt, "Letting the Grandchildren Do It; Environmental Planning during the Ascent of Oil as a Major Energy Source," *Public Historian* 2, no. 4 (Summer 1980): 28–61.

18. Joseph A. Pratt, "Growth or a Clean Environment? Responses to Petroleum-Related Pollution in the Gulf Coast Refining Region," *Business History Review* 52, no. 1 (Spring 1978): 1–29.

19. Joseph A. Pratt, "Creating Coordination in the Modern Petroleum Industry: The American Petroleum Institute and the Emergence of Secondary Organizations in Oil," *Research in Economic History* 8 (1983): 179–215; Pratt, "Growth or a Clean Environment?" 11–16; Hugh S. Gorman, *Redefining Efficiency: Pollution Concerns, Regulatory Mechanisms, and Technological Change in the U.S. Petroleum Industry* (Akron, Ohio: University of Akron Press, 2001), 118–33.

20. Department of the Interior, U.S. Bureau of Mines, *Report on Pollution by Oil of the Atlantic and Gulf Coast Waters, Appendix D* (Washington, D.C.: Government Printing Office, 1923), 440; for Galveston and pollution from the Brazos River, see appendix A, 127–28.

21. For conditions along the Texas coast, see ibid., appendix D, 435–44.

22. This motivation was particularly strong in the mid-1920s, since the Oil Pollution Act of 1924 called for further surveys of oil pollution to be used to pass stricter controls if necessary.

23. The process within the API involved the creation of committees of industry experts to survey conditions in the industry and to publish a list of best practices, which individuals within the industry could then adopt if they so chose. See, for example, American Petroleum Institute, *Waste Water Containing Oil,* vol. 1 of *Manual on the Disposal of Refinery Wastes* (New York: American Petroleum Institute, 1930). Such reports were updated by the standing committees when conditions changed.

24. For an example of the oil industry's lobbying strategy on these issues, see Bronson Batchelor, *Stream Pollution: A Study of Proposed Federal Regulation and Its Effect on the Oil Industry* (New York: American Petroleum Institute, 1937).

25. Pratt, *Growth of a Refining Region,* 22, 84.

26. For a general overview of the expansion of the Houston economy, see Walter L. Buenger and Joseph A. Pratt, *But Also Good Business: Texas Commerce Banks and the Financ-*

ing of Houston and Texas, 1886–1986 (College Station: Texas A&M University Press, 1986); Marvin Hurley, *Decisive Years for Houston* (Houston: Houston Magazine, 1966); and Pratt, *Growth of a Refining Region,* 127–51.

27. Gorman's book provides a detailed and thorough examination of the oil industry's efforts at self-regulation in this era. See Gorman, *Redefining Efficiency,* 137–268.

28. Advisory Committee on Tetraethyl Lead to Surgeon General of Public Health Service, *Public Health Aspects of Increasing Tetraethyl Lead Content in Motor Fuel* (Washington, D.C.: Public Health Service, 1959), xiii.

29. See Pratt, "Letting the Grandchildren Do It," 41–52; and Gerald Markowitz and David Rosner, *Deceit and Denial: The Deadly Politics of Industrial Pollution* (Berkeley: University of California Press, 2002), 17–26.

30. Pratt, *Growth of a Refining Region,* 71.

31. The impact on the health of workers of the emissions from the early generations of chemical and petrochemical plants has been little studied. Although environment, health, and safety laws passed during and after the 1960s created more awareness of the air emissions from refineries and petrochemical plants, many workers had almost a quarter of a century exposure to such emissions before the coming of stricter standards. See, for example, Pratt, "Letting the Grandchildren Do It," 52–61; see also Markowitz and Rosner, *Deceit and Denial,* 195–286.

32. Pratt, "Letting the Grandchildren Do It," 57–58. For the letter itself, see Willard Denno (medical director, Standard of New Jersey) to C. W. Mitchell (surgeon, Bureau of Mines), October 11, 1922, file 022.17, General Classified Files, 1921, Records of the Bureau of Mines.

33. Pratt, "Coordination in the Modern Petroleum Industry."

34. For an overview of these developments, see Joseph A. Pratt, Tyler Priest, and Christopher J. Castaneda, *Offshore Pioneers: Brown & Root and the History of Offshore Oil and Gas* (Houston: Gulf, 1997), 1–33.

35. A more nuanced discussion of this key problem is found in Craig Colten, "Texas v. the Petrochemical Industry, Contesting Pollution in an Era of Industrial Growth," Center for Hazards and Environmental Geography, Department of Geography and Planning, Southwest Texas State University (now Texas State University), May 1998.

36. The first dependable supply of natural gas was piped into Houston in 1927. After World War II, a major new business, the transportation of natural gas from the southwestern United States to markets in the Midwest and Northeast, emerged with the construction of cross-country pipelines. This new industry was centered in Houston.

37. For an overview of California's early responses to air pollution, see James E. Krier and Edmund Ursin, *Pollution and Policy: A Case Essay on California and Federal Experience with Motor Vehicle Air Pollution, 1940–1975* (Berkeley: University of California Press, 1977).

38. Gorman, *Redefining Efficiency,* 242–44; Markowitz and Rosner, *Deceit and Denial,* 156–67.

39. W. B. Hart, an oil industry specialist who wrote extensively in the mid-twentieth century on the control of oil-related pollution, was a strong believer in self-regulation, but he noted that industry had often waited too long to find effective means for controlling the discharge of wastes. W. B. Hart, "Controlled Disposal of Wastes Versus Pollution," *Oil and Gas Journal,* May 14, 1936, 234. For a collection of his technical articles, see W. B. Hart, *Industrial Waste Disposal for Petroleum Refineries and Allied Plants* (Cleveland: Petroleum Processing, 1947).

40. Herbert C. McKee, *Air Pollution Survey of the Houston Area, 1964–66, Final Report*

of Project No. 21-1582 to Houston Chamber of Commerce (Houston: Southwest Research Institute, October 1966), vi.

41. For examples of the sensational coverage of pollution in this era, see Jim Curran, "Pollution Getting Worse in Houston," *Houston Chronicle*, June 23, 1970; Jim Rice and Jim Curran, "Severity of Air Pollution Here Called 'Unbearable,'" *Houston Chronicle*, June 24, 1970.

42. On the national level, see Robert Engler, *The Politics of Oil, Private Power and Democratic Directions* (New York: Macmillan, 1961). For Texas, see Green, *Establishment in Texas Politics*. For Houston, see Feagin, *Free Enterprise City*. See also Pratt, "8F and Many More," 2–7, 31–44.

43. Gorman, *Redefining Efficiency*, 267–365; Anthony E. Kopp, *Regulating Competition in Oil: Government Intervention in the U.S. Refining Industry, 1948–1975* (College Station: Texas A&M University Press, 1976).

44. Of the many good books on the environmental laws in this era, see Samuel P. Hays, *A History of Environmental Politics since 1945* (Pittsburgh: University of Pittsburgh Press, 2000); Richard N. L. Andrews, *Managing the Environment, Managing Ourselves: A History of American Environmental Policy* (New Haven, Conn.: Yale University Press, 1999); and Scott Hamilton Dewey, *Don't Breathe the Air: Air Pollution and U.S. Environmental Politics* (College Station: Texas A&M University Press, 2000).

45. For a useful history of the trans-Alaska pipeline, see Peter Coates, *The Trans-Alaskan Pipeline Controversy: Technology, Conservation, and the Frontier* (Bethlehem, Pa.: Lehigh University Press, 1991).

46. For views of the costs of refinery-related pollution and the refiners' compliance with new regulations, see Joan Norris Booth, *Cleaning Up: The Costs of Refinery Pollution* (New York: Council on Economic Priorities, 1975). See also John Durham and Margaret Blair, "Pollution Control May Cost $18 Billion," *Houston Chronicle*, May 27, 1975. This newspaper article reports a national estimate of the cost of business expenditures to meet pollution control requirements in the year 1974 as about $6 billion to $8 billion. The same article reports that interviews with people at nineteen industries on the Houston Ship Channel produced an estimate of more than $180 million in expenditures to meet pollution control requirements in the years from 1965 to 1975.

47. For an oil industry view of this problem, see National Petroleum Council, *Petroleum Refining in the 1990s—Meeting the Challenges of the Clean Air Act* (Washington, D.C.: National Petroleum Council, 1991).

48. Although representatives of the oil industry have made a strong argument that environmental regulations have made it most difficult to build a new "greenfield" refinery in the United States since the 1970s, considerable new and improved refining capacity has been built within existing refineries around the nation.

49. In January 2006, the *Houston Chronicle* ran a five-day series of articles on pollution in the Houston region entitled "In Harm's Way." These articles were as follows: "Dangers in the Air We Breathe," *Houston Chronicle*, January 16, 2005; Dina Cappiello, "Safeguards-Barely," *Houston Chronicle*, January 17, 2005; Leigh Hopper, "The Burden of Proof," *Houston Chronicle*, January 18, 2005; Dina Cappiello, "Unseen Dangers," *Houston Chronicle*, January 19, 2005; Dina Cappiello and Dan Feldstein, "In the Buffer Zone," *Houston Chronicle*, January 20, 2005. For an earlier report, see Jim Yardley, "Houston, Smarting Economic Ally from Smog, Searches for Remedies," *New York Times*, September 24, 2000.

50. For the response in Texas to oil spill preparedness after the *Exxon Valdez* spill, see Garry Mauro, Texas General Land Office, *Oil Spill Prevention and Response Act of*

1991: Recent History and New Solutions to a Major Threat on the Texas Gulf Coast (Austin: Texas General Land Office, May 1991). For background on the law and its impact on the oil industry, see National Petroleum Council, *The Oil Pollution Act of 1990: An Interim Report of the National Petroleum Council* (Washington, D.C.: National Petroleum Council, December 1993).

51. For an account of the building and impact of the first freeway in Houston, see Tom Watson McKinney, "Superhighway Deluxe: Houston's Gulf Freeway" (Ph.D. diss., University of Houston, 2006). For a recent account of the continuing spread of Houston, see Nancy Sarnoff, "Houston's Final Frontier: How Far Out Will We Go?" *Houston Chronicle,* June 5, 2005.

52. John Durham, "EPQA Is Apparently Abandoning Its Big Auto Cut Plan Here," *Houston Chronicle,* May 29, 1975. This story is one of many in this newspaper about the ongoing battle between local authorities and the EPA over the enforcement of the agency's plans to regulate auto-related air pollution in the region.

53. Clayton Forswall and Kathryn Higgins, *Clean Air Act Implementation in Houston: An Historical Perspective, 1970–1995* (Houston: Rice University, Shell Center for Sustainability, 2005). For an interview with Dr. Eleanor MacDonald, a pioneer in the study of the impact of air pollution on cancer risks, see Hopper, "Burden of Proof."

54. For discussions of green oil companies from different perspectives, see Joseph A. Pratt, *Prelude to Merger: A History of Amoco Corporation, 1973–1998* (Houston: Hart, 2000), 255–84; Neil Cunningham, Robert A. Kagan, and Dorothy Thornton, *Shades of Green: Business, Regulation, and Environment* (Stanford, Calif.: Stanford University Press, 2003); and Andrew Hoffman, *From Heresy to Dogma* (San Francisco: New Lexington Press, 1997).

Chapter 2.
The Houston Ship Channel

1. U.S. Army Corps of Engineers, U.S. Waterway Data, Waterborne Commerce of the United States, www.iwr/usace.army.mil/ndc.

2. Some economists argue that federal pollution control law has had no effect on air quality; see Indur Goklany, *Clearing the Air: The Real Story of the War on Air Pollution* (Washington, D.C.: Cato Institute, 1999).

3. Steven Klineberg, "Houston Attitudes Concerning the Environment," paper presented at "An Environmental History of Houston and the Gulf Coast," Houston, March 14–15, 2003; see also Pratt, "Growth or a Clean Environment?"

4. National Center for Policy Analysis, Brief Analysis No. 322, April 25, 2000. In 1999, Houston's fifty-two days with ozone violations surpassed Los Angeles's forty-two days.

5. "Houston Air Is Focus of $20 Million Study," *Environment* 78 (November 6, 2000): 32–33.

6. Houston Ship Channel TMDL Project, Texas Commission on Environmental Quality, www.tceq.state.tx.us; Oliver A. Houck, *The Clean Water Act TMDL Program: Law, Policy, and Implementation* (Washington, D.C.: Environmental Law Institute, 1999).

7. Hugh S. Gorman, "Brownfields in Historical Context," *Environmental Practice* 5, no. 1 (2003): 21–24.

8. R. M. Farrar, *The Story of Buffalo Bayou and the Houston Ship Channel* (Houston: Chamber of Commerce, 1926).

9. Sibley, *Port of Houston,* 12–30.

10. Kelly F. Himmel, *The Conquest of the Karankawas and Tonkawas, 1821–1859* (College Station: Texas A&M University Press, 1999).

11. Lyle Saxon and E. H. Suydam, *Lafitte, the Pirate* (New Orleans: Crager, 1930).

12. Sibley, *Port of Houston*, 18–59.

13. Harris County, Texas, County Level Data, 1850 U.S. Decennial Census.

14. Sibley, *Port of Houston*, 60–78.

15. Ibid., 52.

16. Ibid., 79–120.

17. Martin V. Melosi, *The Sanitary City: Urban Infrastructure in America from Colonial Times to the Present* (Baltimore: Johns Hopkins University Press, 2001).

18. Years of Historic Houston, 1900–1910, http://houstonhistory.com/decades/history5g.htm.

19. Sibley, *Port of Houston*.

20. Harold F. Williamson and Arnold R. Daum, *The American Petroleum Industry: The Age of Illumination, 1859–1899* (Evanston, Ill.: Northwestern University Press, 1959); Pratt, *Growth of a Refining Region*.

21. Craig Thompson, *Gulf: A Human Story of Gulf's First Half Century* (Pittsburgh: Gulf Oil, 1951); Texas Company, *Texaco* (Houston: Texas Company, 1931); August W. Giebelhaus, *Business and Government in Industry: A Case Study of Sun Oil, 1876–1945* (Greenwich, Conn.: JAI Press, 1980); Pratt, "Growth or a Clean Environment?"

22. Farrar, *Story of Buffalo Bayou*.

23. *Map of the Houston Ship Channel*, reduction of "The Authorized Channel Map," Port Commission, 1925; Port of Houston Industrial Map, 1945, Industrial Maps, Box 16, J. Russell Wait Collection, Woodson Research Center, Rice University, Houston.

24. J. G. Burr to House, Rivers and Harbors Committee, January 19, 1921, Pollution of Waters, Commerce Papers, Herbert Hoover Presidential Library, West Branch, Iowa; W. W. Moore to A. E. Amerman, Mayor, City of Houston, December 11, 1920, J. S. Cullinan Papers, Folder 4, Box 77, J. S. Cullinan Collection, Houston Metropolitan Research Center, Houston Public Library, Houston.

25. A. E. Amerman, Mayor, City of Houston, to W. W. Moore, December 13, 1920, Folder 4, Box 77, J. S. Cullinan Collection.

26. Benjamin Casey Allin III, *Reaching for the Sea* (Boston: Meador, 1956), 102.

27. Gorman, *Redefining Efficiency*, 13–33.

28. Pratt, "Letting the Grandchildren Do It"; Douglas Drake, "Herbert Hoover, Ecologist: The Politics of Oil Pollution Control, 1921–1926," *Mid-America* 55 (1973): 207–28.

29. Statement of Van H. Manning, Senate Committee on Commerce, *Pollution of Navigable Waters*, 58th Cong., 1st sess., January 9, 1924, 93–94; the numbers he gave were on the optimistic side of being accurate.

30. For a general examination of the efficiency ethic, see Samuel P. Hays, *Conservation and the Gospel of Efficiency: The Progressive Conservation Movement, 1890–1920* (Cambridge, Mass.: Harvard University Press, 1959).

31. U.S. Bureau of Mines, *Pollution by Oil of the Coast Waters of the United States*, preliminary report (Washington, D.C.: Bureau of Mines, 1923); Oil Pollution Act of 1924, 43 Stat. 604.

32. American Petroleum Institute, *Waste Water Containing Oil;* American Petroleum Institute, *Waste Gases or Vapors*, vol. 2 of *Manual on the Disposal of Refinery Wastes* (New York: American Petroleum Institute, 1931); American Petroleum Institute, *Chemical Wastes*, vol. 3 of *Manual on the Disposal of Refinery Wastes* (New York: American Petroleum Institute, 1935).

33. U.S. Corps of Engineers, *Pollution Affecting Navigation or Commerce on Navigable*

Waters, H. Doc. 417, Report of the Secretary of War to the House of Representatives, 69th Cong., 1st. sess., 1926, 3; Vladimir A. Kalichevsky and Bert A. Stagner, *Chemical Refining of Petroleum: The Action of Various Refining Agents and Chemicals on Petroleum and Its Products* (New York: Chemical Catalog, 1933).

34. Norman Hurd Ricker, "The Autobiography of Norman Hurd Ricker," unpublished ms., Box 23, Norman Hurd Ricker Papers, Woodson Research Center.

35. Gorman, *Redefining Efficiency*, 137–268.

36. John Ise, *The United States Oil Policy* (New Haven, Conn.: Yale University Press, 1926); American Bar Association, *Section of Mineral Law, Legal History of Conservation of Oil and Gas* (Chicago: American Bar Association, 1938).

37. Hugh S. Gorman, "Efficiency, Environmental Quality, and Oil Field Brines: The Success and Failure of Pollution Control by Self-Regulation," *Business History Review* 73 (Winter 1999): 601–40.

38. Federal Oil Conservation Board, *Complete Record of Public Hearings, February 10 and 11* (Washington, D.C.: Government Printing Office, 1926).

39. "Refinery Locations and Crude-Oil Capacities," supplement to the *Oil and Gas Journal* 48 (September 21, 1950); Standard Oil (N.J.) Marine Department, "Oil Refineries and Principle [*sic*] Oil Fields of the United States," January 1, 1927.

40. Hugh S. Gorman, "Manufacturing Brownfields: The Case of Neville Township, Pennsylvania, 1899–1989," *Technology and Culture* 38 (July 1997): 539–74; Craig E. Colten, *Industrial Wastes in the Calumet Area, 1869–1970: An Historical Geography* (Champaign: Illinois Department of Energy and Natural Resources, 1985).

41. Frank J. Metyko, "Industrial Effluents and Marine Pollution," address to the Marine Laboratory of the Texas Fish and Game Commission, Rockport, Tex., October 27, 1949.

42. Port of Houston Industrial Map, 1945, Industrial Maps, Box 16, J. Russel Wait Collection.

43. *Directory of Oil Refineries and Field Processing Plants* (Tulsa, Okla.: Oil and Gas Journal, 1952), 7–8.

44. W. A. Quebedeaux to Jean Paul Bradshaw, December 7, 1950, Folder 7, Box 7, Dr. W. A. Quebedeaux Environmental Collection, Houston Metropolitan Research Center; W. A. Quebedeaux, "Air and Stream Pollution Control in Harris County, Texas," *Public Health Reports* 69 (September 1954): 836–40.

45. Comments of John Latchford (Texas Water Quality Board), "The Houston Ship Channel," Folder 8, Box 5, Citizens' Environmental Coalition Collection, Houston Metropolitan Research Center.

46. Air Pollution Control Association, Directors, *Journal of the Air Pollution Control Association* 5 (August 1956): 114.

47. Dewey, *Don't Breath the Air*; Krier and Ursin, *Pollution and Policy*.

48. Joel A. Tarr, *The Search for the Ultimate Sink* (Akron, Ohio: University of Akron Press, 1996), 227–61.

49. Los Angeles County Air Pollution Control District et al., *Proceedings of the Southern California Conference on Elimination of Air Pollution, November 10, 1955* (Los Angeles: Los Angeles County Air Pollution Control District, 1955).

50. W. A. Quebedeaux, "Comments on Particulate Pollutants in the Air of the United States," *Journal of the Air Pollution Control Association* 10 (April 1960): 144, 171.

51. Quebedeaux, "Air and Stream Pollution Control."

52. Remarks of Mayor Louie Welch, Manufacturing Chemists Association Workshop

on Air Pollution Control, 1967, Box 4, Series III, CMDA Records, Beckman Center for the History of Chemistry, Philadelphia.

53. For example, in 1961, when one member of the Texas Railroad Commission retired after two decades of service, his associates honored him with a party held on a boat that sailed down the ship channel. See "My Trip Down the Houston Ship Channel," photograph album, Box 3S28, Olin Culberson Papers, Center for American History, University of Texas at Austin.

54. "Before It's Too Late—Make the Ship Channel Safe," *Houston Press*, February 18, 1960.

55. W. A. Quebedeaux, "Prosecution of Air Pollution Cases under Common Law Nuisance," *Journal of the Air Pollution Control Association* 12 (April 1962): 187–91, 242; W. A. Quebedeaux, "The Relationship of Water Pollution to Conservation," January 20, 1961, Folder 11, Box 3, Dr. W. A. Quebedeaux Environmental Collection; W. A. Quebedeaux, "Active Prosecution as the Key to Air Pollution Control," paper presented at the 1957 meeting of the Air Pollution Control Association, Folder 10, Box 3, Dr. W. A. Quebedeaux Environmental Collection. See also Harold W. Kennedy, "Fifty Years of Air Pollution Law," *Journal of the Air Pollution Control Foundation* 7 (August 1957): 125–40.

56. John C. Esposito and Larry J. Silverman, *Vanishing Air: The Ralph Nader Study Group on Air Pollution* (New York: Grossman, 1970): 200–203.

57. "How Humble Combats Water and Air Pollution," *Oil and Gas Journal* 64 (March 28, 1966): 132–36.

58. Terrence Kehoe, *Cleaning Up the Great Lakes: From Cooperation to Confrontation* (De Kalb: Northern Illinois University Press, 1997).

59. Darryl Randerson, "A Study of Air Pollution Sources as Viewed by Earth Satellites," *Journal of the Air Pollution Control Association* 18 (April 1968): 249–53.

60. Senate Subcommittee on Air and Water Pollution, *Hearings on Water Pollution,* 89th Cong., 1st sess., June 23 and June 24, 1965, 912; Robert Martin and Lloyd Symington, "A Guide to the Air Quality Act of 1967," in *Air Pollution Control,* ed. Clark C. Havighurst (Dobbs Ferry, N.Y.: Oceana, 1969), 43–78.

61. The organization is Hobil, and its philosophy is "It's the bicycles, stupid." Hobil removed its "ten reasons" Web pages in April 2004 in response to policy changes implemented by the mayor.

Chapter 3. "Bad Science"

1. Janice Harper, "Air Pollution and Asthma in the Private City," paper presented at the Environmental History Conference, University of Houston, March 14–15, 2003.

2. EPA report cited in Rick Abraham, *The Oil and Chemical Dependency of George W. Bush* (Houston: Mainstream, 2000), 22.

3. Texas Natural Resources and Conservation Commission, "Periodic Emissions Inventory," 1999, Texas Archival Resources Online, www.lib.utexas.edu/taro.

4. Raj Chhikaraand Floyd Spears, "Toxic Air Pollution Evaluation Patterns in the Houston Area," in *Environmental Institute of Houston Annual Report* 1998, www.eih.uh.edu/publications/98annrep/chhikara.htm.

5. Texas Natural Resource Conservation Commission, "Ozone Air Pollution Facts," February 2001, 1, Texas Archival Resources Online, www.lib.utexas.edu/taro.

6. Texas Natural Resource Conservation Commission, "Ozone Forecast Accuracy Based on Measured Ozone Levels," www.tnrcc.state.tx.us/air/monops/ozonestats.html.

7. Ibid.; GHASP, John D. Wilson, "Public Outcry Over Threat to Houston Clean Air Plan," May 31, 2002, News Release.

8. Mothers for Clean Air, Texas Air Quality Study 2000, 1, www.mothersforclearair.org.

9. Dewey, *Don't Breathe the Air,* 8.

10. Mothers for Clean Air, "Air Quality Basics," 2, www.mothersforclearair.org.

11. Rafael Gerlein and Jerry Wood, "Air Quality," 2003, www.centerforhoustonsfuture.org.

12. Hays, *History of Environmental Politics,* 109. Hays also sees this subject as a research area of critical importance and surprising neglect in the field.

13. Jacqueline V. Switzer, *Environmental Politics* (Belmont, Calif.: Wadsworth, 2004); Hays, *History of Environmental Politics.*

14. Hays, *History of Environmental Politics,* 149.

15. Dewey, *Don't Breathe the Air,* 9–10.

16. Switzer, *Environmental Politics,* 49.

17. Gregory Squires, "Partnership and the Pursuit of the Private City," in *Urban Life in Transition,* ed. M. Gottdeiner and C. G. Pickvance (Newbury Park, Calif.: Sage 1991), 197. Earlier, Sam Bass Warner Jr. referred to it as the culture of the American "private city": "What the private market could do well American cities have done well; what the private market did badly, or neglected, our cities have been unable to overcome." Sam Bass Warner, *The Private City: Philadelphia in Three Periods of Growth* (Philadelphia: University of Pennsylvania Press, 1968). For a cultural perspective on Houston's privatism, see P. Lopate, "Pursuing the Unicorn: Public Space in Houston," *Cite* (Winter 1984): 19–22.

18. This meant developing public-sector programs to support private ventures, such as the public-sector dredging of the Houston Ship Channel during the First World War, while keeping the costs of doing business to a minimum, especially for the oil industry so fundamental to the city's economy. Public provision of services was not intended as part of the political culture, except when urban elites sought it, such as after 1982 when the price of oil collapsed and sent the city (and the state) into a downward spiral. Then the private sector realized it needed explicit public partnerships to achieve its ends. For example, see P. Lupsha and W. Siembieda, "The Poverty of Public Services in the Land of Plenty," in D. Perry et al., *The Rise of the Sunbelt Cities* (Thousand Oaks, Calif.: Sage, 1998), 169–91; and R. Thomas, and Richard Murray, *Progrowth Politics: Change and Governance in Houston* (Berkeley: Institute of Government Studies, University of California, 1991).

19. Don Carleton, *Red Scare!: Right Wing Hysteria, 50s Fanaticism, and Their Legacy in Texas* (Austin: Texas Monthly Press, 1985).

20. A. Schaffer, "The Houston Growth Coalition in 'Boom' and 'Bust,'" *Journal of Urban Affairs* 11, no. 1 (1989): 21–38. The Greater Houston Partnership was formally created in 1989 when the Houston Chamber of Commerce merged with the Houston World Trade Association and the Houston Economic Development Council.

21. Robert Fisher, "Organizing in the Private City: The Case of Houston, Texas," in *Black Dixie: Afro-Texan History and Culture in Houston,* ed. Howard Beeth and Cry D. Wintz (College Station: Texas A&M University Press, 1992), 253–76.

22. Herbert McKee, "How Can We Improve Air Pollution Control?" October 2001, 42, www.capmagazine.org. McKee is a former director of Houston's environmental regulatory programs.

23. Harris County Health Department, "History," 1, www.harriscountyhealth.com.

24. Jane Hamilton, "History of the Clean Air Act in Houston," unpublished paper, 1999.

25. McComb, *Houston: The Bayou City,* 150.

26. Dewey, *Don't Breathe the Air,* 10.

27. Ibid., 228, 244.

28. Ibid., 229

29. Switzer, *Environmental Politics.*

30. This was the opposite argument that business leaders used when Houston became the worst ozone city in the nation in 1999. Then they proposed it was due to Houston's stagnant weather patterns, something out of their control.

31. McComb, *Houston: The Bayou City,* 151.

32. Ibid.

33. Ibid.

34. Dewey, *Don't Breathe the Air,* 246.

35. Environmental Literacy Council, "The Clean Air Act," www.enviroliteracy.org.

36. Switzer, *Environmental Politics,* 233; McKee, "How Can We Improve Air Pollution Control?" 42.

37. Environmental Protection Agency (EPA) press release, "Transportation Controls Established in Major Urban Areas to Lower Air Pollution Levels," October 15, 1973, www.epa.gov/history/topics/caa70/10.htm.

38. One weakness of the legislation was to heavily or completely exempt or "grandfather" approximately 40 percent of industrial sources of pollution.

39. Doyle Pendleton, "Environmental Health Issues," *Environmental Health Perspectives* 1995 103/S6, http://ehp.niehs.nih.gov/members/1995/Suppl-6/pendelton-full.html.

40. Harris County Health History, ibid.

41. "Off the Kuff," November 16, 2002. www.offthekuff.com/mt/archives/001048.html.

42. The one-hour design values index measures the fourth-highest one-hour value over three consecutive years with sufficient data capture and calculated site by site.

43. John Wilson, personal communication, November 1, 2004. The Houston Area Research Center (HARC), for example, proposes that ozone exceedance days declined in Houston from a high of eighty in the 1980s to forty in the early 1990s. www.harc.edu/4site/4siteFCRair.html. Using Ozone Trend Data, the Coalition for Clean Air Progress noted a 37 percent reduction between 1980 and 1999, which translates to a reduction of 23.3 exceedance days from an average 62.3 days in 1980–82 to one of 39.0 days in 1997–99. www.cleanairprogress.org/top_20/number13.asp.

44. McComb, *Houston: The Bayou City.*

45. McKee, "How Can We Improve Air Pollution Control?"

46. EPA press release, "Transportation Controls Established in Major Urban Areas."

47. Bill Dawson, "Air Quality Efforts Hit a New Level," *Houston Chronicle,* February 19, 1995.

48. Alfonso Holguin et al., "The Effects of Ozone on Asthmatics in the Houston Area," in *Evolution of the Scientific Basis for Ozone/Oxidant Standard,* ed. S Lee, Proceedings of the Air Pollution Control Association, International Specialty Conference, Houston, November 1984.

49. Gerlein and Wood, "Air Quality."

50. Hamilton, "History of the Clean Air Act."

51. Switzer, *Environmental Politics,* 235.

52. McKee, "How Can We Improve Air Pollution Control?" 43; Marilu Hastings, "Sorting Out the Topic of Air Pollution," March 1, 2000, HARC press release, www.harc .edu/harc/content/newsevents/shownews.aspx/146.

53. Bill Dawson, "Economy, Environment Clash," *Houston Chronicle,* April 15, 1990. This position was even repudiated by the Ozone Task Force.

54. Ibid.

55. Ron Embry, Exxon spokesperson, memo cited in Harper, "Air Pollution and Asthma."

56. Dawson, "Economy, Environment Clash."

57. Ibid.

58. V. Godines, "Levels Prompt City Health," *Houston Chronicle,* August 11, 1993.

59. Dewey, *Don't Breathe the Air,* 95.

60. Bill Dawson, "Business Leaders Try to Gain Delay on Smog Alerts," *Houston Chronicle,* March 4, 1994.

61. Ibid.

62. Bill Dawson, "Smog-Program Postponed," *Houston Chronicle,* August 27, 1994.

63. Bill Dawson, "Smog Alert Program Disperses," *Houston Chronicle,* May 27, 1995.

64. Ibid.

65. Bill Dawson, "Task Force Says Smog Forecasts Could Help Protect Public Health," *Houston Chronicle,* June 16, 1995.

66. Bill Dawson, "Lanier Approves Smog Alerts for City," *Houston Chronicle,* August 4, 1995.

67. W. Stockwell et al., "The Scientific Basis of NOAA's Air Quality Forecasting Program," *EM* (December 2002): 20.

68. Ibid.

69. Ibid., 27.

70. For the same period, it was nineteen days in 1994, twelve in 1993, twenty-six in 1992, and twenty-seven in 1991.

71. Bill Dawson, "Dog Day, Smog Day," *Houston Chronicle,* September 12, 1995.

72. Bill Dawson, "City's Smog Data to Go Nationwide," *Houston Chronicle,* September 27, 1977.

73. Texas Natural Resource Conservation Commission, "Ozone Forecast Accuracy Based on Measured One-Hour Ozone Levels," www.tnrcc.state.tx.us/air/monops/ozone stats.html.

74. Robert Fisher, "Houston's Air: It's More Than You Expected," *Houston Chronicle,* November 16, 1997.

75. Abraham, *Oil and Chemical Dependency.*

76. P. Crimmins, "TNRCC Sets Position on EPA's Proposed Air Quality Standards," Texas Natural Resource Conservation Commission media release, March 6, 1997.

77. Blackburn, cited in Harper, "Air Pollution and Asthma."

78. Judith Garber and Robyne Turner, "Growth, Decline, and Planning: Urban Political Economy in Houston and Edmonton," paper presented at the Shaping the Urban Future conference, Bristol, England, July 10–13, 1994.

79. Michael King, "Who's Poisoning Texas?: Voluntary Grandfathers and Big Dirty Secrets," *Texas Observer,* May 8, 1998.

80. Gerlein and Wood, "Air Quality."

81. Texas Chemical Council, "Our Industry," 1, http://209.235.208.145/cgi-bin/websuite/tcsAssnWebSuite.

82. P. O'Driscoll, "Checking Ozone Levels Becoming Routine for Many," *USA TODAY,* August 20, 2001.

83. American Lung Association, *State of the Air, 2003,* http://lungaction.org/reports/SOTA03_stateozone.html.

84. Lambeth, cited in "Off the Kuff," November 16, 2002, www.offthekuff.com/mt/archives/001048.html. See also Gerlein and Wood, "Air Quality."

85. GHASP, "Ozone Smog," 2001, www.ghasp.org/issues/smog.html.

86. GHASP, "How Bad Is Houston's Smog," January 2002, 1, www.ghasp.org.

87. Bill Dawson, "Smoggy Air Apparent," *Houston Chronicle,* November 7, 1997.

88. O'Driscoll, "Checking Ozone Levels."

89. Brian Lambeth of the Texas Natural Resource Conservation Commission concurs that weather heavily contributes to the ups and downs of Houston's annual ozone pollution. www.offthekuff.com/mt/archives/001048.html.

90. GHASP, "How Bad Is Houston's Smog," 1.

91. Harper, "Air Pollution and Asthma."

92. Wilson, personal communication, November 1, 2004.

93. Greater Houston Partnership press release, "Partnerships Receives EPA Award: Work on Air Quality Improvement Plan Cited," October 16, 2001, 1, www.houston.org/mediaRelations/pressReleases/101601.html.

94. Texas Chemical Council, "Our Industry," 1.

95. Foundation for Clean Air Progress, "Top 20 Improved Areas," www.cleanairprogress.org/top_20/number13.asp; "Ozone Linked to Premature Deaths in Urban Areas," *Hartford Courant,* November 17, 2004.

96. Houston Area Research Council, "Houston Environmental Foresight: Seeking Environmental Improvement," www.harc.edu/4site/4siteFCRair.html; Greater Houston Partnership, "Straight Talk about Houston," www.houston.org/aboutHouston/thefacts.htm. Another document states that the pollution in the northwestern part of the city results primarily from the prevailing summer winds. Gerlein and Wood, "Air Quality."

97. Dawson, "Air Quality Efforts Hit a New Level."

98. John Wilson, personal communication, July 23, 2004.

99. See, for example, Thomas Coles, *No Color Is My Kind: The Life of Eldrewey Stearns* (Austin: University of Texas Press, 1997). For a splendid look at Houston as a not atypical city that sought modern urban innovations just like other major cities, see Platt, *City Building in the New South.*

100. Brad Tyer, "Harris County Hijacking: How Big Business Pirated Enviro Funds to Fund Its Own Non-Compliance," *Texas Observer,* May 10, 2002, www.texasobserver.org/show/Article.asp?ArticleI.

101. Greater Houston Partnership, "Federal Ozone Standard," 2002, www.houston.org/governmentrelations/federalIssues2002/federalOzoneStandard.htm.

102. Ibid.

103. Dewey, *Don't Breathe the Air,* 252.

104. Juan Lozano, "Group Says Smog Was Particularly Bad This Week in Houston," October 1, 2005, Associated Press Wire, http://web.lexis-nexus.com/universe/document?_m=b4d.

105. Tyer, "Harris County Hijacking."

106. Ibid.

107. John D. Wilson, "Clear Houston's Air of 131,000 Tons of Pollution," *Houston Chronicle,* June 4, 2002.

108. Tyer, "Harris County Hijacking."

109. Ibid.

110. Greater Houston Partnership, "Federal Ozone Standard," 2002.

111. Kenneth Chilton and Christopher Boerner, "Health and Smog: No Cause for Alarm," *Regulation* 18, no. 3 (1995), www.cato.org/pubs/regulation/reg18n3f.html.

112. Wilson, personal communication, July 23, 2004.

Chapter 4. "The Air-Conditioning Capital of the World"

1. George Fuermann, *Houston: Land of the Big Rich* (Garden City, N.Y.: Doubleday, 1951), 198.

2. For a discussion of interior space and modern landscapes, see "Introduction: In Search of Nature," in *Uncommon Ground: Rethinking the Human Place in Nature,* ed. William Cronon (New York: W. W. Norton, 1996).

3. For a collection of essays regarding the Sun Belt, see Raymond Mohl, ed., *Searching for the Sunbelt: Historical Perspectives on a Region* (Knoxville: University of Tennessee Press, 1990).

4. For comparisons of temperatures and relative humidity between different cities in the United States, see National Climatic Data Center, "United States Climate Normals, 1971–2000: Normal Daily Mean Temperatures, Degrees F," http://www.ncdc.noaa.gov/oa/climate/online/ccd/nrmavg.txt; and "United States Climate Normals, 1971–2000: Average Relative Humidity," http://www.ncdc.noaa.gov/oa/climate/online/ccd/relhum.txt.

5. National Climatic Data Center, "United States Climate Normals, 1971–2000: Mean Number of Days Maximum Temperature 90 Deg. F or Higher," http://www.ncdc.noaa.gov/oa/climate/online/ccd/mxge90.txt.

6. Fuermann, *Houston,* 201.

7. James C. Cobb, "The Sunbelt South: Industrialization in Regional, National, and International Perspective," in Mohl, *Searching for the Sunbelt,* 32.

8. Houston City Planning Commission, *Houston Year 2000: Report and Map* (Houston: Houston City Planning Commission, 1980), V-13; Marguerite Johnson, *Houston: The Unknown City, 1836–1946* (College Station: Texas A&M University Press, 1991), 131.

9. Department of Commerce, Bureau of the Census, *U.S. Census of Population and Housing, 1990.*

10. William Henry Kellar, *Make Haste Slowly: Moderates, Conservatives, and School Desegregation in Houston* (College Station: Texas A&M University Press), 43–44.

11. Martin V. Melosi, "Dallas–Fort Worth: Marketing the Metroplex," in *Sunbelt Cities: Politics and Growth since World War II,* ed. Richard M. Bernard and Bradley R. Rice (Austin: University of Texas Press, 1983), 178–79; Feagin, *Free Enterprise City.*

12. Kellar, *Make Haste Slowly,* 77; "Minister Charges Life Threatened Over Segregation," *Houston Chronicle,* July 29, 1954.

13. Department of Commerce, Bureau of the Census, *U.S. Census of Population and Housing, 1970.*

14. Henry Miller, *The Air-Conditioned Nightmare* (New York: New Directions Books, 1945); Frank Putnam, "Houston, Texas, an Inland Seaport," *New England Magazine* (1907): 368, in George Fuermann Collection (GFCUH), Box 1, Folder 5, Special Collections, University of Houston.

15. Bernard Kalb, "Air-Conditioned Life Brings Change," *New York Times,* May 15, 1955.

16. Raymond Arsenault, "The End of the Long Hot Summer: The Air Conditioner and Southern Culture," *Journal of Southern History* 50 (November 1984): 610.

17. Advertisement for "Famous Rheem 1954 Model Air Conditioners" in *Houston Chronicle,* August 3, 1954.

18. "ACRMA Gives Sales Figures on Room Conditioners," *Heating, Piping and Air Conditioning* 25 (January 1953): 176; "Room Air-Conditioner Sales Hit 1.3 Million," *Heating, Piping and Air Conditioning* 27 (November 1955): 121.

19. Gail Cooper, *Air-conditioning America: Engineers and the Controlled Environment, 1900–1960* (Baltimore: Johns Hopkins University Press, 1998), 147.

20. Marsha E. Ackermann, *Cool Comfort: America's Romance with Air-Conditioning* (Washington, D.C.: Smithsonian Institution Press, 2002), 110.

21. Stanley Walker, "Houston: Coolest Spot in the U.S.," *New York Times,* May 15, 1955.

22. Cooper, *Air-conditioning America,* 146–47.

23. George Fuermann, "Post Card" (ca. 1957), quoting an anonymous Houston minister: "It's kind of hard to preach here. Hell just doesn't have many terrors for anyone who lives through one of these Houston summers," GFCUH, Box 39, Folder 6.

24. Fuermann, *Houston,* 166.

25. Samuel R. Lewis, "Church Air Conditioning Solves Many Problems," *Heating, Piping and Air Conditioning* 24 (December 1952): 107; Morgan M. Miles, "A Minister Tells Why Air Conditioning for Churches Is Necessary," *Heating, Piping and Air Conditioning* 27 (July 1955): 104–5; David C. Pfeiffer, "Storage Air Conditioning System Boosts Attendance at Big Church," *Heating, Piping and Air Conditioning* 22 (February 1950): 73.

26. Walker, "Houston."

27. Kalb, "Air-Conditioned Life Brings Change."

28. Guy Furgielle, "Air Condition Bank of the Southwest," *Heating, Piping and Air Conditioning* 24 (April 1957): 104–6; Cooper, *Air-conditioning America,* 159; "News of the Month: Houston," *Heating and Ventilating* 51 (May 1952): 122.

29. "Big Building Air Conditioning to Boom," *Heating, Piping and Air Conditioning* 28 (March 1956): 120.

30. "Texas Air Conditioners Strike," *New York Times,* July 28, 1954; "Air-Condition Men Picket 10 Buildings," *Houston Chronicle,* July 28, 1954.

31. Kalb, "Air-Conditioned Life."

32. "Air Cooling Strike Off," *Houston Chronicle,* July 30, 1954.

33. Houston City Planning Commission, *Comprehensive Plan-Houston Urban Area* (Houston: Houston City Planning Commission, 1958), 100; David Goldfield, *Region, Race, and Cities: Interpreting the Urban South* (Baton Rouge: Louisiana State University Press, 1997), 295–96.

34. Department of Commerce, Bureau of the Census, *U.S. Census of Population and Housing, 1960.*

35. Houston City Planning Commission, *Comprehensive Plan,* 78.

36. For a brief history of the Houston wards, see Jeannie Kever, "Pride Lives On in City's Six Historical Wards," *Houston Chronicle,* September 6, 2004.

37. "Water, Power Use Hit All-Time High," *Houston Chronicle,* July 28, 1954.

38. "The Ebey Story," *Houston Post,* July 12, 1954; "Rush to Suburbs Just Starting: Sensational Growth Ahead in Next 20 Years," *U.S. News and World Report,* March 2, 1956, 37; "History of the City of West University Place, Texas," (1959), 5, GFCUH, Box 38, Folder 9.

39. "Texas Suburban Shopping," *Texas Business Review* 29, no. 8 (August 1955): 1, GFCUH, Box 38, Folder 5.

40. Ibid., 106.

41. Houston City Planning Commission, *Houston Year 2000,* II-10.

42. Melosi, *Effluent America,* 195–96.

43. Richard Rutter, "Air Conditioning Wooing Markets," *New York Times,* August 2, 1964.

44. *Houston Post,* September 14, 1961.

45. For an excellent historiography regarding suburban malls and city growth, see Kenneth T. Jackson, "All the World's a Mall: Reflections in the Social and Economic Consequences of the American Mall," *American Historical Review* 101 (October 1996): 1111–21.

46. Department of Commerce, Bureau of the Census, *U.S. Census of Population and Housing, 1960.*

47. Walker, "Houston."

48. "How's the Climate?" (1952), from "a pamphlet about Houston published by the Prudential Insurance Company of America in 1952 for guidance of employees being transferred from New Jersey to Southwestern Home Office here," GFCUH, Box 39, Folder 6.

49. "Factory Air Conditioning for Worker Comfort is Definite Trend," *Heating, Piping and Air Conditioning* 18 (October 1956): 89.

50. National Advisory Commission on Civil Disorders, *Report of the National Advisory Commission on Civil Disorders* (New York: E. P. Dutton, 1968), 22.

51. Department of Commerce, Bureau of the Census, *U.S. Census of Population and Housing, 1970.*

52. Ernest Bailey, "Closer-In Living on Rise?" *Houston Post,* May 24, 1964; "Twenty Sixteen Main, Houston's First Downtown High Rise in Thirty-Seven Years," advertisement for the opening of a deluxe, air-conditioned high rise in *Houston Chronicle,* April 11, 1965.

53. *Houston Post,* July 7, 1967.

54. For a good discussion of "law and order" politics, see Dan T. Carter, *The Politics of Rage: George Wallace, the Origins of the New Conservatism, and the Transformation of American Politics* (Baton Rouge: Louisiana State University Press, 1995), 349.

55. "Hoover Terms Black Panthers 'Most Dangerous,'" *Houston Chronicle,* July 14, 1970.

56. Martha E. Ackermann, *Cool Comfort: America's Romance with Air-Conditioning* (Washington, D.C.: Smithsonian Institution Press, 2002), 122.

57. "Houstonians Sizzle with 100 Reading," *Houston Chronicle,* July 10, 1969.

58. Rutter, "Air Conditioning Wooing Markets."

59. Ibid.; Cooper, *Air-conditioning America,* 60; "Houston's Climate: Ideal the Year 'Round," from the Houston Chamber of Commerce, date unknown (most likely late 1960s per its place in the archival collection), GFCUH, Box 39, Folder 6.

60. Department of Commerce, Bureau of the Census, *U.S. Census of Population and Housing, 1970.*

61. Barry J. Kaplan, "Houston: The Golden Buckle of the Sunbelt," in Bernard and Rice, *Sunbelt Cities,* 197–98; Stuart E. Jones and William Albert Allard, "Houston: Prairie Dynamo," *National Geographic* 132, no. 3 (September 1967), 349.

62. Jim Schefter, "NASA Moves Control of Gemini from Cape Kennedy to Houston," *Houston Chronicle,* April 11, 1965.

63. "'Team Effort,'" *Newsweek,* July 7, 1969, 55–56.

64. "Record 48,145 Jampack Dome," *Houston Chronicle,* April 11, 1965.

65. Arthur Daley, "Indoors and Out," *New York Times,* April 14, 1965.

66. "Astrodome: Houston, Texas," promotional flier for the 1965 Houston Astros season, George Kirksey Papers, Box 7, Folder 19, 1965, HAS-Astrodome, Special Collections, University of Houston.

67. Mayor Louie Welch, in *Inside the Astrodome: Eighth Wonder of the World,* 6, George Kirksey Papers, Box 7, Folder 20, 1965.

68. Billy Graham, in "Astrodome."

69. "Astrodome."

70. Jim Maloney, "Study Aimed at Staving Off Blight in Downtown," *Houston Post,* September 30, 1962, GFCUH, Box 31, Folder 8; Len Billingsley, "Economist Is Secretary of Development Project," *Houston Post,* July 21, 1963.

71. Barrie Zimmelman, in the pamphlet "A Report to the Downtown Houston Banks on the Need for a Central Houston Association" (July 1968); excerpt from *Texas Magazine,* January 9, 1966, in Fuermann Collection, Box 31, Folder 8.

72. *Houston Downtown Tunnel System* (Houston: Houston City Planning Department, 1977), 1, 8–9; George Fuermann, "Post Card—Tuesday, October 3, 1961," Fuermann Collection, Box 31, Folder 8; "Downtown Tunnel Network," *Houston Chronicle,* October 29, 1967, in Fuermann Collection, Box 31, Folder 8.

73. Ralph Dodd, "Downtown Slum Area Is Bought," *Houston Post,* January 24, 1964, Fuermann Collection, Box 31, Folder 8.

74. Christopher J. Cantaneda and Joseph A. Pratt, *From Texas to the East: A Strategic History of Texas Eastern Corporation* (College Station: Texas A&M University Press, 1993), 178–87, 226–35.

75. Melosi, *Effluent America,* 197–98.

76. William K. Stevens, "Poor and Elderly People in Sun Belt Cities Suffer an Unremitting Misery," *New York Times,* July 21, 1980.

77. Marylin Bender, "Shell's Move Two Years and $35-Million Later," *New York Times,* January 16, 1972.

78. Patrick Horsbaugh, "Blight: A Foretold Affliction," *Texas Architect* 16, no. 5 (April 1966), in Fuermann Collection, Box 29, Folder 3.

79. Planning Commission, *Houston Year 2000: Report and Map,* II-3; James P. Sterba, "Houston Tangles with the Problems of Success," *New York Times,* December 16, 1977.

80. Melosi, *Effluent America,* 197–98.

81. Ray Waldrep, "Parents Try to Avoid Integration by Turning to Private Schools," *Houston Chronicle,* July 12, 1970; "Segregation in Private Schools Not Tax-Exempt," *Houston Chronicle,* July 13, 1970.

82. Kevin Fox Gotham, *Race, Real Estate, and Uneven Development: The Kansas City Experience, 1900–2000* (Albany: State University of New York Press, 2002), 10.

83. James P. Sterba, "Infectious Mosquitoes Plaguing Houston," *New York Times,* August 28, 1976.

84. The growth of activism against suburban sprawl took shape across the country beginning in the late 1950s. See Adam W. Rome, *The Bulldozer in the Countryside: Suburban Sprawl and the Rise of American Environmentalism* (Cambridge: Cambridge University Press, 2001).

85. Morris Ketchum Jr., "Blueprint for the Future," *Texas Architect* 16, no. 5 (May 1966), GFCUH, Box 29, Folder 3; Judy Carmack York, "An Architect for the People," *Houston Chronicle*, July 12, 1970.

86. Sterba, "Problems of Success."

87. Department of Commerce, Bureau of the Census, *U.S. Census of Population and Housing, 1980*.

88. Stevens, "Poor and Elderly People."

89. J. Murray Mitchell Jr., "Warmer Climate and City Growth Spur Air Conditioning," *Heating, Piping and Air Conditioning* 28 (November 1956): 118–20.

90. J. Marshall Shepherd and Steven J. Burian, "Detections of Urban-Induced Rainfall Anomalies in a Major Coastal City," *Earth Interactions* 7 (2003): 2; Eric Berger, "Hot? Blame the Pavement," *Houston Chronicle*, June 7, 2003.

91. Leigh Hopper, "Heat Wave Heralds a Deadly Season," *Houston Chronicle*, June 1, 2003.

92. Ibid.

93. Houston is not unique in dealing only haphazardly with dangerous heat waves. Regarding Chicago, see Eric Klinenberg, *Heat Wave: A Social Autopsy of Disaster in Chicago* (Chicago: University of Chicago Press, 2002); for Paris, see Jean de Kervasdoué, "Jarring Lessons of a Tragedy," *World Press Review* 50 (November 2003); and Elaine Sciolino, "France Addresses Issue of Another Hot Summer," *New York Times*, June 13, 2004.

94. Clifford Pugh, "It's Still Lonely Being Green," *Houston Chronicle*, February 16, 2003.

95. Kathy Huber and Dina Cappiello, "'Green' Roof a Cool Hybrid of Ecology, Engineering," *Houston Chronicle*, November 29, 2003.

96. Jim Yardley, "Last Innings at a Can-Do Cathedral," *New York Times*, October 3, 1999.

97. Eric Hanson, "Field of Memories," *Houston Chronicle*, October 2, 1999.

98. Claudia Feldman, "Hot Energy Tips Help You Keep Your Cool and More of Your Money," *Houston Chronicle*, July 18, 2001.

99. Jeffrey Ball, "Air Conditioners Get New Rules on Energy Use," *Wall Street Journal*, March 18, 2004.

100. Stevens, "Poor and Elderly."

Chapter 5. Houston's Public Sinks

I would like to acknowledge Tom Watson McKinney for gathering important data, especially for the post-1945 sections of the chapter.

1. Material on Houston's sanitary services prior to 1945 was drawn from Martin V. Melosi, "Sanitary Services and Decision-Making in Houston, 1876–1945," *Journal of Urban History* 20 (May 1994): 365–406.

2. Joe R. Feagin's *Free Enterprise City: Houston in Political and Economic Perspective* best embodies this interpretation of Houston.

3. See Melosi, *Sanitary City*, 74, 117–48. For Galveston's water supply system, see Janice Ranee Clark, "Fresh Water for Galveston: The Development of a Supply System from the Texas Mainland" (Master's thesis, University of Houston, 2000).

4. Bud A. Randolph, "The History of Houston's Water Supply," *Texas Commercial News* (June 1927): 43.

5. Planning and Development Department, *Public Utilities Profile for Houston*, III-15.

6. Charles D. Green, *Fire Fighters of Houston, 1838–1915* (Houston: Dealy-Adey, 1915), 13, 21; Houston Fire Museum, *The History of the Houston Fire Department, 1838–1988*, text

by Thomas A. McDonald and F. Scott Mellott (Houston: Taylor, 1988), 8; H. H. Page, comp., *Houston and Harris County Facts* (Houston: Facts, 1939), 98.

7. "First Survey of Houston's Water Supply Was Prompted by 1869 Yellow Fever Epidemic," *Houston Chronicle,* October 26, 1933; Page, *Houston and Harris County Facts,* 98.

8. Platt, *City Building in the New South,* 67.

9. T. Lindsay Baker, "Houston Waterworks: Its Early Development," *Southwest Water Works Journal* 56 (July 1974): 37; Andrew Morrison, *The City of Houston* (St. Louis: Englehardt, 1891), 15–17; City of Houston, Water Department, *Report of Director for Year 1942,* 2; "History of the Houston Water System," in Department of Utilities, City of Houston, *Water Service in Houston,* prep. Edna D. Wood (October 1953), 19; Platt, *City Building in the New South,* 67; McComb, *Houston: The Bayou City,* 87–89.

10. "History of the Houston Water System," 19; Green, *Fire Fighters,* 126.

11. See Morrison, *City of Houston,* 16; Platt, *City Building in the New South,* 67; Randolph, "Houston's Water Supply," 43; Houston Fire Museum, *Houston Fire Department,* 13; McComb, *Houston: The Bayou City,* 88–89.

12. McComb, *Houston: The Bayou City,* 89.

13. *Directory of the City of Houston, 1890–91,* 5.

14. "History of the Houston Water System," 19; Water Department, *Report of Director for Year 1942,* 3; City of Houston, *Public Water Supply System* (January 1948), 2; Baker, "Houston Waterworks," 37; Platt, *City Building in the New South,* 197–99.

15. See Houston, "City Council Minutes," Book M, June 16, 1902, 52–57; January 12, 1903, 336–37; Book N, April 5, 1904, 426–27; April 25, 1904, 458, 462–63; May 23, 1904, 537; April 17, 1905, 372–73; "Report of Chief of Fire Department," City of Houston, *Annual Report, 1903,* 105; Green, *Fire Fighters,* 126.

16. The trend toward municipal ownership was well under way in the United States in the 1870s; by 1899, 205 waterworks changed from private to public systems. See Letty D. Anderson, "The Diffusion of Technology in the Nineteenth Century American City: Municipal Water Supply Investments" (Ph.D. diss., Northwestern University, 1980), 104, 106; Joel A. Tarr, "The Evolution of the Urban Infrastructure in the Nineteenth and Twentieth Centuries," in *Perspectives on Urban Infrastructure,* ed. Royce Hanson (Washington, D.C.: National Academy Press, 1984), 30–31; and B. M. Wagner, "The Acquisition of Private Water Plants by Municipalities," *Journal of the American Water Works Association* 2 (March 1915): 25–41.

17. "Mayor's Message," City of Houston, *Annual Report, 1905,* 8–9.

18. Quoted in Water Department, *Report of Director for Year 1942,* 6.

19. "Report of Water, Light and Health Committee," City of Houston, *Annual Report, 1908,* 23; "Water Supply and Works," *Progressive Houston* 1 (August 1909): 1.

20. See "Report of Water Committee," City of Houston, *Annual Report, 1912,* 24.

21. Houston Fire Museum, *Houston Fire Department,* 28; Green, *Fire Fighters,* 127.

22. City of Houston, *Public Water Supply System* (January 1948), 4.

23. Fugate, "Houston's Water Supply," 1768–69.

24. "Report of Water Committee," City of Houston, *Annual Report, 1909,* 24.

25. See Melosi, *Sanitary City,* 123–26.

26. Department of Utilities, *Water Service in Houston,* 20; Water Department, *Report of Director for Year 1942,* 7–8; "Report of Water Committee," City of Houston, *Annual Report, 1909,* 25–31; "Report of Water Committee," City of Houston, *Annual Report, 1910,* 19, 22; "Report of Water Committee," City of Houston, *Annual Report, 1912,* 18–21, 23; *Progressive Houston* 4 (June 1912).

27. "Report of Water Commissioner," *City Book of Houston, 1914,* 106; "Mayor's Message," *City Book of Houston, 1914,* 80–81; *City Book of Houston, 1925,* 41; Randolph, "Houston's Water Supply," 122.

28. In 1900, the incorporated area covered nine square miles; in 1937, seventy-three square miles. By the 1940s, the Houston public water supply consisted of six interconnected water systems, each having its own wells and storage reservoirs, supplying a particular portion of the distribution network. See Clyde R. Harvill et al., "Maintenance of Chlorine Residual in the Distribution System," *Journal of the American Water Works Association* 34 (December 1942): 1797–98.

29. The decision grew out of the *Allred* case. James V. Allred, Texas attorney general, refused to approve a 1933 bond issue passed by the Houston City Council because he believed it violated state law and an earlier bond agreement entered into by the city in 1926. The city brought suit against the attorney general to compel him to approve the bonds, but the supreme court sided with Allred. See also Clyde R. Harvill, "The Houston Water System," *Southwest Water Works Journal* (October 1935): 20–21; Department of Utilities, *Water Service in Houston,* 20; Water Department, *Report of Director for 1942,* 9–10.

30. Department of Utilities, *Water Service in Houston,* 21. See also Page, *Houston and Harris County Facts,* 99. Regular chlorination of the water supply began in 1929; chloramination was instituted in 1933.

31. Fugate, "Houston's Water Supply," 1770–71.

32. The National Board of Fire Underwriters supported continuation of groundwater withdrawal, but its report stressed the "very poor" quality of the overall water system because of inadequate fire protection capability and deficiency in mains and fire hydrants in many large residential areas. See "Spending $4,000,000 to $6,000,000 on Water System Urged," *Houston Chronicle,* September 2, 1937.

33. Fugate, "Houston's Water Supply," 1772–74; "Water Supply for City Said to be Ample," *Houston Chronicle,* October 17, 1932; George B. Waters, "Controversy Over River Water and Well Supply Rages," *Houston Press,* June 8, 1938; "City Urged to Tap San Jacinto to Insure Adequate Water Supply," *Houston* (March 22, 1938): 1; Committee Report to Houston Water Board, September 12, 1938.

34. Alvord, Burdick, & Howson, "Report on an Adequate Water Supply for the City of Houston, Texas," Chicago, February 1938, 1–3, 75–76.

35. "New Plan for Water Supply to Be Offered," *Houston Chronicle,* February 3, 1939. See also Fugate, "Houston's Water Supply," 1175–78.

36. J. M. Nagle, "Houston Gets Needed Water," *American City* 60 (February 1945): 77.

37. The district was created by the legislature in 1937, empowered to develop the San Jacinto watershed and its tributaries under supervision of the Texas Board of Water Engineers. Harris County, in which most of Houston was located, was omitted from the district because of possible conflicts with the authority of the Harris County Flood Control District. See William W. McClendon, "The San Jacinto River Conservation and Reclamation District's Proposed Plan of Full Scale Development," *Slide Rule* (March 1945): 11; Water Department, *Report of Director for Year 1942,* 12–16; "Water Supply Dam to Be Built on San Jacinto," *Houston Post,* July 14, 1942; City of Houston, Utilities Department, *Engineering Report for Water Works Improvements,* January 17, 1944. The war years also exposed weaknesses in the existing water supply system, and the growing population made effective service for "fringe populations" more difficult. See Water Department, *Report of Director for Year 1942,* 26–27.

38. Lake Houston, the site of the dam, is a reservoir on the San Jacinto River located

twenty-five miles east-northeast of downtown Houston. Water Department, *Report of Director for Year 1942,* 16; McComb, *Houston: The Bayou City,* 146; Nagle, "Houston Gets Needed Water," 77; "Houston's Greater Water Supply Near," *Houston* 23 (August 1952): 8–9; "Its First Water Filter Plant," *American City* 67 (March 1952): 102; Water Supply and Conservation Committee, Houston Chamber of Commerce, *Water for the Houston Area* (December 1954), 3; U.S. Department of the Interior, U.S. Geological Survey, "Characteristics of Water-Quality Data for Lake Houston, Selected Tributary Inflows to Lake Houston, and the Trinity River near Lake Houston, August 1983–September 1990," *Water-Resources Investigations Report 99-4129* (1999), 2, 4; "'Water . . . Water . . . Everywhere!'" *Houston* 37 (May 1966): 75.

39. Melosi, *Sanitary City,* 90–99; Joel A. Tarr, "Sewerage and the Development of the Networked City in the United States, 1850–1930," in *Technology and the Rise of the Networked City in Europe and America,* ed. Joel A. Tarr and Gabriel Dupuy (Philadelphia: Temple University Press, 1988), 159–66.

40. Platt, *City Building in the New South,* 40.

41. "Harris County's Flooding History," Harris County Flood Control District, http://www.hcfcd.org/hcfloodhistory.html.

42. Ibid. See also "History: Houston and Brays Bayou" and "History: Flooding and Flood Policies," Rice University/Texas Medical Center, Brays Bayou Flood ALERT System, January 12, 2006, http://www.floodalert.org/BraysFAS/index.php?sPageID=History&sRadar=KHGX.

43. Elisabeth O'Kane, "'To Lift the City Out of the Mud': Health, Sanitation and Sewerage in Houston, 1840–1920," paper presented at the American Society for Environmental History Conference, Houston, 1991; Platt, *City Building in the New South,* 40; "Report of the City Engineer," City of Houston, *Annual Report, 1902,* 40.

44. McComb, *Houston: The Bayou City,* 89.

45. See O'Kane, "'To Lift the City,'" 11–13; McComb, *Houston: The Bayou City,* 89.

46. Joel A. Tarr, "The Separate vs. Combined Sewer Problem: A Case Study in Urban Technology Design Choice," *Journal of Urban History* 5 (May 1979): 308–12, 317–30; Melosi, *Sanitary City,* 153–61.

47. *Directory of the City of Houston, 1892–93,* 7; O'Kane, "'To Lift the City,'" 15–18.

48. Cost savings often was a consideration of the city council. See, for example, Houston, "City Council Minutes," Book J, November 9, 1896, 164.

49. An evaluation of the work to be done in several sewer districts indicated little uniformity. The 1902 *Annual Report,* 52, showed the following distribution of sewer lines in the city (in linear feet): First Ward (northwest of downtown): Sanitary, 9,750; Second Ward (east of downtown): Sanitary, 15,300, Combined, 4,000, Storm, 5,100; Third Ward (downtown): Sanitary, 77,600, Combined, 1,734, Storm, 12,500; Fourth Ward (west of downtown): Sanitary, 35,100, Storm, 3,400; Fifth Ward (north of downtown): Storm, 16,200, Force Main, 9,500; Sixth Ward (west of downtown): Sanitary, 2,000, Combined, 5,000.

50. Houston, "City Council Minutes," Book N, September 14, 1903, 101–2.

51. *Directory of the City of Houston, 1912,* 3; City of Houston, *First Annual Report of the City Engineer on Sewage Disposal* (February 1, 1916), 11–12; "Report of the City Engineer," City of Houston, *Annual Report, 1902,* 38–43; "Mayor's Message," City of Houston, *Annual Report, 1904,* 11; "Report of the City Engineer," City of Houston, *Annual Report, 1904,* 76; "Report of Sewer Inspector," City of Houston, *Annual Report, 1904,* 90–91; "Report of the Water, Light, and Health Committee," City of Houston, *Annual Report,*

1905, 24–25; "Report of the Water, Light, and Health Committee," City of Houston, *Annual Report, 1906*, 22–23; "Report of the City Sewer Inspector," City of Houston, *Annual Report, 1906*, 127; "Report of City Engineer," City of Houston, *Annual Report, 1908*, 73; "Engineering Department," *Municipal Book of the City of Houston, 1922*, 109.

52. "Report of City Engineer," City of Houston, *Annual Report, 1902*, 40–41; "Report of City Engineer," City of Houston, *Annual Report, 1909*, 110; "Report of Consulting Engineer," City of Houston, *Annual Report, 1909*, 135; "The Austin Street Reinforced Concrete Sewer," *Progressive Houston* 1 (December 1909); "Austin Street Sewer," *Progressive Houston* 1 (May 1909).

53. McComb, *Houston: The Bayou City*, 89–91; O'Kane, "'To Lift the City,'" 18–26.

54. See Tarr, "Sewerage and the Development of the Networked City," 169–70.

55. The Willow Street Pump Station, completed in 1902, was the first lift station in Houston. See Mark Rothfeld, "History of the Willow Street Pump Station," in Stephen Cruse, Michael Moore, Mark Rothfeld, and Debbie Winikates, *Willow Street Pumping Station Preservation Project Report*, unpublished report, Institute for Public History, University of Houston, 1993, III-3–III-4; G. L. Fugate, *Division of Sewage Pumping and Disposal Plants: Report of Operation to December, 1923* (Houston: City of Houston, Engineering Department, Division of Sewage, 1924), 7–9.

56. McComb, *Houston: The Bayou City*, 89–91; "Engineering Department," *Municipal Book of the City of Houston, 1922*, 110; "Engineering Department," City of Houston, *Annual Report, February 28, 1915*, 14.

57. The Houston plants were uncontested until 1925, when Milwaukee constructed a plant that treated forty-five mgd using the activated sludge process. See Melosi, *Sanitary City*, 172, 249, 253–54; and Debbie Winikates, "A History of Houston Bayou Technology," in Cruse et al., *Willow Street Pumping Station*, 1-1, 1-2, 1-8, 1-9, 1-24.

58. The manufacture of fertilizer in Houston did not move much beyond the experimental stage until the 1930s. See Charles N. Tunnell, "24 Hours from Sewage to Fertilizer," *American City* 46 (January 1932): 70–71; Tarr, "Sewerage and the Development of the Networked City," 170; O'Kane, "'To Lift the City,'" 28–29; *Municipal Book, City of Houston, 1928*, 63; Thomas N. Hildreth, "Remodeling a Sewage-Treatment Plant in Houston, Texas," *American City* 41 (October 1929): 117; "Report of the City Engineer," City of Houston, *Annual Report, March 1, 1929*, 18; *Quarterly Municipal Review, 1927*, 3.

59. "The Houston Sewerage Program," *American City* 62 (February 1947): 90.

60. E. E. Rosaire, "Engineer's Council Raps at City Water Pollution," *Slide Rule* (February 1947), 8; Gordon H. Turrentine, *Clean Air and Water—A Trust; A Review of Houston Chamber of Commerce Efforts in Behalf of Effective Abatement of Pollution, 1941–1972* (Houston: Houston Chamber of Commerce, 1972), 1–3, 6–8, 10, 15.

61. Houston, "Minutes of City Council," Book H, October 10, 1892, 93; December 12, 1892, 100; September 30, 1895, 628; Book J, June 7, 1897, 286; December 27, 1897, 382; January 30, 1899, 608.

62. U.S. Department of the Interior, Census Office, *Report on the Social Statistics of Cities, Tenth Census, 1880*, comp. George E. Waring Jr. (Washington, D.C., 1886).

63. "Mayor's Message," City of Houston, *Annual Report, 1902*, 10; "Report of City Engineer," City of Houston, *Annual Report, 1902*, 43; "Report of Water, Light and Health Committee," City of Houston, *Annual Report, 1905*, 22–23; "Report of the Water, Light and Health Committee," City of Houston, *Annual Report, 1906*, 23–24; "Report of Water, Light, and Health Committee," City of Houston, *Annual Report, 1908*, 26.

64. "Report of the City Health Officer," City of Houston, *Annual Report, 1904*, 145–46.

65. "Report of the Health Officer," City of Houston, *Annual Report, 1912,* 141; "Health Department," *The City Book of Houston, 1925,* 127; *Quarterly Municipal Review, 1926,* 8, 30; Dr. A. H. Flickwir, "Health Department," *Municipal Book, City of Houston, 1928,* 69, 71. See also James B. Speer Jr., "Pestilence and Progress: Health Reform in Galveston and Houston during the Nineteenth Century," *Houston Review* 2, no. 3 (1980): 122, 124–28.

66. See *Municipal Book, City of Houston, 1928,* 63; Health Department, City of Houston, *Annual Report, 1930,* 1–2; Ed J. Ryan, "Cost of Garbage Collection by Contract," *American City* 44 (June 1931): 97–98; "Health Department," *City Book of Houston, 1925,* 127.

67. For a more complete discussion of siting disposal facilities in Houston, see Robert Bullard's chapter in this volume. See also Robert Bullard, "Endangered Environs: The Price of Unplanned Growth in Boomtown Houston," *California Sociologist* 7 (Summer 1984): 94–97; Robert Bullard, "Solid Waste Sites and the Black Houston Community," *Sociological Inquiry* 53 (Spring 1983): 277–81; and Robert Bullard, *Dumping in Dixie: Race, Class, and Environmental Quality* (Bolder, Colo.: Westview Press, 1990), 6–7, 50–54.

68. Martin V. Melosi, "The Viability of Incineration as a Disposal Option: The Evolution of a Niche Technology, 1885–1995," *Public Works Management and Policy* 1 (July 1996): 32–37. See also Martin V. Melosi, *Garbage in the Cities: Refuse, Reform and the Environment,* rev. ed. (Pittsburgh: University of Pittsburgh Press, 2005), 185–87; for more information on sanitary landfills, see 182–84.

69. Melosi, "Community and the Growth of Houston," 193–95.

70. Water Supply and Conservation Committee, *Water for the Houston Area,* I–II.

71. A survey indicated that seventeen of thirty-four private water systems within the city had been connected to the Houston system by the end of 1950. Four separate systems also were purchased from the Texas Water Company. See Department of Utilities, City of Houston, *Consolidated Progress Report of the Department of Utilities* (1950), 32.

72. Metropolitan Harris County, *A Report of the Harris Home Rule Commission* (1957), 59–60; Regional Service Systems Task Force, Water Supply System Task Group, Houston Chamber of Commerce, *A Water Supply System Plan for the Greater Houston Region* (October 1983), 2; Metropolitan Water Systems, Division of Public Works, *City of Houston, Texas, Resume* (1979), 1.

73. The project on the San Jacinto called for half of the supply reserved for municipal use and half allocated to industries. The total yield of water was approximately 150 mgd. See Water Supply and Conservation Committee, *Water for the Houston Area,* 4.

74. Turner, Collie & Braden, Consulting Engineers, for the City of Houston, Department of Public Works, Water Division, *Fifth Quadrennial Engineering Report on Physical Condition and Adequacy of the Water System* (November 1970), 6; Metropolitan Water Systems, *City of Houston, Texas, Resume,* 3; "Welcome to Lake Houston" (n.d.), 2; City of Houston, *Public Water Supply System* (January 1948), 17; Water Supply and Conservation Committee, *Water for the Houston Area,* 3; Nagle, "Houston Gets Needed Water," 77.

75. The CIWA Conveyance System consists of a pumping station on the Trinity River about seven miles downstream from Liberty, Texas, in the headwaters of the Wallisville reservoir; a canal system terminating at a point north of the Houston Ship Channel at Lynchburg; and a distribution system to the industrial area along the ship channel. See Metropolitan Water Systems, *City of Houston, Texas, Resume,* 9.

76. "Water . . . Water . . . Everywhere!" 75; Coastal Industrial Water Authority, *The CIWA Waterway: Houston's New River* (n.d.); Alison Hart Hill, "Water Supply in Houston," unpublished paper, University of Houston, 1993, 9–10; "What Is Houston's Real Water

Picture?" *Houston* 28 (April 1957): 27; "Ground Water . . . A Rich Resource," *Houston* 32 (September 1961): 15; Regional Service Systems Task Force, *Water Supply System Plan for the Greater Houston Region,* 1, 5; U.S. Department of the Interior, "Characteristics of Water-Quality Data for Lake Houston," 2–3.

77. City of Houston, *Public Water Supply System* (January 1948), 9; Turner, Collie & Braden, *Fourth Quadrennial Engineering Report on Physical Condition and Adequacy of the Water System* (April 1966), 2–3.

78. Metropolitan Harris County, *A Report of the Harris Home Rule Commission* (1957), 55.

79. Turner, Collie & Braden, *Fourth Quadrennial Engineering Report,* 9.

80. Virginia Marion Perrenod, *Special Districts, Special Purposes: Fringe Governments and Urban Problems in the Houston Area* (College Station: Texas A&M University Press, 1984), 14–34.

81. Metropolitan Harris County, *A Report of the Harris Home Rule Commission* (1957), 57, 59, 62; Metropolitan Water Systems, Division of Public Works, *City of Houston, Texas, Resume* (1979), 11; Turner, Collie & Braden, *Fourth Quadrennial Engineering Report,* 7, 24–26; Tax Research Association, *Water Districts in Harris County: A Survey of Water District Finances* (Houston: Tax Research Association, February 1969), 4, 6–7, 9–11.

82. City of Houston, Water Department, *Report of Director for Year 1942,* 26–27; Nagle, "Houston Gets Needed Water," 77; Water Supply and Conservation Committee, *Water for the Houston Area,* I–II, 9.

83. See Regional Service Systems Task Force, *Water Supply System Plan for the Greater Houston Region,* 3; "Houston," in *Twentieth Century Cities,* 121–24; W. H. Gaines, A. G. Winslow, and J. R. Barnes, *Water Supply of the Gulf Coast Region,* Bulletin 5101 (Washington, D.C.: Geological Survey, U.S. Department of the Interior, January 1951), 5; Water Supply and Conservation Committee, *Water for the Houston Area,* 6–8; Water Supply and Conservation Committee, Ground Water Division, Houston Chamber of Commerce, *Ground Water Resources of the Metropolitan Houston Area* (June 1961), 16–17; "Water for Houston: Blessing and Challenge," *Houston* 39 (May 1968): 81–82; Metropolitan Water Systems, Division of Public Works, *City of Houston, Texas, Resume* (1979), 4; Citizens' Environmental Coalition, *2004 Environmental Resource Guide* (Houston: Citizens' Environmental Coalition, 2004), 19; Harris-Galveston Coastal Subsidence District, *Subsidence '81,* 1–2, 7, 10–11; Perrenod, *Special Districts,* 87–98.

84. U.S. Department of the Interior, "Characteristics of Water-Quality Data for Lake Houston," 2; Regional Service Systems Task Force, *Water Supply System Plan for the Greater Houston Region,* I, 2, 6–7; Planning and Development Department, *Public Utilities Profile for Houston,* III-5, III-16, III-20.

85. Ken Kramer, "Water," in *Houston 2001: A Livable City?* (Houston: Institute of Labor and Industrial Relations, University of Houston, 1977), 71–72, 75, 79.

86. Citizens' Environmental Coalition, *2004 Environmental Resource Guide,* 19. For documentation on some early pollution control efforts, see City of Houston, *Public Water Supply System* (January 1948): 13–14; Department of Utilities, *Consolidated Progress Report of the Department of Utilities,* 21; and City of Houston, Surface Water Supply Department, *1962 Annual Report,* 4.

87. Bonnie B. Pendergrass, "Water for Houston: The Wallisville Case," unpublished paper funded by the Galveston District of the U.S. Army Corps of Engineers, 1987, 1–2; Robert J. Dacey, *Record of Decision for Wallisville Lake, Texas* (February 25, 1984), 1.

88. Frank Fuller, prep., *Subsidized Destruction: The Wallisville Lake Project and Galveston Bay* (Austin: Texas Center for Policy Studies, October 1995), i.

89. Pendergrass, "Water for Houston," 1–2.

90. Robert W. McFarlane, "Wallisville Dam—The Persistent Folly" (n.d.), 1; Gary Cartwright, "Holy Trinity," *Texas Monthly* 31 (October 2003), www.texasmonthly.com; Fuller, *Subsidized Destruction,* i–iv.

91. Pendergrass, "Water for Houston," 1.

92. Ibid., 2–4; Dacey, *Record of Decision,* 1.

93. Pendergrass, "Water for Houston," 4.

94. Ibid., 5; John A. Tudela, *Wallisville Lake—Abstract* (Galveston, Tex.: U.S. Army Corps of Engineers, Galveston District, 1987).

95. Planning and Development Department, *Public Utilities Profile for Houston,* III-26–III-27; "TRA Makes First Payment on Wallisville Saltwater Barrier Project," *Intra: Newsletter of the Trinity River Authority of Texas* (December 2003 / January 2004), 1, 7; Bill Dawson, "Federal Funds Soon Available for Reservoir, *Houston Chronicle,* November 15, 1989; Harold Scarlett, "Panel's Vote Appears to Be Setback for Wallisville Reservoir," *Houston Post,* February 3, 1989.

96. "TRA Makes First Payment," 1; B. C. Robison, "Wallisville: The Scam Lives On," *Houston Post,* May 23, 1987; B. C. Robison, "Wallisville Dam: A Scandal in Disguise," *Houston Post,* October 19, 1986.

97. See Planning and Development Department, *Public Utilities Profile for Houston,* V-5–V-6.

98. Rainfall and flood-flow data collection began as early as 1936 at Buffalo, Brays, and White Oak bayous. David E. Winslow, "Flood Control in Houston: A Historical Perspective," *LJA Insight* 7 (Summer 2002), www.ljaengineering.com; "History: Flooding and Flood Policies"; "Harris County's Flooding History."

99. See Planning and Development Department, *Public Utilities Profile for Houston,* V-3; "History: Flooding and Flood Policies."

100. "History: Flood Policies and Sources"; "Flood Policy 1980 to Present."

101. In this instance, more than 65 percent of the areas flooded from Allison were outside the 100-year floodplain—an area with a 1 percent chance of being flooded in any given year, which obviously broke all the rules of a more typical flood event. See Tropical Storm Allison Recovery Project, *Off the Charts: Tropical Storm Allison Public Report* (Federal Emergency Management Agency and Harris County Flood Control District, June, 2002), 1, 4, 8, 10; "Harris County's Flooding History."

102. "Harris County's Flooding History."

103. Planning and Development Department, *Public Utilities Profile for Houston,* V-3, V-8, V-10. On flooding, see also Regional Service Systems Task Force, Drainage and Flood Control Task Group, Houston Chamber of Commerce, *Final Review Draft: Drainage and Flood Control System Plan for the Greater Houston Region* (August 1983), 1, 5.

104. Regional Service Systems Task Force, *Final Review Draft,* 12; see also 1, 3.

105. "The Houston Sewage Program," *American City* 62 (February 1947): 90–91.

106. Regional Service Systems Task Force, *Final Review Draft,* 12.

107. The reality was that Houston often annexed areas where the treatment system had aged to the point where the city had to invest heavily to replace or improve the system. See ibid.

108. Roger Moehlman, Phillip W. Young, and Meryl L. Olson, "County-Wide Planning for Sewage Treatment," *Public Works Magazine* (June 1957), 1.

109. Metropolitan Harris County, *A Report of the Harris Home Rule Commission* (1957), 65–66; Moehlman, Young, and Olson, "County-Wide Planning," 1–3.

110. Moehlman, Young, and Olson, "County-Wide Planning," 5.

111. U.S. Environmental Protection Agency, *Draft Environmental Impact Statement for District 47 Regional Wastewater Facilities, City of Houston* (Dallas: Office of Grants Coordination, Region VI, Environmental Protection Agency, October 1974), 5–8. See also Houston-Galveston Area Council, *Gulf Coast State Planning Region Waste Treatment Management Study, Appendix II* (May 1975); Harold Scarlett, "Sewage Woes," *Houston Post,* October 29, 1972; Harold Scarlett, "Sewage Solutions Expensive," *Houston Post,* September 23, 1974; City of Houston, Planning Department, *Sewer Plant/Population Study* (November 1975), 20.

112. Regional Service System Task Force, *Final Review Draft,* 18; Niall Q. Washburn, "Houston Sewers and Storm Drains: Looking for Service Equity at Both Ends of the Pipe," unpublished paper, May 5, 1991, 19; Thomas Beck, "Preservation of the Buffalo: Modern Wastewater Treatment in Houston," unpublished paper, December 10, 1990, 2.

113. Houston-Galveston Area Council, *Gulf Coast State Planning Region, Executive Summary, Areawide Waste Treatment Management Study* (May 1975), 31–32.

114. Turner, Collie & Braden, *A Regional Wastewater Facility Plan for the City of Houston Extraterritorial Jurisdiction, Draft Report* (July 1989), I-1–I-3. See also Texas Department of Water Resources, *Water Planning Information: Southeast Texas and Upper Gulf Coast Region with State Summary Data* (Austin: Texas Department of Water Resources, June 1983). In 1983, an enormous sewage treatment plant was completed to relieve overloads at the Northside Plant. The 69th Street Wastewater Treatment Plant, which serves the central business district and major shopping centers and industrial/commercial districts to the north, had a capacity of 200 million gallons of wastewater in the 1990s. A principal element of an extensive capital improvement campaign, the complex was meant to double the city's sewage treatment capabilities. See "69th St. Wastewater Treatment Plant," www.hal-pc.org/~dphan/69.html; and Washburn, "Houston Sewers and Storm Drains," 12, 15.

115. Planning and Development Department, *Public Utilities Profile for Houston,* IV-3.

116. Ibid., IV-3, IV-10.

117. City of Houston, Public Works Department, *Draft Final Report* (June 13, 1991).

118. In addition, the growing city had to contend with larger amounts of sedimentation-tank sludge and excess amounts of activated sludge from two large treatment plants.

119. Houston dried raw activated sludge was sold under the registered name Hou-Actinite. See J. G. Turney, Samuel A. Greeley, and Paul E. Langdon, "Houston Sludge Disposal Plant Profits from Use of Existing Facilities," *Civil Engineering* 21 (May 1951): 26–29; W. E. White, "From Sewage Sludge Houston Produces Marketable Fertilizer," *Texas Municipalities* 8 (March 1961): 12–14; "Houston Sewerage Program," *American City* 62 (February 1947): 90–91; and "City to Receive Federal Grant for Sludge Disposal Facility," *Houston Chronicle,* August 24, 1974.

120. E. E. Rosaire, "Engineer's Council Raps at City Water Pollution," *Slide Rule* (February 1947): 8.

121. "The Pollution Problem," *Houston Chronicle,* June 28, 1948; "City Attacks Sewer Stink, but Slowly," *Houston Chronicle,* February 1, 1952.

122. Turrentine, *Clean Air and Water,* 1a.

123. Ibid., 2–3, 6–8, 12–16, 35.

124. Arthur Hill, "Cleanup Proposal for Ship Channel Hits Resistance," *Houston*

Chronicle, August 13, 1974; Tom Kennedy, "Court Orders City to Act 'Without Delay' on Sewage Problem," *Houston Post,* August 21, 1974.

125. Planning and Development Department, *Public Utilities Profile for Houston,* IV-10.

126. Jim Morris, "'Benign Neglect' Begets City Filth," *Houston Chronicle,* June 9, 1991.

127. Planning and Development Department, *Public Utilities Profile for Houston,* IV-10; Bob Burtman, "Below Standards," *Houston Press,* April 2–8, 1998.

128. Houston Chamber of Commerce, Regional Service Systems Task Force, Solid Waste System Task Group, *Final Report: Solid Waste System Management Plan for the Greater Houston Region* (September 1983), i.

129. "Houston, Texas, Builds Dual Incinerator," *American City* 63 (January 1948): 98; W. G. Wilson to F. N. Baldwin, July 1, 1949, Texas Room, Houston City Library.

130. League of Women Voters of Houston, "Your Garbage and Mine: Waste Incineration Issues in the 1980's," *Focus* 66 (November 1986): 2. See also "Incinerator Shut Down Indefinitely," *Houston Chronicle,* June 8, 1969.

131. For more details on sanitary landfills, see Melosi, *Garbage in the Cities.*

132. Bernard Johnson Engineers, *Comprehensive Master Plan for Collection and Disposal of Solid Waste for the City of Houston, Texas—Report Number One* (May 1971), 5-1.

133. Citizens' Environmental Coalition, *2004 Environmental Resource Guide,* 28.

134. Private waste collectors offer service to commercial and residential customers who are not served by the city. See Solid Waste System Task Group, *Final Report,* ii, 19, 35–36.

135. "Houston's 'Round-the-Clock Landfill," *Waste Age* 12 (March 1981): 29. The McCarty landfill was owned and operated by Browning-Ferris Industries, a Houston-based agglomerate and one of the country's largest private solid waste companies. Despite the fact that the city of Houston was its largest single client, the city and the company were often at loggerheads. See Alison Cook, "Trashy Business," *Texas Monthly* 12 (November 1984): 264.

136. Bill Dawson, "Concern Mounts Over Disposal of Garbage," *Houston Chronicle,* July 26, 1992.

137. Citizens' Environmental Coalition, *2004 Environmental Resource Guide,* 28.

138. A separate "garbage department" was established for the city in 1971. Since that time, the structure and organization of the Houston Solid Waste Management Department—as it is known today—have changed several times.

139. W. G. Wilson to Frank N. Baldwin, January 29, 1951, Texas Room, Houston Public Library.

140. For about two decades, however, the director of the solid waste department has been traditionally an African American. Such visibility for minority officials clearly had a political dimension.

141. Paul Burka, "Why Is Houston Falling Apart?" *Texas Monthly* 8 (November 1980): 190; "Whitmire 'Slaps Wrists' of Strikers," *Informer* (September 6, 1986).

142. For a more complete discussion of siting disposal facilities in Houston, see Robert Bullard's and Elizabeth Blum's chapters in this volume.

143. Dina Cappiello and Kristen Mack, "Old Landfills Pose Risk for Residents," *Houston Chronicle,* October 12, 2003; Citizens' Environmental Coalition, *2004 Environmental Resource Guide,* 28.

144. In 1967, a composting plant was built, operated by Metropolitan Waste Conversion Company, but eventually it was closed. See League of Women Voters of Houston, "Your Garbage and Mine," 2.

145. Edward T. Chen, "Recycling in Houston—Plans and Progress," *Public Works* 123 (May 1992): 38–41; Edward T. Chen, "Houston Recycles with Foresight," *Waste Age* 22 (February 1991): 81–82; Mary Dodd Dubbert, *Why the Subdivision of Meyerland Should Establish a Recycling Program* (Houston: Meyerland Community Improvement Association, January 1990), 7–8; Planning and Development Department, *Public Utilities Profile for Houston*, II-14–II-17. See also Lee Latham and Steve Reynolds, *A Brief History of the City of Houston's Recycling Program* (Houston: Institute for Public History, December 1997).

146. Citizens' Environmental Coalition, *2004 Environmental Resource Guide*, 28–29; Melosi, *Garbage in the Cities*, 222; Matt Stiles, "A Waste of Time or a Waste of Worth?" *Houston Chronicle*, September 13, 2005.

147. Stiles, "A Waste of Time?"

148. Rachel Graves, "In Need of Repairs," *Houston Chronicle*, July 8, 2001. See also Matt Schwartz, "Public Work Faces Big Woes in Big City," *Houston Chronicle*, July 9, 2001.

Chapter 6. Superhighway Deluxe

1. *United States Statutes at Large Containing the Laws and Concurrent Resolutions Enacted during the Second Session of the Seventy-eighth Congress of the United States of America, 1944, and Proclamations, Treaties, and Other International Agreements Other Than Treaties, Compiled, Edited, Indexed, and Published by Authority of Law under the Direction of the Secretary of State*, Vol. 58, Part 1, *Public Laws* (Washington, D.C.: Government Printing Office, 1945), 838.

2. Ibid.

3. Ibid., 840.

4. "Texas Roads and Highways," Texas Legislative Council, October 1952, Nos. 52–53, 26, Austin.

5. Ibid., 79–80.

6. *United States Statutes*, 842.

7. Edward Weiner, *Urban Transportation Planning in the United States: An Historical Overview* (Westport, Conn.: Praeger, 1999), 27.

8. "Texas Roads and Highways," 24.

9. Richard Morehead, *Dewitt C. Greer: King of the Highway Builders* (Austin, Tex.: Eakin Press, 1984), 46.

10. Remarks of Division Engineer J. A. Elliot, Division Six 1949 Bureau of Public Roads Administration meeting, Files of the United States Bureau of Public Roads, RG 30, "Records of Division Six, January–March 1949," Box 1782, Folder 3, National Archives, College Park, Md.

11. "$2,000,000 per Mile Gulf Freeway Is the 'Cheapest Highway Ever Built,'" *Houston Chronicle*, January 20, 1952.

12. Ibid.

13. "Texas Roads and Highways," 25.

14. "Early in World War II, it was thought that all possible energies should be devoted to national war production. To place this new policy in effect, the national government suspended federal aid and discouraged all except vitally-needed programs. These moves virtually suspended highway construction from 1942 through 1945." Ibid., 94–95.

15. Ibid., 24.

16. Ibid., 95.

17. D. W. Loutzenheiser to J. Barnett, December 12, 1949, United States Bureau of Public Roads, RG 30, "Records of Correspondence FAS-Texas, July–December 1949," Box 2985–86, Folder 1, National Archives, College Park, Md.

18. Remarks of Division Engineer J. A. Elliott.

19. Texas Transportation Institute, *A Mass Transportation Concept for Metropolitan Houston: Immediate Action, Flexibility, Economy* (College Station: Texas Transportation Institute, 1970), 6.

20. Harris County Home Rule Commission, "Metropolitan Harris County; A Report of the Harris County Home Rule Commission" (Houston: Harris County Home Rule Commission, 1957), 4.

21. Ibid., 20.

22. "Expressway Work Slated in January," *Houston Chronicle,* October 3, 1945.

23. "Here's How Money Went for Freeway," *Houston Chronicle,* April 29, 1952.

24. It is important to note that the state of Texas was unable to condemn land for right-of-way purposes until after 1951.

25. Sam Weiner, "Van London's Roller-Coaster," *Parade Magazine*, February 5, 1950.

26. "Freeway Was Dream of Two Men," *Houston Chronicle,* August 1, 1952.

27. Ibid.

28. "Highway of Tomorrow," *Rotogravure Magazine,* August 31, 1947, 5–7.

29. Slotboom, *Houston Freeways,* 145, 147.

30. "The Houston Expressway," *American City* 63 (November 1948): 116–17.

31. Dick Tate, "$28,000,000 Gulf Freeway 'Will Pay For Itself in Four Years in Time Saved,'" *Houston Chronicle,* August 1, 1952.

32. Weiner, "Van London's Roller-Coaster."

33. "Opening of Our Splendid Freeway," *Houston Chronicle,* August 3, 1952.

34. "Expressway Work Slated in January."

35. "Making Way for Expressway," *Houston Chronicle,* September 8, 1946.

36. "Expressway Work Slated in January."

37. Ibid.

38. "Moving Homes for Expressway Is Problem," *Houston Chronicle,* February 1, 1946.

39. "Houston-Galveston Superhighway Construction Job Is Progressing," *Houston Chronicle,* March 16, 1947.

40. "Land Acquisition for Superhighway Pushed," *Houston Chronicle,* July 1, 1947.

41. Ibid.

42. Ibid.

43. For a more complete description of how the highway impacted this area, see Amy L. Bacon, "From West Ranch to Space City: A History of Houston's Growth Revealed through the Development of Clear Lake" (Master's thesis, University of Houston, 1996).

44. "Gulf Freeway Siege Flares into Battle," *Houston Chronicle,* March 29, 1949.

45. Ibid.

46. See Tom Lewis, *Divided Highways* (New York: Penguin Books, 1997); Robert B. Fairbanks, *For the City as a Whole: Planning, Politics, and the Public Interest in Dallas, Texas, 1900–1965* (Columbus: Ohio State University Press, 1998); and John Lauritz Larson, *Internal Improvement: National Public Works and the Promise of Popular Government in the Early United States* (Chapel Hill: University of North Carolina Press, 2001).

47. See "3rd Ward Group Protests Plan to Expand Freeway," *Houston Chronicle,* February 22, 1973, *Houston Chronicle* Morgue, Houston Metropolitan Research Center.

48. "Land Owned by Mayor and Deats Taken for Freeway," *Houston Chronicle,* July 4, 1949.

49. "Progress Reported on Extension of Freeway," *Houston Chronicle,* February 18, 1949.

50. McComb, *Galveston*, 1.

51. See *Progressive Houston* 4 (February 1913): 8–9.

52. "Mayor Throws Switch to Light the Freeway," *Houston Chronicle*, October 1, 1948, *Houston Chronicle* Morgue.

53. "Superhighway Will Be Known as Gulf Freeway," *Houston Chronicle*, December 17, 1948, *Houston Chronicle* Morgue.

54. "Engineer to Make Check of Traffic on Gulf Freeway," *Houston Chronicle*, February 11, 1949, *Houston Chronicle* Morgue.

55. Willier's formula used two cents per minute for passenger vehicles and five cents per minute for commercial vehicles. "Willier Reports Gulf Freeway Is 'A Paying Proposition,'" *Houston Chronicle*, July 28, 1949.

56. "Gulf Freeway Traffic Gives Engineers Kick," *Houston Chronicle*, August 4, 1949.

57. "Progress Reported on Extension of Freeway," *Houston Chronicle*, February 18, 1949.

58. "Houston to Get Back Its Money," *Houston Chronicle*, August 1, 1952.

59. Sam Weiner, "Some Land Values Up 1,000 Per Cent," *Houston Post*, August 2, 1952.

60. For a more complete exploration of the value of freeway-fronting land, see Joel Garreau, *Edge City: Life on the New Frontier* (New York: Anchor Books, 1991).

61. Thomas E. Willier and Eugene Maier, *Economic Evaluation of the Gulf Freeway* (Houston: City of Houston, Department of Traffic and Transportation, July 1949), introduction.

62. Ibid., 17, 21.

63. Ibid., 25.

64. Ibid., acknowledgments.

65. "Willier Reports Gulf Freeway Is 'A Paying Proposition.'"

66. METRO introduced Park and Ride in 1977, and it was so successful that METRO's lots were filled beyond capacity. Metropolitan Transit Authority, *Regional Transit Plan* (Houston: METRO, 1978), 21.

67. There are many works on this subject. See Garreau's *Edge City* for one of the best known. His chapter on Houston's Galleria area is particularly relevant.

68. For information regarding the development of this trend, see "The Suburban Office Market After the Shakeout," *Houston* 46, no. 12 (January 1976): 21–24, 74; and Dmitry Mesyanzhinov, "Suburbanization of Office Space: A Case Study of Houston, Texas" (Ph. D. diss., Louisiana State University, 1997).

69. Jimmie Brough, "Gulf Freeway Proves to Be Big Money-Saver," *Houston Chronicle*, September 19, 1949.

70. "87% in Houston Area Drive Car to Work," *Houston Chronicle*, December 12, 1973.

71. Houston Urban Office, Texas State Department of Highways and Public Transportation, *Cost Effectiveness Analysis of Alternatives for Gulf Freeway Busway (I-45) from FM 1959 (Near City Limits) to Downtown Houston, Texas* (Houston: Houston Urban Office, Texas State Department of Highways and Public Transportation, 1979), i.

72. "Work Due in '50 on Freeway Extension to Buffalo Drive," *Houston Chronicle*, December 7, 1949, *Houston Chronicle* Morgue.

73. "Freeway May Span City in Next Two Years," *Houston Chronicle*, January 16, 1950, Progress Edition.

74. Ibid.

75. Mark Morrison, "Gulf Freeway Completion Reportedly Years Away," *Houston Post*, May 28, 1973.

76. Rick Barrs, "The Gulf Freeway—Job That's Never Done," *Houston Post,* February 7, 1982.

77. Ibid.

78. "Ceremony Held," *Houston Chronicle,* August 11, 1950.

79. "Gulf Freeway No Longer Most Congested," *Houston Post,* February 7, 1982.

80. Houston Planning Commission, *A Study of Thoroughfare Development in the Southeast Area of Metropolitan Houston and Harris County* (Houston: Houston Planning Commission, 1963), 2.

81. Ibid., 5.

82. *Texas State Almanac and Industrial Guide 1940–1960* (Dallas: Belo, 1940–60).

83. Paul Alejandro Levengood, "For the Duration and Beyond: World War II and the Creation of Modern Houston, Texas" (Ph.D. diss., Rice University, 1999), 313–14.

84. Ibid., 322.

85. Norris & Elder Consulting Engineers, *A 15-Year Study of Land Values and Land Use along the Gulf Freeway in the City of Houston, Texas* (Houston: Norris and Elder, 1956), 15–16, 38–40.

86. The original date of the opening was moved from July 19, 1952, as heavy weather shut down the project. "Rains Delay Opening of Freeway until Aug. 2," *Houston Chronicle,* June 8, 1952.

87. Chester Rogers, "5000 See Freeway Dedicated," *Houston Chronicle,* August 3, 1952.

88. "Freeway Opening Saturday," *Houston Chronicle,* July 31, 1952.

89. George A. Seel, "Modern Highway System Today Connects Island with Rest of State," *Galveston Daily News,* April 11, 1942.

90. C. E. McClelland, "The Editor Writes," *Galveston Daily News,* August 2, 1952.

91. Ibid.

92. "Civic Leaders Say Freeway Brings New Era," *Galveston Daily News,* August 2, 1952.

93. "Gulf Freeway," *Houston Press* Collection, Vertical Files, Box 3, Houston Metropolitan Research Center.

94. Steven M. Baron, *Houston Electric: The Street Railways of Houston, Texas* (Lexington, Ky.: Steven M. Baron, 1996), 54.

95. Orland O. Dodson, "Galveston Eyes New Business Horizon Opened by Freeway," *Houston Post,* August 2, 1952.

96. Ibid.

97. "Pulling the Two Cities Closer Together," *Houston Chronicle,* August 2, 1952.

98. "Houston Is Now the South's Largest City; Recent Election Adds 110,000," *Houston* 20 (January 1950): 5.

99. John W. Yeats, "South Houston Says Freeway Boon to Safety, Home-Owning," *Houston Post,* August 2, 1952.

100. Ibid.

101. Sam Bass Warner Jr., *Streetcar Suburbs: The Process of Growth in Boston (1870–1900),* 2nd ed. (Cambridge, Mass.: Harvard University Press, 1978).

102. "Friendswood Now Closer to Channel," *Houston Post,* August 2, 1952.

103. Ibid.

104. For a better description of this process, see the other chapters in this work.

105. "Getting the Big Picture on Houston's Air Pollution," in NASA Life on Earth, http://www.nasa.gov/vision/earth/everydaylife/archives/HP_ILP_Feature_03.html.

106. See Erik Slotboom, *Houston Freeways: A Historical and Visual Journey* (Houston: Oscar Slotboom, 2003), 24–29.

107. Marvin Hurley, *Decisive Years For Houston* (Houston: Houston Magazine, 1966), 169–70.

108. Ibid., 170.

109. David G. McComb, *Houston: A History* (Austin: University of Texas Press, 1981), 142–43.

110. See Bacon, "From West Ranch to Space City."

111. Ralph Ellifrit, quoted in Bacon, "From West Ranch to Space City," 91.

112. E. L. Wall, "Freeway Studied as 'Model,'" *Houston Chronicle,* March 23, 1951.

113. Ibid.

114. "The Houston Expressway," *American City* 63 (November 1948): 116–17.

115. Texas Highway Department, "Houston Urban Expressways: Gulf Freeway, Houston, Texas" (Houston: Texas Highway Department, 1949).

116. "Chamber of Commerce 1952 Annual Report," *Houston* 23, no. 11 (December 1952): 10–36.

117. "Holcombe Asks More Freeways," *Houston Post,* August 2, 1952.

Chapter 7. Urban Sprawl and the Piney Woods

1. Robert E. Lange, *Edgeless Cities: Exploring the Elusive Metropolis* (Washington, D.C.: Brookings Institution, 2002).

2. Department of Commerce, Bureau of the Census, 2004, www.census.gov.

3. Patrick E. Miller and Andrew J. Hartsell, *Forest Statistics for East Texas Counties—1992, Resource Bulletin SO-173* (December 1992), New Orleans, U.S. Department of Agriculture Forest Service, Southern Forest Experiment Station.

4. Ibid.

5. See, for example, Robert Fishman, *Bourgeois Utopias: The Rise and Fall of Suburbia* (New York: Basic Books, 1987); Garreau, *Edge City;* Andres Duany, Elizabeth Plater-Zyberk, and Jeff Speck, *Suburban Nation: The Rise of Sprawl and the Decline of the American Dream* (New York: North Point Press, 2000); and Lang, *Edgeless Cities.*

6. G. Galster, R. Hanson, M. R. Ratcliffe, H. Wolman, S. Coleman, and J. Freihage, "Wrestling Sprawl to the Ground: Defining and Measuring an Elusive Concept," *Housing Policy Debate,* 12, no. 4 (2001): 681–717.

7. Citizens' Environmental Coalition, *2002 Environmental Resource Guide* (Houston: Citizens' Environmental Coalition, 2000), 8.

8. David Schrank and Tim Lomax, *The Urban Mobility Report* (College Station: Texas Transportation Institute, 2002), 67.

9. Citizens' Environmental Coalition, *2002 Environmental Resource Guide,* 24.

10. Mary Sanger and Cyrus Reed, *Texas Environmental Almanac,* 2nd ed. (Austin: University of Texas Press, 2000).

11. Diane C. Bates, "Suburban Sprawl and Deforestation: Community Responses in a Southeast Texas Watershed," paper presented at the Eighty-second Annual Meeting of the Southwestern Sociological Association, New Orleans, March 27–30, 2002. See also George T. Morgan Jr. and John O. King, *The Woodlands: New Community Development, 1964–1983* (College Station: Texas A&M University Press, 1987).

12. James F. Rosson Jr., *Forest Resources of East Texas, 1992, Resource Bulletin SRS-53* (2000), 3–4, New Orleans, U.S. Department of Agriculture Forest Service, Southern Research Station.

13. Robert P. Schultz, *Loblolly Pine: The Ecology and Culture of the Loblolly Pine* (Pinus taeda L.*)*, Agricultural Handbook 713 (Washington, D.C.: U.S. Department of Agriculture, Forest Service, 1997), 3–15.

14. Daniel W. Moulton and John S. Jacob, *Texas Coastal Wetlands Guidebook* (Bryan: Texas Sea Grant College Program, n.d.), 24.

15. American Forests, *Urban Ecosystem Analysis for the Houston Gulf Coast Region: Calculating the Value of Nature* (Houston: Houston Green Coalition, 2000).

16. D. Beach, *Coastal Sprawl: The Effects of Urban Design on Aquatic Ecosystems in the United States* (Arlington, Va.: Pew Oceans Commission, 2002).

17. Christof Spieler, "METRO: What's Next?" *Cite* 61 (2004): 14–15.

18. Bates, "Suburban Sprawl."

Chapter 8. A Tale of Two Texas Cities

1. Gary Cartwright, *Galveston: A History of the Island* (New York: Atheneum, 1991), 1; Edna Ferber, *A Kind of Magic* (London: Victor Gollancz, 1963), 102.

2. There are several sensationalist accounts of the 1900 storm, including Paul Lester, *The Great Galveston Disaster* (Philadelphia: Globe, 1900); and Murat Halstead, *Galveston: The Horrors of a Stricken City* (Chicago: American Publishers Association, 1900). For more balance, see John Edward Weems, *A Weekend in September* (College Station: Texas A&M University Press, 1980); and Herbert Molloy Mason Jr., *Death from the Sea* (New York: Dial Press, 1972). The best general history is McComb, *Galveston.*

3. Citizens of Texas, "Memorial of Citizens of Texas Praying that the Galveston and Red River Railway Company may be granted the privilege of extending their railroad through the territory of the United States to the Pacific ocean," January 24, 1849, Center for American History, Austin.

4. Caleb G. Forshey, "Texas Rail Roads," *Galveston Daily News,* July 13, 1866.

5. Lucy Weston Shaw, Lucy Weston Shaw Letters, February 24 and April 21, 1839, Letters from Galveston, Texas, to Jane Weston of Eastport, Maine, Manuscripts and Typescripts, Rosenberg Library, Galveston.

6. Shaw, Lucy Weston Shaw Letters, August 25, 1850, and September 28, 1839.

7. Shaw, Lucy Weston Shaw Letters, February 24 and March 22, 1839.

8. Land and Thompson, *Galveston: The Commercial Metropolis and Principal Seaport of the Great Southwest* (Galveston, Tex.: Land and Thomson, 1885), 9–10.

9. Ibid., 56, 20.

10. McComb, *Galveston,* 49, 51–53, 60, 119. See also Earle B. Young, *Galveston and the Great West* (College Station: Texas A&M University Press, 1997); and Earle B. Young, *Tracks to the Sea: Galveston and Western Railroad Development, 1866–1900* (College Station: Texas A&M University Press, 1999).

11. Recent books emphasizing the storm as the root of Galveston's decline include Erik Larson, *Isaac's Storm: A Man, a Time, and the Deadliest Hurricane in History* (New York: Crown, 1999); Casey Greene and Shelly Henley Kelly, eds., *Through a Night of Horrors: Voices from the 1900 Galveston Storm* (College Station: Texas A&M University Press, 2000); and Patricia Bellis Bixel and Elizabeth Hayes Turner, *Galveston and the 1900 Storm* (Austin: University of Texas Press, 2000).

12. Edward King, *The Great South* (1875; reprint, New York: Arno Press, 1969), 103–4.

13. Ibid., 109.

14. Cronon, *Nature's Metropolis.*

15. King, *Great South,* 111, 112, 115.

16. McComb, *Houston: A History,* 76–77.

17. King, *Great South,* 115, 114.

18. On the slogans Houston boosters used, see Frances Dressman, "Visions for Houston: Booster Literature, 1886–1926," *Houston Review* 9, no. 3 (1987): 138–39.

19. Sibley, *Port of Houston,* 92–101; McComb, *Houston: A History,* 33–35.

20. Sibley, *Port of Houston,* 111–26; McComb, *Houston: A History,* 65–67.

21. See Daniel Yergin, *The Prize: The Epic Quest for Oil, Money, and Power* (New York: Simon and Schuster, 1991), 11–12, 87, 95; and Martin V. Melosi, *Coping with Abundance: Energy and Environment in Industrial America* (Philadelphia: Temple University Press, 1985), 50. See also Seth S. McKay and Odie B. Faulk, *Texas After Spindletop* (Austin, Tex.: Steck-Vaughn, 1965), 7–10; and Paul N. Spellman, *Spindletop Boom Days* (College Station: Texas A&M University Press, 2001), 11–12, 49.

22. Melosi, *Coping with Abundance,* 43–46; McComb, *Houston: A History,* 78–80.

23. Pratt, *Growth of a Refining Region,* 34.

24. Yergin, *Prize,* 88, 410.

25. Harry Van Demark, "What's Happening in Houston," *Texas Monthly* 3 (May 1929): 598–608. See also Pratt, *Growth of a Refining Region,* 77–83; McKay and Faulk, *Texas After Spindletop,* 10–16; and McComb, *Houston: A History,* 78–81.

26. South Texas National Bank, *Houston, Texas: The Manchester of America, Where 17 Railroads Meet the Sea, the Site of the World's Next Big City, Nearest the Panama Canal* (Dallas: Southwestern Folder, 1910). Bird's-eye view titled "Houston—a Modern City" produced by Hopkins and Motter, 1912. See Carol Johnson, "Watching Houston Grow: Maps in the Houston Public Library's Collection," *Houston Review* 16, no. 1 (1994): 54.

27. McComb, *Houston: A History,* 84.

28. *Houston Press,* April 3, 1924.

29. *Houston Chronicle,* February 21, 1923.

30. *Houston Chronicle,* March 20, 1923.

31. Paul W. Brown, "A Traveler in Texas—Part VII—Houston and Its Ship Channel," *America at Work,* April 20, 1923, 17–18, Pamphlets, Rosenberg Library.

32. Houston Natural Gas Company, *Civics for Houston* 1, no. 11 (December 1928): 33. On the natural gas industry, see Christopher J. Castaneda, *Invisible Fuel: Manufactured and Natural Gas in America, 1800–2000* (New York: Twayne, 1999).

33. Peter C. Papademetriou, "Urban Development and Public Policy in the Progressive Era: 1890–1940," *Houston Review* 5, no. 3 (1983): 116.

34. McKay and Faulk, *Texas After Spindletop,* 99.

35. Van Demark, "What's Happening in Houston," 601–2. See also McComb, *Houston: A History,* 75.

36. P. J. R. MacIntosh, "Concrete Roads to Houston," *Texas Monthly* 5 (May 1930): 395.

37. Papademetriou, "Urban Development," 123.

38. Houston Electric Company, *Civics for Houston* 1, no. 8 (September 1928): 12; Katherine Pollard, "Houston Provides Shopping on Wheels," *Civics for Houston* 1, no. 12 (January 1929): 6–7.

39. Archie Henderson, "City Planning in Houston, 1920–1930," *Houston Review* 9, no. 3 (1987): 117–27.

40. Ibid., 129–36; McComb, *Houston: A History,* 96–99, 123, 157–58; Papademetriou, "Urban Development," 122–31.

41. Houston Electric Company, *Civics for Houston,* 29; Houston Gas & Fuel Company, *Civics for Houston* 1, no. 7 (August 1928): 32.

42. River Oaks, *Civics for Houston* 1, no. 6 (June 1928): 38; Galveston Houston Electric Railway, *Civics for Houston* 1, no. 4 (April 1928): 1; McComb, *Houston: A History,* 97–99, 156–59.

43. Matilda C. H. Houston, *Texas and the Gulf of Mexico; Or, Yachting in the New World* (1843; reprint, Austin, Tex.: W. Thomas Taylor, 1991), 98.

44. King, *Great South,* 102.

45. On vacationing in nineteenth-century America, see Cindy S. Aron, *Working at Play: A History of Vacations in the United States* (New York: Oxford University Press, 1999).

46. McComb, *Galveston,* 64–65, 171–72.

47. Ibid., 151, 156–66, 174–87.

48. Steven M. Baron, "Streetcars and the Growth of Houston," *Houston Review* 16, no. 2 (1994): 86–89; McComb, *Houston: A History,* 73–75.

49. MacIntosh, "Concrete Roads," 393, 394, 403.

50. Ibid., 403.

51. On historic preservation in large cities, see David Hamer, *History in Urban Places: The Historic Districts of the United States* (Columbus: Ohio State University Press, 1998).

52. Howard Barnstone, *The Galveston That Was* (Houston: Museum of Fine Arts, 1966). On historic preservation and George Mitchell, see Cartwright, *Galveston,* 301–11; and Ray Miller, *Ray Miller's Galveston* (Houston: Cordovan Press, 1983), 216–26.

53. On modern U.S. consumption, see Lizabeth Cohen, "Is There an Urban History of Consumption?" *Journal of Urban History* 29, no. 2 (January 2003): 87–106; and Lizabeth Cohen, *A Consumers' Republic: The Politics of Mass Consumption in Postwar America* (New York: Alfred A. Knopf, 2003). On Las Vegas, see Hal Rothman, *Neon Metropolis: How Las Vegas Started the Twenty-first Century* (New York: Routledge, 2002).

54. James A. Clark and Michel T. Halbouty, *Spindletop* (New York: Random House, 1952), xv–xvi. A monument built for the fiftieth anniversary used similar language. The inscription on the Spindletop Monument, Beaumont, Texas, reads: "On this Spot on the tenth day of the Twentieth Century a New Era in Civilization Began."

Chapter 9. Dumping on Houston's Black Neighborhoods

1. Henry Allen Bullock, *Pathways to the Houston Negro Market* (Ann Arbor, Mich.: J. N. Edwards, 1957), 60–61.

2. Robert D. Bullard, *Invisible Houston: The Black Experience in Boom and Bust* (College Station: Texas A&M University Press, 1987), chapter 2.

3. The author served as an expert witness on the *Bean v. Southwestern Waste Management, Inc.* lawsuit and as a researcher on the case while employed as an untenured assistant professor at Texas Southern University (1979–85).

4. R. D. Bullard, "Leveling the Playing Field through Environmental Justice," *Vermont Law Review* 23 (Spring 1999): 454–78.

5. See Bullard, *Dumping in Dixie;* R. D. Bullard and G. S. Johnson, "Environmental Justice: Grassroots Activism and Its Impact on Public Policy Decision Making," *Journal of Social Issues* 56 (2000): 555–78.

6. See R. D. Bullard and B. H. Wright, "The Politics of Pollution: Implications for the Black Community," *Phylon* 47 (March 1986): 71–78.

7. R. D. Bullard, "Dismantling Environmental Racism in the USA," *Local Environment* 4 (1999): 5–19.

8. Luke W. Cole and Sheila R. Foster, *From the Ground Up: Environmental Racism and the Rise of the Environmental Justice Movement* (New York: New York University Press, 2001);

Laura Westra and Bill E. Lawson, eds., *Faces of Environmental Racism: Confronting Issues of Global Justice,* 2nd ed. (Lanham, Md.: Rowan and Littlefield, 2001).

9. See Robert D. Bullard, ed., *Confronting Environmental Racism: Voices from the Grassroots* (Boston: South End, 1993); Robert D. Bullard, "The Threat of Environmental Racism," *Natural Resources and Environment* 7 (Winter 1993): 23–26; Bunyan Bryant and Paul Mohai, eds., *Race and the Incidence of Environmental Hazards* (Boulder, Colo.: Westview Press, 1992); Regina Austin and Michael Schill, "Black, Brown, Poor and Poisoned: Minority Grassroots Environmentalism and the Quest for Eco-Justice," *Kansas Journal of Law and Public Policy* 1 (1991): 69–82; Kelly C. Colquette and Elizabeth A. Henry Robertson, "Environmental Racism: The Causes, Consequences, and Commendations," *Tulane Environmental Law Journal* 5 (1991): 153–207; and Rachel D. Godsil, "Remedying Environmental Racism," *Michigan Law Review* 90 (1991): 394–427.

10. Devon Pena, *The Terror of the Machine: Technology, Work, Gender and Ecology on the U.S.-Mexico Border* (Austin: Center for Mexican American Studies, University of Texas, Austin, 1996); Davis Naguib Pellow, *Garbage Wars: The Struggle for Environmental Justice in Chicago* (Cambridge, Mass.: MIT Press, 2002); Ike Okonta and Oronto Douglas, *Where Vultures Feast: Shell, Human Rights and Oil* (New York: Verso, 2003); Mario Murillo, *Island of Resistance: Vieques, Puerto Rico, and U.S. Policy* (New York: Seven Stories Press, 2001).

11. Bullard, *Invisible Houston,* 72–73.

12. Ruth Rosen, "Who Gets Polluted: The Movement for Environmental Justice," *Dissent* (Spring 1994): 223–30; Robert D. Bullard, "Environmental Justice: It's More Than Waste Facility Siting," *Social Science Quarterly* 77 (September 1996): 493–99.

13. See Robert D. Bullard, "Endangered Environs: The Price of Unplanned Growth in Boomtown Houston," *California Sociologist* 7 (Summer 1984): 84–102; Richard Babcock, "Houston Unzoned, Unfettered, and Mostly Unrepentent," *Planning* 48 (1982): 21–23; and Joe R. Feagin, "The Global Context of Metropolitan Growth: Houston and the Oil Industry," *American Journal of Sociology* 90 (May 1985): 1204–30.

14. Bullard, "Solid Waste Sites," 273–88.

15. Ibid.

16. "Council Awards Garbage Transfer Station Contract," *Houston Chronicle,* March 3, 1983.

17. Bullard, "Solid Waste Sites," 273–88.

18. Mike Snyder, "Neglected Neighborhoods," *Houston Chronicle,* November 19, 2002.

19. Matt Schwartz, "Though City Touts Neighborhood Initiatives, Many Have Yet to Bring Substantial Change," *Houston Chronicle,* September 19, 2002.

20. Bullard, *Invisible Houston,* 60–75.

21. Kristen Mack, "Acres Homes as Impoverished Today as It Was 10 Years Ago," *Houston Chronicle,* November 19, 2002.

22. Ibid.

23. The neighborhood descriptions that follow are derived from Bullard, *Invisible Houston.*

24. Armando Villafranca, "APV Demolition Nearing End," *Houston Chronicle,* June 4, 1997.

25. Robert D. Bullard, J. Eugene Grigsby II, and Charles Lee, *Residential Apartheid: The American Legacy* (Los Angeles: UCLA Center for African American Studies Publication, 1984).

26. Bullard, *Invisible Houston,* 110–11.

27. National Advisory Commission on Civil Disorders, *Report,* 40–41.

28. Bullard, *Invisible Houston,* 111.

29. Bullard, *Dumping in Dixie.*

30. Lori Rodriguez, "Dump Plan Would Hurt Area, Residents Tell Court," *Houston Chronicle,* November 16, 1979.

31. Ibid., quote from Patricia Reaux.

32. For a detailed account of this dispute, see Bullard, *Invisible Houston,* chapter 6; *Houston Chronicle,* November 8, 11, 15, 22, 1979; December 15, 22, 1979; June 19, 1980; and *Houston Post,* December 15, 1981.

33. Lori Rodriguez, "School Superintendent Testifies," *Houston Chronicle,* November 22, 1979.

34. Ibid.

35. "Judge Denies Request to Halt Dump Opening," *Houston Chronicle,* December 22, 1979.

36. Ibid.

37. Raul Reyes, "Council Resolution Condemns Sanitary Landfill Site," *Houston Chronicle,* June 19, 1980.

Chapter 10. The Gunfighters of Northwood Manor

This essay has been adapted from several sources, and I am deeply indebted to numerous people for assisting with the refinement and discussion of the ideas contained herein. I began to examine this issue in my dissertation, and for a more in-depth look, see Elizabeth D. Blum, "Pink and Green: A Comparative Study of Black and White Women's Environmental Activism in the Twentieth Century" (Ph.D. diss, University of Houston, 2000). Some of the ideas have also been hammered out in paper presentations to the American Society for Environmental History (ASEH), the American Historical Association (AHA), and, of course, "An Environmental History of Houston and the Gulf Coast" conference held in March 2003. As a basic definition for my framework within this article, I will define environmental justice in the following way: The environmental justice movement fights against the unequal distribution of pollution, polluting sites, and concurrent city services affecting poor and minority communities. Essentially, it hopes to distribute the burdens of an industrial society as well as the benefits of a wealthy one to all equally.

1. Complaint, Civ. Action No. H-79-2215, October 26, 1979, *Bean v. Southwestern Waste Management* Case file, Civil Action No. H-79-2215, FRC Accession No. 021-94-0483, FRC Location No. I-230-1215, Agency Box No. 112–13, National Archives and Records Administration–Southwest Region, Fort Worth, Texas [hereinafter referred to as Bean Case Records].

2. Now Benjamin Muhammad, Chavis went on to lead the NAACP briefly and in 2001 became the president and CEO of the Hip-Hop Summit Action Network. See Elizabeth Blum, "Muhammad, Benjamin Franklin Chavis," in *Encyclopedia of Southern Culture,* forthcoming.

3. For a more detailed account of NACW activism during the Progressive Era, see Elizabeth Blum, "Protecting Home and Remolding Stereotypes: National Association of Colored Women Environmental Rationale and Activism during the Progressive Era," in *"To Love the Wind and the Rain": African Americans and Environmental History,* ed. Dianne Glave and Mark Stoll (Pittsburgh: University of Pittsburgh Press, 2005), 77–92. For other perspectives on African American women's environmental activities during the Progressive Era, see Dianne Glave, "'A Garden So Brilliant with Colors, So Original in Its Design': Rural African American Women, Gardening, Progressive Reform, and the Foundation

of an African American Environmental Perspective," *Environmental History* 8, no. 2 (July 2003): 395–411.

4. The history of the NACW is also discussed in Deborah Gray White, *Too Heavy a Load: African-American Women in Defense of Themselves, 1894–1994* (New York: W. W. Norton, 1999), 21–141; and Daphne Spain, *How Women Saved the City* (Minneapolis: University of Minnesota Press, 2001). For a discussion of the African American club women's movement, see Darlene Clark Hine, ed., *African-American Women in United States History* (Brooklyn, N.Y.: Carlson, 1990). For an "inside" history of the NACW up to the 1930s, see Elizabeth Lindsay Davis, *Lifting as They Climb* (New York: Humanity Books, 2003); for autobiographical information on some of the NACW's prominent leaders, see Mary Church Terrell, *A Colored Woman in a White World* (New York: G. K. Hall, 1996); and Ida Wells-Barnett, *Crusade for Justice: The Autobiography of Ida B. Wells,* ed. Alfreda M. Duster (Chicago: University of Chicago Press, 1970). For an excellent source of primary records, see Lillian Serece Williams, consulting ed., *Records of the National Association of Colored Women's Clubs, 1895–1992* (Bethesda, Md.: University Publications of America, 1993), microform [hereinafter cited as *NACW Records*].

5. "Names of the Clubs of the National Association of Colored Women," *National Association Notes,* April 1890, 1, *NACW Records;* Blum, "Protecting Home," 79.

6. NACW Resolution, Meeting Minutes of 1895, 11, *NACW Records.*

7. Mary Church Terrell, "A Reply to Hannibal Thomas," *National Association Notes,* May 1901, 3, *NACW Records;* Blum, "Protecting Home," 81.

8. Dr. Mary F. Waring, "Clean Up," *National Association Notes,* July 1913, 17, *NACW Records;* Blum, "Protecting Home," 87–88.

9. Mary Fitsbutler Waring, M.D., "Sanitation," *National Association Notes,* April–May, 1917, 8, *NACW Records;* Blum, "Protecting Home," 89.

10. Waring, "Clean Up," 17; Blum, "Protecting Home," 88.

11. M. S. Pearson, "The Home," *National Association Notes* 12, *NACW Records.*

12. *National Association Notes,* September 1898, *NACW Records.*

13. The project also was known as the Community Improvement Contest and the Home and Neighborhood Improvement Contest.

14. Irene Gaines, president of the organization and prime mover on the project, came from a background of activism in housing problems. In 1932, she received an appointment from President Herbert Hoover to sit as a member of his Conference on Housing. A social worker by training and education, she also served as a caseworker for the Cook County (Illinois) Bureau of Public Welfare. Gaines visited and gathered information from over 2,000 homes during her tenure, later testifying before both the city council and state legislature to push for passage of a slum clearance bill and funding. In addition to her testimony, Gaines also served on numerous housing groups, including the Southside Housing and Planning Council of Chicago and the Chicago Women's Joint Committee on Adequate Housing. Gaines information sheet, n.d., *NACW Records.*

15. Irene Gaines, 1956 President's Address, 33, *NACW Records.*

16. Irene Gaines, "Dear Clubwomen," Community Project Contest, 1956–57 Brochure, 2, *NACW Records.*

17. Gaines, 1956 President's Address, 35.

18. "Project Suggestions," Community Project Contest Brochure, 4, *NACW Records.*

19. Ibid., 4–6.

20. "Five Won and No One Lost," *National Notes,* Summer ed., 1957, 4, *NACW Records.*

21. Ibid., 6–7.

22. "Neighborhood Improvement Contests," *NACWC—What You Should Know About It* (Washington, D.C.: National Association of Colored Women's Clubs, 1959), n.p., *NACW Records.*

23. Henry Bullock, quoted in Bullard, *Invisible Houston,* 23, 24–26.

24. Bullard, *Invisible Houston,* 19.

25. Answers of Defendant Southwestern Waste Management to Plaintiff's Interrogatories, May 23, 1980, 9, Bean Case Records. The area had been investigated previously as a possible site for a landfill in 1971, although county commissioners ultimately voted against the project.

26. Jack C. Carmichael, Texas Department of Health Brief to John Richards, May 1, 1978, 3, Bean Case Records.

27. Answers of Defendant Southwestern Waste Management to Plaintiff's Interrogatories, 1–2, Bean Case Records.

28. Answers of Defendant Southwestern Waste Management to Plaintiff's Interrogatories, 8; Answers of Defendant Browning Ferris Industries to Plaintiff's Interrogatories, 1, 8–9, Bean Case Records; Michael Lawlor, Preliminary Injunction Hearing, 4–6, Bean Case Records.

29. Answers of Defendant Southwestern Waste Management to Plaintiff's Interrogatories, 8; Answers of Defendant Browning Ferris Industries to Plaintiff's Interrogatories, 1, 8–9; Lawlor, Preliminary Injunction Hearing, 4–6.

30. Although some residents appeared at the hearing on the permit, many remained confused or under different impressions as to the nature of the construction at the site. One resident, Louise Black, learned from different sources that the construction was not a landfill. "I learned of it when I saw the construction coming up and immediately . . . who I asked was the guy that sold me my house. He was still in the area selling homes. . . . And he said, well, I've heard that it was going to be a shopping center." Louise Black, Transcript of Court Trial, July 11, 1984, 85, Bean Case Records. The site itself failed to give uninformed residents clues to the nature of the construction. Landscaping professionals left trees standing on the site, on either side of the entrance to the landfill. In addition, construction efforts, as required, included an "earthen berm" to "bar visual site of the operation." At the entrance to the site, the only sign was from Teal Construction, without any indication of the type of facility, or Browning Ferris Industry's involvement. Lawlor, Preliminary Injunction Hearing, November 21, 1979, 6–13, Bean Case Records. Evidence at trial reinforced this view. Judge Gabrielle McDonald who presided over the initial injunction hearings, found that pictures introduced as evidence "clearly showed that a passerby would not be able to discern that a land fill was under construction." Judge Gabrielle McDonald, Memorandum Opinion and Order, December 21, 1979, 3, footnote 3, Bean Case Records.

31. McDonald, Memorandum Opinion and Order, 3, footnote 3.

32. Charles Bean, interview by Elizabeth D. Blum, April 25, 2000, tape recording, Houston, in author's collection; Margaret Lair, interview by Elizabeth D. Blum, April 27, 2000, tape recording, Houston, in author's collection; Patricia Reaux, interview by Elizabeth D. Blum, April 27, 2000, tape recording, Houston, in author's collection; Mildred Douglas, interview by Elizabeth D. Blum, May 2, 2000, tape recording, Houston, in author's collection; Louise Black, interview by Elizabeth D. Blum, May 4, 2000, tape recording, Houston, in author's collection.

33. Lair, interview.

34. Patricia Reaux, Court Trial Transcript, July 11, 1984, 66, Bean Case Records.

35. Ibid., 85, 86.

36. Reaux, interview.

37. Reaux, Lair, Black, Douglas, interviews.

38. Lair, interview.

39. Douglas, interview.

40. Patricia Reaux, Class Motion Hearing, 37, Bean Case Records.

41. Dorothy Grimes, Class Motion Hearing, 11, Bean Case Records.

42. Ibid., 12.

43. Reaux, interview.

44. John Singleton, Memorandum and Order, February 7, 1985, 2–3, Bean Case Records. The permit was actually approved on both occasions by the Texas Department of Health, although the earlier attempt at a landfill never materialized.

45. Grimes, Class Motion Hearing, 12.

46. Reaux, Class Motion Hearing, 37.

47. Reaux, Court Trial Transcript, 74.

48. Reaux, interview.

49. Reaux, Court Trial Transcript, 66.

50. Reaux, interview.

51. Reaux, Bean, Black, interviews.

52. "Stops Dump . . . Temporarily," *Houston Forward Times,* November 10, 1979; Plaintiffs' Post Trial Brief, December 10, 1979, 2, Bean Case Records.

53. Estimates on the number of participants vary. Mildred Douglas estimated the crowd turnout at about 60, while Patricia Reaux remembered between 200 and 300. These estimates are not necessarily contradictory, for both remember picking up additional support from people who were driving by and decided to stop and become a part of the protest spontaneously. Douglas, Reaux, interviews.

54. Reaux, Lair, Douglas, Black, Bean, interviews.

55. Bean, interview.

56. Reaux, Lair, Bean, Douglas, Black, interviews.

57. The African American women involved in the lawsuit came from working- or middle-class backgrounds. Only two of the eight women in the final group of named plaintiffs listed themselves as housewives. Margaret Bean and Patricia Reaux both worked for Baker Packers; Dorothy Jean Grimes was an "invoice price specialist" for Aramco Services Company. Louise Black described herself as an accountant with Strachan Shipping Co., while Mildred Douglass owned and operated her own business with her family, Douglas Tire Repair. All of the women had completed high school, and most had at least some college or postsecondary education. See Answers of Plaintiffs to Defendant Browning Ferris' Interrogatories, December 15, 1980; and Answers of Plaintiffs to Defendant Southwestern Waste Management's Interrogatories, December 15, 1980, Bean Case Records.

58. Complaint, October 26, 1979, 1, Bean Case Records. McKeever Bullard eventually filed three amended complaints to the original. The last, the Third Amended Complaint, was filed December 31, 1981.

59. Complaint, paragraph 13, Bean Case Records.

60. McDonald, Memorandum Opinion and Order, 2, 3. Quote is from Tex. Rev. Civ. Stat. Ann., art. 4477-7, Sec. 4(e)(8).

61. *Yick Wo v. Hopkins,* 118 US 356, 30 L.Ed 220, 68 S.Ct. 1064 (1896); *Gomillion v. Lightfoot,* 364 US 339, 5 L.Ed 2d 110, 81 S.Ct. 125 (1960); *Washington v. Davis,* 96 US 229, 48

L.Ed2d 597, 96 S.Ct. 2040 (1976); *Arlington Heights v. Metro Housing Corp.*, 429 US 252, 50 L.Ed 2d 450, 97 S.Ct. 555 (1977).

62. Robert D. Bullard, Vita, filed with Final Pretrial Order, May 23, 1984, Bean Case Records.

63. For more information about the proceedings of the case itself, see Blum, "Pink And Green," 303–34.

64. McDonald, Memorandum Opinion and Order, 1.

65. Bean, interview.

66. Charles Stradit, Plaintiffs' Hearing on Motion for Temporary Restraining Order, November 4, 1981, 94–98, Bean Case Records.

67. Bean, Lair, Reaux, interviews.

68. Ed Shannon, memo to Charles Moore, "List of Northside Residents/Browning Ferris Situation," October 8, 1980, with Hearing on Motion for Temporary Restraining Order exhibits, Bean Case Records.

69. Reaux, interview.

70. Charles Moore, "Narrative," December 10, 1980, with Hearing on Motion for Temporary Restraining Order exhibits, Bean Case Records.

71. Specifically, Singleton agreed with the defense's presentation that Bullard's definition of neighborhood was too subjective. In looking only at neighborhoods with churches and schools, he discounted purely residential subdivisions, often nonminority in nature. For example, through cross-examination, the court discovered that "the [West University and Bellaire landfill] sites were actually outside the boundaries of the Riceville community and closer to the sites was a small subdivision of residents occupied exclusively by whites." Singleton also took issue with Bullard's definition of schools that were "near" to a landfill. He "did not define near," Singleton began, and, "his definition of near only accounted for predominantly African-American schools located in predominantly African-American neighborhoods near a landfill. . . . [He] did not look at all schools near a landfill." If a more "scientific" definition of "near" were adopted, Singleton noted, such as drawing a circle around a site, "evidence disclosed that the landfills were surrounded by more non-African American schools than African-American schools." Singleton, Memorandum and Order, 7, 8.

72. Appeal Decision, January 22, 1986, 2, Bean Case Records.

73. W. Battle, letter to the editor, *Birmingham Post Herald*, 1994, quoted in "Introduction," in Laura Westra and Peter S. Wenz, eds., *Faces of Environmental Racism: Confronting Issue of Social Justice* (Lanham, Md.: Rowan and Littlefield, 1995), xvii.

74. Carl Anthony, "A Place at the Table," *Sierra* 78, no. 3 (May/June 1993): 57.

75. The "Big Ten" includes the National Parks and Conservation Association, the National Wildlife Federation, the Izaak Walton League, the National Audubon Society, the Sierra Club, the Wilderness Society, the Natural Resources Defense Council, the Environmental Defense Fund, the Environmental Policy Center, and the Friends of the Earth. Robert Gottlieb, *Forcing the Spring: The Transformation of the American Environmental Movement* (Washington, D.C.: Island Press, 1993), 118.

76. Ibid., 117–61.

77. Dorceta Taylor, "Environmentalism and the Politics of Inclusion," in Bullard, *Confronting Environmental Racism*, 57.

78. Martin V. Melosi, "Environmental Justice, Political Agenda Setting, and the Myths of History," *Journal of Policy History* 12 (2000): 52–56, 54–55 (quotes); see also Martin Melosi, "Equity, Eco-Racism, and Environmental History," in *Out of the Woods:*

Essays in Environmental History, ed. Char Miller and Hal Rothman (Pittsburgh: University of Pittsburgh Press, 1997), 194–211. For another critique of the environmental justice movement, see Christopher H. Foreman, *The Promise and Peril of Environmental Justice* (Washington, D.C.: Brookings Institute, 1998).

79. R. D. Bullard, "Anatomy of Environmental Racism and the Environmental Justice Movement," in Bullard, *Confronting Environmental Racism,* 30; Taylor, "Environmentalism," 57.

80. Judi Bari and Judith Kohl, "Environmental Justice: Highlander After Myles," *Social Policy* 21, no. 3 (Winter 1991): 73, 75.

81. For more revealing, complicated stories, see Pellow, *Garbage Wars*; and Colin Crawford, *Uproar at Dancing Rabbit Creek: The Battle over Race, Class and the Environment in the New South* (Reading, Mass.: Addison-Wesley, 1996).

Chapter 11. "To Combine Many and Varied Forces"

1. Samuel P. Hays, in collaboration with Barbara D. Hays, *Beauty, Health, and Permanence: Environmental Politics in the United States, 1955–1985* (Cambridge: Cambridge University Press, 1987), 62.

2. Douglas caught the imagination of the naturalists and concerned citizens across the country in seven editions printed between 1947 and 1962. Marjorie Stoneman Douglas, *The Everglades: River of Grass* (New York: Rinehart, 1947).

3. Hal Rothman's treatment of the Echo Park Dam controversy is another example of a regional issue attracting national attention. In the early 1950s, the Bureau of Reclamation sited a dam project in Echo Park in Dinosaur National Park, straddled between Colorado and Utah. State and national conservation groups used direct mail, a documentary film, a book of essays and photos, and newspapers to present the controversy to the national public. Appealing to the public rather than to government or elected officials, conservationists achieved a response in letters to Congress, eighty to one against the dam. See Hal Rothman, *Saving the Planet: The American Response to the Environment in the Twentieth Century* (Chicago: Ivan R. Dee, 2000).

4. Jim Blackburn, *The Book of Texas Bays* (College Station: Texas A&M University Press, 2004); Geoff Winningham, *Along Forgotten River* (Austin: Texas State Historical Association, 2003).

5. U.S. Congress, House Committee on Resources, "Buffalo Bayou National Heritage Area Study Act (to accompany H.R. 1776)" (Washington, D.C.: Government Printing Office, 2001, 2002).

6. A comprehensive listing of Houston's environmental groups is not the purpose of this chapter; rather it presents a preliminary exploration of the origins, spirit, and temper of local environmental activism. See the Web site for the Citizens' Environmental Coalition, www.cechouston.org, for information about current groups concerned with environmental issues in Houston.

7. McComb, *Houston: A History.*

8. Henderson, "City Planning in Houston," 107–36.

9. Don E. Carleton and Thomas H. Kreneck, "Houston, Back Where We Started," *Houston City Magazine* (1979).

10. Barry J. Kaplan and Charles Orson Cook, "Civic Elites and Urban Planning: Houston's River Oaks," in *Houston: A Twentieth Century Urban Frontier,* ed. Francisco A. Rosales and Barry J. Kaplan (Port Washington, N.Y.: Association Faculty Press, 1983), 22–33.

11. Bruce J. Weber, "Will Hogg and the Business of Reform" (Ph. D. diss., University of Houston, 1979), 163–73.

12. Ibid., 176; Robert Fisher, "'Be on the Lookout': Neighborhood Civic Clubs in Houston," *Houston Review* 6, no. 3 (1984): 108; "Original Covenants and Restrictions in Country Club Estates Additions: River Oaks Section Adjoining River Oaks Country Club, Houston, Texas" (1924).

13. In 1912, Houston had five parks totaling 112 acres. In his plan for Houston, Arthur Comey calculated that Houston should have one acre of park space for every 110 people; hence, in 1912 Houston should have had 800 acres of park space. See Arthur Coleman Comey, *Houston; Tentative Plans for Its Development; Report to the Houston Park Commission* (Boston: Press of Geo. H. Ellis, 1913), 21, 27.

14. Charles N. Burris, *Study of Parks and Greenspace in Houston* (Houston: University of Houston, Institute for Public History, 1993), 21–25.

15. Stephen Fox, "Big Park, Little Plans: A History of Hermann Park—Part I," *Archives-Hermann Park-Writings,* n.d., http://www.georgekessler.org.

16. Comey, *Houston,* 35.

17. http://www.houstontx.gov/municipalgolf/hermann/index.html. By the late 1980s, Friends of Hermann Park rescued a rundown Hermann Park damaged by insufficient public funds and overuse. In 2004, under the new name of the Hermann Park Conservancy, the organization committed to permanent stewardship of Hermann Park. See http://www.hermannpark.org/about.html.

18. Inspired by the Civic Arts Movement, such ventures occurred across the country. See http://www.ci.houston.tx.us/municipalart/samhouston.htm.

19. Ilona B. Benda, "Pine Trees at Camp Logan Sing of Past, *Houston Chronicle,* July 15, 1923; Robert V. Haynes, *A Night of Violence: The Houston Riot of 1917* (Baton Rouge: Louisiana State University Press, 1976).

20. Sarah H. Emmott, *Memorial Park: A Priceless Legacy* (Houston: Herring Press, 1992), 23.

21. Ibid., 24; "Woman with Purpose," *Houston Sportsman News and Views,* 1971, n.p., in J. M. Heiser Jr. Environmental Collection, Houston Metropolitan Research Center.

22. Along with Ima Hogg, Mike and Will were the three civic-minded children of Texas two-term governor (1891–95) James Stephen Hogg. A fourth son did not share the family's philanthropic impulses. Mike graduated from University of Texas law school and oversaw the family estate. Ima did not participate in the family businesses but lent her energies to community efforts such as establishing the Houston Symphony. With her brother Mike, she designed their home, Bayou Bend, now a museum for American antiques and material culture. Weber, "Will Hogg," 5–11.

23. Emmott, *Memorial Park,* 24

24. Burris, "Study of Parks," 26. Historically, the success of preservation initiatives in Houston was limited by the inherent circularity of business interests, local government, and philanthropy, because the large fortunes were earned by business elites who dominated local government.

25. Terry Hershey, personal communication, March 17, 2006; Emmott, *Memorial Park,* 23.

26. Terry Hershey suggested that it was most likely Ima Hogg who encouraged her brothers to sell the land to the city to establish the acreage for Memorial Park. Terry Hershey, personal communication, March 17, 2006.

27. Andrew Sansom, "And the Things That Belong," *Horizons* (Texas Nature Conservancy, San Antonio) 9, no. 2 (1984): 3.

28. *Outdoor Nature Club History, 1923–2003,* Slide Presentation Script, 2004, slide 5; David Thomas Radcliff. "The of Houston's Environmental Organizations" (Master's thesis, University of Houston, 1997); Terry Hershey, inscription in Stephen R. Fox, *John Muir and His Legacy: The American Conservation Movement* (Boston: Little, Brown, 1981), J. M. Heiser Jr. Environmental Collection.

29. "Biographical Note," J. M. Heiser Jr. Environmental Collection.

30. Radcliffe, "Houston's Environmental Organizations," 65–66.

31. Ibid., 68–70; *Outdoor Nature Club History,* slide 12.

32. Martin V. Melosi, "Environmental Reform in the Industrial Cities," in Melosi, *Effluent America,* 211–24.

33. Garden Club of Houston, 1985–86, "History of the Garden Club of Houston" and photocopied page, n.d., Marguerite Johnston Barnes Research Collection, Box 10, Folder 19, Rice University.

34. Johnson, *Houston,* 243.

35. One of Hogg's objectives was the construction of a civic center, for which he bought the land and promoted the development. See Bruce J. Weber and Charles Orson Cook, "Will Hogg and Civic Consciousness: Houston Style," *Houston Review* 2, no. 4 (1980): 22–25.

36. Paul S. Boyer, "The Ideology of the Civic Arts Movement in America, 1890–1920," *Houston Review* 2, no. 4 (1980): 3–19; Hays, *Conservation and the Gospel of Efficiency.*

37. John Avery Lomax, *Will Hogg, Texan* (Austin: University of Texas Press, 1956). Hogg's distribution of plants such as crepe myrtle trees irritated the editor of the *Forum* journal because she believed that activity was best left to the garden clubs. Weber and Cook, "Will Hogg," 33–34

38. Weber and Cook, "Will Hogg," 33.

39. "A Forum of Civics for Houston," Houston, May 15, 1926, J. M. Heiser Jr. Environmental Collection; and Forum of Civics: "An organization designed to stimulate civic pride and to combine many and varied forces for the betterment and beautification of our city and county," *Report of the City Planning Commission,* Houston, December 1929, title page.

40. The Outdoor Nature Club was a member of the Forum of Civics. See Hester Scott to J. M. Heiser, March 13, 1928, J. M. Heiser Jr. Environmental Collection. On April 22, 1956, Heiser wrote a letter to the editor of the *Houston Chronicle* expressing his concern over pressing civic problems and the need for a public forum designed along the same lines that a "far-sighted Will Hogg" had planned.

41. Kendrick A. Clements, *Hoover, Conservation, and Consumerism: Engineering the Good Life* (Lawrence: University Press of Kansas, 2000).

42. The Garden Club of Houston and the River Oaks Garden Club also continued their work throughout the Depression and World War II and still remain vital parts of an environmental network.

43. The Houston Audubon Society was not established until 1969.

44. *Outdoor Nature Club History,* slides 23–26.

45. Jonathan J. Keyes, "A Place of Its Own: Urban Environmental History," *Journal of Urban History* 26, no. 3 (2000): 380–90; see also D. B. Botkin and C. E. Beveridge, "Cities as Environments," *Urban Ecosystems* 1 (1997): 3–19.

46. Fox, *John Muir.* Joseph Petulla, in *American Environmentalism* (College Station: Texas A&M University Press, 1980), mapped the history and philosophies behind basic belief systems that underpin the conflict between Muir and Pinchot. The biocentric perspective privileges nature as an end in itself, a value to be set aside from human exploitation. The economic perspective advocates the production and management of resources for human use. The ecological perspective would enter the debate later in the century. An ecological perspective recognizes balance within and among natural systems that are self-sustaining if allowed to self-regulate, with humans included in natural systems.

47. Fox, *John Muir, 333–57*; Andrew Hurley, *Environmental Inequalities: Class, Race, and Industrial Pollution in Gary, Indiana, 1945–1980* (Chapel Hill: University of North Carolina Press, 1995).

48. Gottlieb, *Forcing the Spring.*

49. Melosi, "Environmental Reform," 211–24.

50. Hays, *Beauty,* 2–10.

51. Urban environmental activism from the 1930s through the 1950s is not covered in detail here. Urban watershed management emerged as a salient issue, and the Harris County Flood Control District began serving the area in 1937. Civic associations arose in subdivisions around the city. Their mission was beautification as well as racial segregation. See Fisher, "'Be on the Lookout,'" 108. Interwar conservation history is still underexamined.

52. Hans W. Baade, "Roman Law in the Water, Mineral and Public Land Law of the Southwestern United States," *American Journal of Comparative Law* 40, no. 4 (1992): 865–77.

53. *John G. and Marie Stella Kenedy Memorial Foundation v. Dewhurst,* Tex. Sup Ct 08-29-2002.

54. Barbara Karkabi, "Naturalist Citizens: Living a Legacy of Respect for the Earth Is Second Nature for the Emmotts," *Houston Chronicle,* April 21, 1991.

55. Ibid.; *Outdoor Nature Club History,* slide 50.

56. Anella Dexter to Membership of Texas Conservation Council, February 5, 1970, J. M. Heiser Jr. Environmental Collection.

57. Radcliff, "Evolution of Organizations," 79–85

58. Inspired by her mother-in-law, Sarah researched and wrote the history of Memorial Park. Emmott, *Memorial Park.*

59. *Nomination of A. V. Emmott and Sarah H. Emmott for the George Washington Honor Medal for Excellence in a Lifetime of Achievement,* Freedom Foundation at Valley Forge, 1992, n.p.; Houston Sportsmen's Club's Conservation Hall of Fame, *Houston Sportsman,* 7–8, 1982, J. M. Heiser Jr. Environmental Collection.

60. Karkabi, "Naturalist Citizens."

61. Radcliff, "Evolution of Organizations," 81.

62. Eckhardt later went to the U.S. House of Representatives and was the "brains behind the Clean Air Act," according to Texan Molly Ivins. *Free Press,* November 12, 2001, http://www.freepress.org/columns/display/1/2001/646.

63. Karkabi, "Naturalist Citizens."

64. Hays, *Beauty,* 2.

65. The Houston Canoe Club formed in 1964 for recreation, but its members became increasingly aware of environmental concerns by exploring the local bayous in canoes that could navigate the winding, shallow waters inaccessible with other modes of transportation. One argument for the relevance of recreation groups is their potential to

develop environmental concern and activism through environmental awareness derived from firsthand observation of hidden natural wonders in local ecosystems. John Bartos, Chair Conservation Committee, Houston Canoe Club, personal communication, February 28, 2006. Another recreation-based group was Citizens for Hike and Bike organized by Marge Selden and Elizabeth Lankford in 1965. By 1970, due partly to their efforts and partly to support from the 1968 National Trails System Act, Houston's Parks and Recreation Department had planned a network of trails that would connect historical, environmental, and cultural points of interest.

66. Harris County Flood Control District, *Riding the Waves of Change: 60 Years of Service* (Houston: Harris County Flood Control District, 1997).

67. Barker and Addicks reservoirs were created as part of a federal project to control flooding on Buffalo Bayou and protect downtown Houston. The U.S. Army Corps of Engineers completed Barker Dam in 1945 and Addicks Dam in 1948. See http://www.hcfcd.org/watersheds.html; and Alperin, *Custodians of the Coast*. For more about the problems with structural intervention as flood management, see Jared Paul Orsi, "Hazardous Metropolis: Flood and Urban Ecology in Los Angeles" (Ph.D. diss. University of Wisconsin–Madison, 1999).

68. The BPA's first protest against structural intervention along Buffalo Bayou preceded the National Environmental Policy Act of 1969 (NEPA) by three years. NEPA would require public input through mandatory environmental impact statements for federally funded projects. According to Hershey, however, Harris County commissioners were so hostile to suggestions for a restudy and so committed to concretizing the bayou that they vowed to continue work with county money when Congress agreed to a restudy. Terry Hershey, personal communication, March 11, 2005.

69. To expand his dedication to sustainable development, George Mitchell established the Houston Area Research Center (HARC) in 1982 to pursue research in basic, applied, and policy issues funded by contracts, grants, and gifts. See http://www.harc.edu.

70. Charles E. Closmann, *Buffalo Bayou: Past, Present and Future* (Houston: University of Houston, Institute for Public History, 1993).

71. Bayou Preservation Association, untitled document on organization letterhead, describing BPA history, 1–3, 1990, Al Morris Environmental Archives, Bayou Preservation Association, Box 1, Folder 2, Houston.

72. Shirley Pfister, "Terese Hershey: A Flood of Details," *Houston Chronicle,* May 28, 1975: "I believe that when you're fortunate enough to have free time, you ought to give something back."

73. Résumé of Terry Hershey, n.d., in Terry Hershey Papers, unprocessed collection in Hershey's possession.

74. The Wild and Scenic Rivers Act of 1977 offered protection to designated waterways, but Buffalo Bayou has not yet qualified for that protection.

75. Some historians and journalists suggest that *Silent Spring* began the modern environmental movement, but there is considerable evidence that activism here and elsewhere had begun before *Silent Spring* was published in 1962. Another point is that *Silent Spring* encouraged citizens to demand information from government and industry. See Sarah Thomas, "A Call to Action: *Silent Spring,* Civic Activism, and the Origins of the Modern Environmental Movement," American Society for Environmental History, Houston, March 16–19, 2005; Rachel Carson, *Silent Spring* (Boston: Houghton Mifflin; Cambridge, Mass.: Riverside Press, 1962).

76. Hana Ginzbarg, personal communication, February 2005.

77. "Bellaire Woman to Receive Honor for Ecology Work," *Houston Post,* April 29, 1974.

78. James C. Patterson, "The Houston-Galveston Area Council: A Regional History of Intergovernmental Cooperation" (Ph.D. diss., University of Houston, 1990).

79. Ibid.

80. Hana Ginzbarg, "How Armand Bayou Park and Nature Center Came to Be: Setting the Record Straight," transcript of talk for Armand Bayou volunteer organization, October 13, 2005, Armand Bayou Park and Nature Center, Houston. Ginzbarg described the Texas Water Plan as a "$3.5 billion boondoggle [that] involved building dams all along Texas Rivers. It would provide lots of jobs but devastate the environment" (2).

81. Brandt Mannchen, "History of the Houston Sierra Club," *Bayou Banner* (Houston Sierra Club newsletter), n.d.

82. The National Audubon Society supported the wardens in Gulf Coast bird sanctuaries but did not participate in local environmental politics. Winnie Burkett, personal communication, April 6, 2006.

83. Houston Audubon Society, information sheet, May 27, 2005. Presently, the Houston Audubon Society owns and manages 3,000 acres of habitat and includes 3,900 members from thirteen counties.

84. Decisions such as these to breach the insularity of local government happened across the country during the 1960s. For example, see James Lewis Longhurst, "'Don't Hold Your Breath, Fight for It!': Women's Activism and Citizen Standing in Pittsburgh and the United States, 1965–1975" (Ph.D. diss., Carnegie Mellon University, 2004).

85. Terry Hershey, personal communication, June 16, 2005.

86. W. Leo Theiss and Mrs. J. W. Hershey to Mrs. Searcy Bracewell, President, The Heritage Society, August 12, 1970, Terry Hershey Papers.

87. S. W. Dorrell Jr., Chairman Sam Houston Resource Conservation and Development Area, to Dr. Arthur Atkinsson, President CEC, July 20, 1970, Terry Hershey Papers.

88. Minutes of the Land Use Subcommittee Meeting, August 28, 1970; neither the national nor other local Audubon groups had addressed this problem.

89. Terry Hershey to N. Robert Batten, Executive Vice President, Greater Houston Builders Association, October 15, 1970, Terry Hershey Papers.

90. Cover letter from W. Leo Theiss and Mrs. J. W. Hershey to addressee, n.d., Terry Hershey Papers; "Recommendations" document explained in subhead: "The Land Use Planning Task Force of the Citizens' Environmental Coalition, headed by Sam Houston Resource Conservation and Development, recommends the following."

91. "Origin and History of CEC," *CEC Resource Manual* (March 1984), 1.

92. Annella Dexter to members of the Texas Conservation Council, February 5, 1970, J. M. Heiser Jr. Environmental Collection.

93. Tribute, "Armand Yrmategui Memorial Book" (scrapbook), 1980, Pasadena Public Library, Pasadena, Tex.

94. James Herzberg, "Naturalist Armand Yramategui, 1923–1970," *Houston Review* 17, no. 2 (1995): 49–58. The entire environmental community gave considerable thought to the appropriate symbolic recognition of Yramatagui's contribution to Houston's environmental activism. Ideas for bayou trails were rejected as too insignificant.

95. Ginzbarg, "Armand Bayou Park," 3.

96. In her narrative on the events leading to the campaign, Ginzbarg stated that Rachel Carson's *Silent Spring* roused her to action as well as the inspiration from Armand Yramategui and Terry Hershey.

97. "Bellaire Woman to Receive Honor."

98. Friendswood Development Company was a subsidiary of Exxon.

99. http://www.abnc.org/mission.php.

100. "The Park People, Inc.," sheet 1 of a speech draft, The Park People Records Collection, Houston History Archives, Special Collections, University of Houston; Speech from Vernon Henry to President's Commission on Americans Outdoors, December 12, 1985, Terry Hershey Papers.

101. Terry Hershey to Susan Garwood, February 7, 2006, Terry Hershey Papers.

102. "Who Are The Park People?" Historical Summary, May 17, 1982, The Park People Records Collection.

103. In 2006, The Park People produced a countywide trails system map after attempts by Citizens Hike and Bike in the late 1960s and early 1970s and a CEC Trails Committee. Hike and bike trails around the bayous and scenic landmarks have been under construction for most of the twentieth century.

104. The Katy Prairie Conservancy (KPC) is a good example of a site-specific community coalition that addresses the major themes of both wilderness and urban environmental concerns, including prevention of prairie loss to suburban growth; wildlife preservation; maintenance of a sustainable ecosystem, including the wetlands; and preservation of open space and biodiversity. KPC's board of directors includes hunters, environmentalists, and community business leaders. Preservation advocates believe that tourism during annual migrations will bring economic advantages, and that such a natural resource enhances the quality of urban life.

105. Marie D. Hoff, ed., *Sustainable Community Development: Studies in Economic, Environmental, and Cultural Revitalization* (Boca Raton, Fla.: Lewis, 1998).

106. David Crossley, "Gulf Coast Institute Information Sheet," Gulf Coast Institute, Houston, March 17, 2006. These beginnings illustrate a more active participation from city government than enjoyed by CEC in the 1970s.

107. Gulf Coast Institute, *Choices: A Smart Growth Primer for the Houston Gulf Coast Region* (Houston: Gulf Coast Institute, 1999). This primer was published after the conference of the same name.

108. Donella H. Meadows, Dennis L. Meadows, and Jørgen Randers, *Beyond the Limits: Confronting Global Collapse, Envisioning a Sustainable Future* (Mills, Vt.: Chelsea Green, 1992).

109. Chattanooga Institute, http://www.csc2.org.

110. David Crossley, ed., *Connecting the Visions: Creating the Future We Want in the Houston Gulf Coast* (Gulf Coast Institute, 2001); see also the following sites: http://www.gulfcoastinstitute.org/connecting/houstonvisions.pdf; and http://www.blueprinthouston.org/documents/compendium_of_plans.doc.

111. http://www.blueprinthouston.org; http://www.blueprinthouston.org/documents/transportation.doc.

112. As an example, after the CEC conducted its community education program on air quality in 1970, the Citizens Clean Air Advisory Council (CCAAC) emerged in 1977, when the Clean Air Act was amended with new deadlines for the National Ambient Air Quality Standards because many areas had failed to meet the 1975 deadlines established by the Clean Air Amendments of 1970. The CCAAC met until the early 1980s, but its efforts to effect air quality enforcement and to construct channels of citizen input to the Texas Air Control Board were unsuccessful. In 1988, the Galveston-Houston Association for Smog Prevention (GHASP) resumed the work of the CCAAC. GHASP's mission

included advocacy for clean air in the Houston region, education for the public, and engagement with government, community, and industry leaders regarding regional air pollution.

113. Adam Rome, "'Give Earth a Chance': The Environmental Movement and the Sixties," *Journal of American History* 90 (September 2003): 525–54; Longstreet, "'Don't Hold Your Breath.'"

114. Carleton, *Red Scare!*

115. Fletcher Winston, "Repertoires, Rules, and Relationships: A Cultural Analysis of Tactical Choice in the Environmental Movement" (Ph.D. diss., State University of New York at Stony Brook, 2004), 61–62.

116. Ruth Rosen, *The World Split Open: How the Modern Women's Movement Changed America* (New York: Viking, 2000); Linda K. Kerber, *No Constitutional Right to Be Ladies: Women and the Obligations of Citizenship* (New York: Hill and Wang, 1998). Participation in the League of Women Voters started in Houston in 1922. With a strong resource and environmental committee at different periods, Houston's League of Women Voters was an active participant in the coalitions described in this essay and in many other organizations and initiatives.

117. Houston's women activists identified strongly with women's empowerment yet often eschewed formal leadership roles, preferring to assume more traditional personas in organizational politics. Houston was the site of the National Women's Conference in 1977, and the Houston Local Arrangements Committee included volunteers from both feminist and traditional women's and political organizations. See Ellen Pratt Fout, "'A Miracle Occurred!': The Houston Committee of International Women's Year, Houston, 1977," *Houston Review of History and Culture* 1, no. 1 (2003): 4–9.

118. Bayou Preservation Association, for example, had a "Ladies Committee" for several years.

119. Cole, *No Color Is My Kind;* Brian D. Behnken, "On Parallel Tracks: A Comparison of the African-American and Latino Civil Rights Movements in Houston" (Master's thesis, University of Houston, 2001).

120. Barry J. Kaplan, "Race, Income, and Ethnicity: Residential Change in a Houston Community, 1920–1970," *Houston Review* 3, no. 4 (1981): 78–202.

121. Francisco A. Rosales, "The Mexican Immigrant Experience in Chicago, Houston and Tucson: Comparisons and Contrasts," in Rosales and Kaplan, *Houston,* 58–77.

122. Bullard, *Dumping in Dixie;* Beeth and Wintz, *Black Dixie;* Bullard, *Invisible Houston;* Arturo F. Rosales, "Mexicans in Houston: The Struggle to Survive, 1908–1975," *Houston Review* 3, no. 2 (1981): 224–48.

123. Divergences between national and local groups are frequently addressed. For a full treatment of the emergence and professionalization of national groups, see Gottlieb, *Forcing the Spring.*

124. Terry Hershey, personal communication, January 2004; résumé of Terry Hershey.

125. Differences in movement constituencies are well recognized by environmental historians. For a detailed analysis of the environmental movement demographics, see Deborah Lynn Guber, *The Grassroots of Green Revolution: Polling America on the Environment* (Cambridge, Mass.: MIT Press, 2003).

126. Andrew Dobson and Derek Bell, eds., *Environmental Citizenship* (Cambridge, Mass.: MIT Press, 2006).

127. Dave Horton, "Demonstrating Environmental Citizenship: A Study of Everyday Life among Green Activists," in Dobson and Bell, *Environmental Citizenship,* 128–29.

Chapter 12. Voices of Discord

1. "CERCLA Overview," in U.S. Environmental Protection Agency Superfund Program, http://www.epa.gov/superfund/action/law/cercla.htm.

2. Jim Hutton and Jim Henderson, *Houston: A History of a Giant* (Tulsa, Okla.: Continental Heritage, 1976), 115.

3. Ibid., 115.

4. Melosi, *Effluent America*.

5. Brio Site Task Force Community Relations Plan, September 1985, updated May 1989, Book 1. This is a reclamation process that removes or extracts chemicals from waste material, and these recovered chemicals are made into other products that are sold for profit. The oil and chemical companies delivered the waste to the dumpsite since they no longer had a use for this waste.

6. Richard J. Lewis Sr., *Hazardous Chemicals Desk Reference*, 4th ed. (New York: Van Nostrand Reinhold, 1997), 1217–18; Lisa H. Newton and Catherine K. Dillingham, *Watersheds: Classic Cases in Environmental Ethics* (Belmont, Calif.: Wadsworth, 1994), 12.

7. Craig E. Colten and Peter N. Skinner, *The Road to Love Canal: Managing Industrial Waste before EPA* (Austin: University of Texas Press, 1996), preface.

8. Ibid., 5.

9. Hays, *History of Environmental Politics*, 17.

10. Brio Site Task Force Community Relations Plan.

11. Resource Engineering, "Brio Refining Site Remedial Investigation Report," Final Report, November 15, 1986, 4–5.

12. Ibid., 24.

13. "NPL Site Narrative for Brio Refining, Inc.," in EPA Superfund NPL Narrative at Listing, http://www.epa.gov/superfund/sites/npl/nar792.htm; *U.S. EPA, Brio Refining, Inc.*, EPA ID# TXD980625453, April 5, 2000 (Washington, D.C.).

14. Donald Y. Joe and Kenneth G. Orloff, "Health Assessment for Brio Refining, Inc. (Brio) and Dixie Oil Processors (DOP) NPL Sites," Brio Repository at San Jacinto Community College, Friendswood, Tex., February 7, 1989, 3, 10.

15. Resource Engineering, "Brio Refining Site Remedial Investigation Report," 26.

16. Ibid., 29.

17. Ibid. This report was compiled by Resource Engineering, the first group to investigate the Brio site. The report detailed the various chemicals stored at Brio and the various operations utilized throughout its history. The purpose of the report was to guide the EPA and the responsible parties to the best methods for cleaning up the waste while aiding them in understanding the full dilemma and the risks of the chemicals at the site.

18. Brio Site Task Force Community Relations Plan, 3.

19. "NPL Site Narrative for Brio Refining"; *U.S. EPA, Brio Refining*.

20. A potential responsible party (PRP) is a corporation or individual who generated or assumed responsibility for the waste at the site. Frequently, the PRP went out of business, could not be identified, or passed on its legal responsibility to someone else; however, Superfund provides joint, several, and retrospective liability, which maintains that every owner and operator, previously or presently, is responsible for damages and cleanup at the site.

21. Resource Engineering, Brio Refining Site Remedial Investigation Report, 7.

22. Brio Site Task Force, "Brio Site History: Beginning to Superfund Nomination," Brio Site Task Force Community Relations Plan, Book 1, September 1985, updated May 1989, 4.

23. Joe and Orloff, "Health Assessment for Brio Refining," 8–10.

24. Hays, *Environmental Politics,* 25.

25. Gottlieb, *Forcing the Spring,* 208–9.

26. Ibid., 209.

27. Hays, *Environmental Politics,* 3.

28. Andrew Szasz, *Ecopopulism: Toxic Waste and the Movement for Environmental Justice* (Minneapolis: University of Minnesota Press, 1994), 65.

29. Trigg Gardner, "Survey Shows Problem for Southbend Births," *South Belt–Ellington Leader,* April 12, 1990.

30. Rebecca Deaton, "Neighborhood in Transition," *Houston Chronicle,* July 15, 1992.

31. Scott Harper, "Brio Study Finds Health Problems," *Houston Post,* October 7, 1994.

32. Deaton, "Neighborhood in Transition."

33. Gardner, "Survey Shows Problem for Southbend Births."

34. Ruth Rendon, "Chemical Release at Brio Cleanup Spurs List of Evacuation Standards," *Houston Chronicle,* October 15, 1993.

35. Gaynell Terrell, "Brio Initiates More Safeguards in Wake of Aug. 25 Gas Release," *Houston Post,* October 27, 1993.

36. Ruth Rendon, "EPA Report Says More Tests Needed at Brio Hazardous Waste Site," *Houston Chronicle,* November 17, 1993.

37. Ruth Rendon, "State Warning," *Houston Chronicle,* November 19, 1993.

38. Ibid.; Scott Harper, "Tests Point to Brio as Source of Chemicals Clear Creek: Results Include Carcinogens," *Houston Post,* March 11, 1994.

39. Scott Harper, "EPA Urges New Tests at Brio Site," *Houston Post,* April 2, 1994.

40. Harold Scarlett, "Children Caught in Brio Hazardous Waste Fracas," *Houston Post,* March 11, 1990.

41. Ruth Rendon, "For School, End of Week Is End of Line," *Houston Chronicle,* March 26, 1992.

42. Joseph A. Pratt, interview by Kimberly Youngblood, March 2000.

43. Trigg Gardner, "M.U.D. 13 to Fight Incineration," *South Belt–Ellington Leader,* September 14, 1989.

44. Bob Sablatura, "Caught in the Toxic Waste Cross-fire," *Houston Chronicle,* October 24, 1995.

45. Ruth Rendon, "Amended Proposal May Finally Bring End to Brio Waste Site Cleanup," *Houston Chronicle,* March 15, 1999.

46. "Brio Superfund Site Slurry Wall Finished," *South Belt–Ellington Leader,* December 14, 2000.

47. Bruce Nichols, "Community Crusader," *Dallas Morning News,* January 28, 1990.

48. Bruce Piasecki and Peter Asmus, *In Search of Environmental Excellence: Moving Beyond Blame* (New York: Simon and Schuster, 1990), 162. Piasecki actually stated that the environmental movement as a whole is fed by the factors of fear and mistrust.

49. Julie Mason, "Huge Suit by Residents against Monsanto Set," *Houston Chronicle,* October 17, 1989.

50. JoAnn Zuniga, "Brio Site Case Draws Angry Homeowners," *Houston Chronicle,* February 3, 1987.

51. David Ellison, "$39 Million Settles Brio Injury Claims," *Houston Post,* June 16, 1992.

52. Harold Scarlett, "Jury Gave a Warning, Experts Say," *Houston Post,* February 13, 1990.

53. JoAnn Zuniga, "Builders, Developers Settle Lawsuit in Waste Dump Flap," *Houston Chronicle,* May 14, 1987.

54. Debbie Housel, "Settlement on Brio Hits $207 Million," *Houston Post,* June 19, 1992.

55. Jerry Urban, "Settlement Reached between U.S., Brio Stockholder Lowe," *Houston Chronicle,* December 30, 1992.

56. Marie Flickinger, "Brio Site a Gold Mine for Monsanto," *South Belt–Ellington Leader,* March 6, 1997.

57. Michael Brown, *Laying Waste: The Poisoning of America by Toxic Chemicals* (New York: Washington Square Press, 1980), 51.

58. Deaton, "Neighborhood in Transition."

59. Colten and Skinner, *Road to Love Canal,* 147.

60. Bullard, *Dumping in Dixie*, 38.

CONTRIBUTORS

William C. Barnett is assistant professor of history at North Central College, Naperville, Illinois. He received his Ph.D. in history from the University of Wisconsin. Under the direction of William Cronon, he completed a dissertation in 2005 entitled "From Gateway to Getaway: Labor, Leisure, and Environment in American Maritime Cities."

Diane C. Bates is assistant professor in the Department of Sociology and Anthropology at the College of New Jersey. She received her Ph.D. from Rutgers University in sociology. She offers courses in quantitative research methods, environmental sociology, social change in Latin America, and introduction to sociology. Aside from research on urban sprawl, she also has worked on smart growth and topics related to Latin America. She recently codirected a competitive grant for the promotion of undergraduate scholarly activity awarded by the Alice and Leslie E. Lancy Foundation.

Elizabeth D. Blum is associate professor of history at Troy University, and also serves as acting archivist for the university. She received her Ph.D. in history at the University of Houston. At Troy, she teaches courses in environmental history, the history of women, contemporary America, and African American history. She has served on the Rachel Carson Prize Committee. She is completing a book for the University Press of Kansas entitled *Enlarging the Picture at Love Canal: Gender, Race, and Class in Environmental Activism.*

Robert D. Bullard is the Ware Distinguished Professor of Sociology and director of the Environmental Justice Resource Center at Clark Atlanta University. He is the author of twelve books that address sustainable development, environmental racism, urban land use, industrial facility permitting, community reinvestment, housing, transportation, and smart growth. His most recent books include *Just Sustainabilities: Development in an Unequal World* (2003) and *Highway Robbery: Transportation Racism and New Routes to Equity* (2004).

Robert Fisher is a professor of community organization in the School of Social Work and director of the Undergraduate Urban and Community Studies Program at the University of Connecticut. He received his Ph.D. from New York University. His areas of specialization include community organizing, urban policy, social movements, social theory, and social welfare history. He is the author of *Let the People Decide: Neighborhood Organizing in America* (1984) and several important pieces on the history of Houston, among others.

Hugh S. Gorman is associate professor of environmental policy and history in the Department of Social Sciences at Michigan Technological University. He studies the interaction of policy, technology, and the environment and is especially interested in the process by which complex technological systems come to be made more compatible with natural systems and human uses of the environment. He is the author of *Redefining Efficiency: Pollution Concerns, Regulatory Mechanisms, and Technological Change in the U.S. Petroleum Industry* (2001). Currently, he is working on a book that examines the notion of sustainability through society's changing knowledge of and interaction with the nitrogen cycle.

Tom Watson McKinney received his Ph.D. in history from the University of Houston. His dissertation focused on the Gulf Freeway and its influence on growth in Houston. He has completed additional research on railroads and has given several presentations on transportation history at a variety of professional meetings. He is currently completing a study on the financing of infrastructure in Houston for the Houston Council of Engineering Companies.

Martin V. Melosi is Distinguished University Professor of History and director of the Center for Public History at the University of Houston. He is president of the Urban History Association, and past president of the American Society for Environmental History, National Council on Public History, and the Public Works Historical Society. He also held the Fulbright Chair in American Studies at the University of Southern Denmark (2000–2001). He is the author or editor of thirteen books, most of which deal with the urban environment, energy history, and public policy.

Joseph A. Pratt is a historian of the petroleum industry. He holds a joint appointment in history and management as the Cullen Professor of History and Business at the University of Houston and serves as associate director of the

Center for Public History. He also served as chair of the Department of History and for five years as the executive director of the Scholars' Community, the largest undergraduate retention program at the University of Houston. He is author of ten books and numerous articles. His research is primarily in energy history and the history of the Houston region.

Robert S. Thompson is a Ph.D. candidate in history at the University of Houston. He also is assistant editor for the *Encyclopedia of American Environmental History*. His dissertation deals with the relationship between American slavery and the environment.

Terry Tomkins-Walsh is project archivist for the Houston History Archives at the University of Houston and consulting archivist for Houston Endowment Inc. She is finishing her dissertation in the Department of History on the response of local environmental activism to urban watershed management in Houston/Harris County, funded in part by grants from the Environmental Institute of Houston.

Kimberley Youngblood is a Ph.D. candidate in history at the University of Houston. She also serves as managing editor for the *Houston History Magazine*, published by the Center for Public History. Her dissertation topic focuses on the shrimping industry and pollution in Galveston Bay, and she also is engaged in research for litigation support.

INDEX

Page numbers in italic type indicate tables, figures, or illustrations.